The Metric System of Measurement

A. Common prefixes

Name	Symbol	Meaning	Multiplier
mega	M	one million	1 000 000
kilo	k	one thousand	1 000
hecto	h	one hundred	100
deca	da	ten	10
deci	d	one tenth of a	0.1
centi	c	one hundredth of a	0.01
milli	m	one thousandth of a	0.001
micro	μ	one millionth of a	0.000 001

B. Système International (SI) base units of measurement

Quantity	Unit	Symbol
Length	metre	m
Mass	kilogram	kg
Time	second	s
Electric current	ampere	A
Thermodynamic temperature	kelvin	K
Amount of substance	mole	mol
Luminous intensity	candela	cd

Derived units often used in physics

Quantity	Unit	Symbol
Force	newton	N
Pressure	pascal	Pa
Energy	joule	J
Power	watt	W
Frequency	hertz	Hz
Electric resistance	ohm	Ω
Electric voltage	volt	V
Radiation activity	becquerel	Bq

Physics: *A Practical Approach*

The photograph on the front cover shows a pattern of light created by a laser. It is one of many patterns that "dance" to music in laserium shows held in planetariums. See pages 434 and 435. (Courtesy of Laser Images Inc., California)

PHYSICS
A PRACTICAL APPROACH

Alan J. Hirsch

John Wiley & Sons

Toronto · New York · Chichester · Brisbane · Singapore

To the Peel County Board of Education

Design: Maher & Murtagh
Technical illustration:
Acorn Technical Art Inc.
Part title illustrations: Clive Dobson

Canadian Cataloguing in Publication Data

  Hirsch, Alan J.
    Physics: a practical approach

  For use in secondary schools.
  Includes index.
  ISBN 0-471-79967-X

  1. Physics.   I. Title: Physics: a practical
  approach.

  QC21.2.H57   530   C81-094381-6

The metric usage in this text has been
reviewed by the Metric Screening Office of
the Canadian General Standards Board.
Metric Commission Canada has granted
permission for use of the National Symbol
for Metric Conversion.

**Printed and bound in Canada
at John Deyell Company**

10 9 8 7 6 5 4

Acknowledgements

I wish to thank the Peel County Board of Education in
Ontario for giving me continual support and
encouragement, and specifically for awarding me a
professional development leave of absence during which
I worked on this book.
  I also wish to thank the many people who helped
make this a better book. While only I am responsible for
final content of this book, many suggestions for changes
by others were incorporated. I thank the following
teachers and specialists who read and criticised all or
part of the manuscript: Peter Twist, Hugh Bolton, Ferris
Wainwright, Dave Farwell, Ravi Shrivastava, Art
Geddis, John Kinsman, Gordon McKye, Rick Wahrer,
Bruce Robb and Professor Peter Matthews of the
University of British Columbia. Special thanks go to
Judy Evans who assisted with the photography, Doug
Bannister, with whom I taught for several years, and
Bob Sheppard who helped me make the subject more
understandable to the reader. I must thank John
Murtagh for his excellent design and Nick Owocki for
his careful work on the illustrations for the book.
Thanks are also extended to Rene Webber who provided
some of the art work, and to Helen Ross and Lorna
Wreford who worked on the original typing.
  Lastly, I wish to thank the hundreds of students whom
I have had the privilege of teaching in the past years.
They provided the inspiration and rewards that have
made the writing of this book a pleasure.

A.J.H.

Contents

Introducing Physics

If the subject of physics had never developed, our world would be very different. We would have to live without cars, boats and airplanes. We would not be able to watch colour television or listen to stereo sound systems. We would not have computers, thrill rides at fairs, telephones, watches or pin-ball machines. Nor would we have the pollution and energy problems that go along with technology. Indeed, without physics, nearly everything we have in our modern world would not have been developed.

Physics is the study of **matter** and **energy** and the interaction between matter and energy. Everything from the smallest particle you can imagine to the galaxies of stars that make up our universe is made up of matter. (Refer to the photograph.) Physics helps us measure matter and

The great Andromeda galaxy consists of many millions of stars. It is similar in shape to our own Milky Way galaxy. There are thousands of galaxies in the universe.

describe its motion. It also helps us understand the forces that cause motion in matter. Then we are able to apply those forces in the use of automobiles, tools, kitchen appliances and thousands of other devices.

Wherever there is matter, there is energy. Our senses of hearing and sight would be useless without sound and light energy. Machines develop mechanical energy to help us do much of our work. Our food is cooked by using heat energy. Electrical energy has been put to use almost everywhere in the world and nuclear energy is being applied to more uses each year.

To help you study the ideas and applications of matter and energy, this book has been designed in a straight-forward, practical way. You will study theory and how we apply it. You will gain experience by performing experiments and answering questions. This will give you a better understanding of what goes on in the world around you.

Studying physics can be challenging as well as rewarding. May you enjoy both the challenges and rewards as you study **Physics: A Practical Approach**!

Alan Hirsch

Note: Circled question numbers indicate a greater degree of difficulty.

I. Preparing for Physics

1
Preparing for a Course in Physics

GOALS: After completing this chapter you should be able to:
1. Observe and report carefully and completely when you perform experiments.
2. Arrange in order a given set of ages during which the subject of physics developed.
3. Name some important scientists and state their place in history.
4. List the steps of the scientific method.

1.1 Powers of Observation

Imagine that you are asked to write a description of the cover of this book. You could use your sense of sight to determine the colours, patterns and words on the cover. You could use your sense of touch to find the cover's texture and shape. You would probably not bother using your senses of hearing, taste or smell. By using a ruler and other measuring devices, you would gather more details for your description.

To describe an object or event, begin by exercising your powers of observation. Then, if possible, perform measurements. The more observant you are and the more measuring tools you use, the better your description will be.

Close observations, careful measurement and detailed reporting have helped physics advance to its present state. They will also help you understand the topics studied in physics. The two experiments that follow will help develop your powers of observation and your ability to describe what you observe.

PRACTICE
1 Name the five senses that humans use to observe objects or events.

1.2 Experiment 1: Invisible Forces

INTRODUCTION

This experiment will be easier to perform than to explain. One clue to help in your explanation of what happens is that wood fibres expand when they absorb water.

PURPOSE: To observe and explain a physical reaction that occurs because of forces that cannot be seen.

APPARATUS: 5 wooden toothpicks; eye dropper

PROCEDURE
1. Snap but don't break each toothpick at its middle. Arrange the five toothpicks on a hard, smooth surface (such as a lab bench) in a pattern like the one shown in Figure 1-1.
2. Using the eye dropper, add two or three drops of water to the very middle of the toothpick pattern. Observe and describe the results. (A diagram will help your description.)

Figure 1-1 Toothpick arrangement

1.3 Experiment 2: Corn Kernels

INTRODUCTION

This experiment, like the previous one, is easy to perform. However, detailed observations must be made so that your explanation will be logical.

PURPOSE: To observe and explain the action of kernels of corn that are being influenced by their surroundings.

APPARATUS: 400 mL beaker; 6 or 7 kernels of popping corn; tablet of Alka Seltzer or similar product; water

PROCEDURE
1. Place about 200 mL of water into the beaker and add the kernels of corn.
2. While observing carefully, add the Alka Seltzer tablet to the water. Describe what happens.
3. Continue watching for several minutes; then describe what you observe.

1.4 Foundation and Framework

The study of physics might be compared to building a house. The foundation of the house is built first. The framework is added, and then the details are completed step by step. The details can only be completed by using careful measurement and the close observation you practised in Experiments 1 and 2.

Similarly, in the study of physics the foundation was laid first. For thousands of years, up to around 1500 A.D., people slowly built up knowledge.

Then the framework of physics developed for about 400 a (years), to about 1900 A.D. In this period scientists improved their observations and measurements; thus, the framework was built.

Finally, many details of physics have been added during this present century.

In studying physics it is easier to understand the details if we know how they came about. The remainder of this chapter gives a summary of the foundation and framework of the study of physics.

1.5 The Development of Physics: A Historical Summary

In the Beginning
Can you imagine what life was like for people who lived 30 000 a ago? They had to get up from their beds of straw when the sun rose and leave the shelter of caves or bushes to try to survive another day. They spent much of their time picking berries or chasing animals, using wooden or bone clubs as weapons. Some people were able to make a few tools and weapons out of stone. (That is why we call this period of time the **Old Stone Age**.) There were also a few people who discovered they could keep a fire going after lightning had struck a tree or bush. But there was no science to think about—only survival to an average age of about thirty years.

Civilization Begins
When our ancestors finally realized they could live more easily if they helped one another, civilization began. Besides hunting for

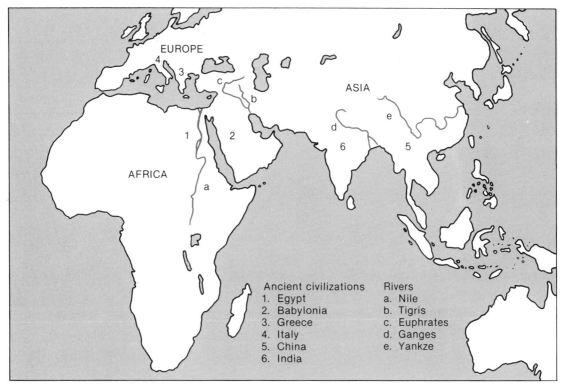

Figure 1-2 Historical regions

food, people started to grow food on land close to rivers and seas. Certain animals were domesticated, and farming began. The stone tools and weapons were sharpened and polished. (We call this period of time the **New Stone Age**.) It took thousands of years for knowledge to grow as information was passed from one generation to the next. Organized knowledge was just beginning.

Civilization Advances

Civilization was much more advanced in some areas of the world than in others. From 5000 B.C. to about 2500 B.C., the most advanced regions were Egypt, Babylonia, India and China. (See Figure 1-2.) People lived along well-known rivers such as the Nile, the Tigris and the Euphrates. Recorded history began about 4000 B.C. with an ancient Egyptian calendar. At that time writing was done on clay tablets. Science grew because information could be kept more permanently and could be used by more people.

Metals Replace Stone

Between 5000 B.C. and 2000 B.C. people, using fire, discovered that metals such as copper, tin and iron could be extracted from

Figure 1-3 "The Charioteer" is a bronze statue fashioned in Delphi, Greece, about 500 B.C. The fact that such a piece of art can last for 2500 a shows the durability of bronze.

ore. They also discovered that a mixture of copper and tin, called bronze, was a useful metal. This discovery was so important that we have named the **Bronze Age** after this versatile alloy. (Refer to Figure 1-3.)

The Greeks

The civilizations that had the greatest direct influence on our Western civilization were those of the Greeks and Romans. When the Greeks first became powerful, they had little scientific knowledge yet they provided a foundation for scientific ideas. That foundation remained unshaken for almost 2000 a. (In fact, some of that foundation exists today. For example, the word physics stems from the Greek word **physis**, which means nature.)

Many Greek architects and philosophers contributed to the advancement of science, especially between 600 B.C. and 200 B.C. To appreciate the development of science, it is worthwhile to mention the ideas of some of the scientific philosophers.

Thales (about 636-546 B.C.), who is sometimes called the first Greek philosopher, thought that water was the origin of all substances. He studied astronomy and was able to predict a solar eclipse. **Pythagoras** (about 582-497 B.C.) founded a religious order that believed in the importance of numbers. The Pythagorean theorem, illustrated in Figure 1-4, originated with him. **Democritus** (about 470-380 B.C.) was the first person to state that matter is made of small particles called atoms.

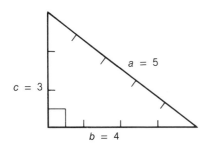

Pythagorus proved that for a right-angled triangle $a^2 = b^2 + c^2$ where a is the hypotenuse, and b and c are the other two sides. In the diagram,

$$5^2 = 4^2 + 3^2$$
$$25 = 16 + 9$$
$$25 = 25$$

Figure 1-4 The Pythagorian Theorem

Aristotle (384-322 B.C.) became the best known and most important scientific philosopher in Greece. He did not conduct experiments, nor did the other Greek philosophers. Their ideas came about by thinking and discussing. Aristotle believed that the earth was the centre of the universe and that matter consisted of four elements—earth, air, fire and water. He thought it logical to assume that heavy objects fell more rapidly toward earth than light objects. These are only a few of Aristotle's many ideas that influenced scientific thought until about 1500 A.D. Some were later shown by experiment to be correct, many others were disproven.

One other scientist, **Archimedes** (about 287-212 B.C.), was a great inventor, mathematician and physicist. Yet, like other people during those times, he tended to accept Aristotle's teachings. (Archimedes' principle is discussed in Chapter 3.)

*Aristotle
(about 384–322 B.C.)*

Thus, scientific knowledge advanced, even though the methods used were not, by our present standards, scientific. Because no experiments were conducted to test ideas, many conclusions these philosophers reached have been proven incorrect. But the foundation of physics was laid.

The Romans

As Greek power diminished, the **Roman Empire** grew. For more than 600 a (from about 200 B.C. to 400 A.D.) the Romans copied and improved ideas from the Greek and other nations. The wide use of the Latin language helped spread scientific knowledge throughout the empire, but the Romans did not create much that was new in science. Although they had powerful armies and interesting architecture, the Romans were held back in science partly because of their number system (Roman numerals). To see how awkward those numbers are, refer to Figure 1–5.

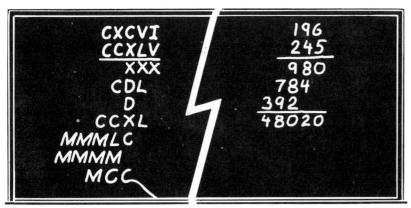

Figure 1–5 Roman numerals are awkward

Islam Fills in the Gaps

During the **Dark Ages**, between the fall of the Roman Empire and 1500 A.D., little inventive work in science occurred in most of Europe, which was greatly influenced by the Roman Catholic Church. Scientists and church leaders accepted Aristotle's claim that the earth was at the centre of the universe.

During this same period of time another religion, Islam, played an important part in unifying scientific knowledge. In the seventh century A.D. the followers of the Islam religion, called Moslems, began to translate information from many nations such as China, India, Egypt and Greece into the Latin language common in Europe. They also improved the system of numbers, called the Arabic or Hindu numbers, which we still use today. Thus, the knowledge of science and the mathematics needed to study it were spreading.

Changing Habits

In the thirteenth century people began to realize that experimentation added greatly to thought and discussion. Centres of learning, called universities, were set up. The printing press was developed in the last half of the fifteenth century, allowing information to be more readily available. These changes set the stage for a new way of solving problems.

New Thinking

In the early 1500s scientists no longer accepted old ideas without question. Even the teachings of the great Aristotle were criticized. Scientists realized the importance of experimenting. (This period in European history is called the **Renaissance**, which is a French word meaning rebirth. The term rebirth is appropriate for the art and architecture that were developed from the classic Greek and Roman styles. Science, however, was not reborn. On the contrary, scientists questioned ancient explanations. Thus, when considering science, the expression "new thinking" applies more aptly than "rebirth".)

One great scientist who influenced the new thinking was an Italian, **Galileo Galilei** (1564-1642). He questioned Aristotle's explanations of motion. In a simple experiment he proved that objects, whether heavy or light, fall to the ground at the same time, assuming no air resistance. You can perform a similar experiment by dropping a 50 g metal mass and a 100 g metal mass from the same height. Galileo constructed the first telescope in 1609. Using the telescope, he provided evidence that the earth revolves around the sun. Later he was forced to deny this fact in order to avoid persecution.

Galileo Galilei
(1564–1642)

The Scientific Method

Although there were many setbacks—people were imprisoned or even executed for their scientific beliefs—scientific knowledge developed rapidly. Experimentation led to what is called the **scientific method**. Using it, the scientist would:

(1) state a **problem**
(2) make a **hypothesis** (a scientific "guess" to answer the problem)
(3) perform a controlled **experiment** to test the hypothesis
(4) come to a logical **conclusion**
(5) offer **predictions** for other situations

This method helped build the framework for the study of physics between 1500 and 1900. It still provides the basis for scientific study today. In fact, we use the scientific method to solve problems in everyday life. If we don't, we often fail to solve the problems. It's a very good method—try it next time something

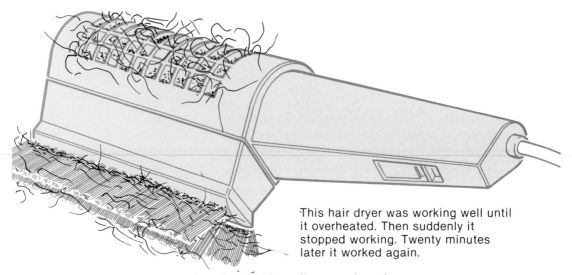

This hair dryer was working well until it overheated. Then suddenly it stopped working. Twenty minutes later it worked again.

Figure 1-6 How would you solve the problem illustrated in the diagram?

stops working, or you spill some ketchup on your sweater. (See Figure 1-6.)

Great Physicists

Thousands of scientists have contributed to the advancements in physics. Besides Galileo Galilei, two other people stand out as especially great contributors. They are Isaac Newton and Albert Einstein.

Sir Isaac Newton (1642-1727) was born in England the year Galileo died. He was a physicist and mathematician who carried on the work of Galileo and other physicists. At the age of twenty-three he was forced, because of the plague, to move away from a crowded city to a small town. He spent his year there studying and experimenting. Among the many subjects he wrote about were motion, force, gravity, light and astronomy. Most of his ideas are still accepted today. In fact, much of the information in this text is based on Newton's physics. His laws of motion are found in Chapter 7.

Albert Einstein (1879-1955) was a German physicist and mathematician who carried Newton's ideas into the realm of high-speed motion. Einstein, who moved to the United States before World War II, is renowned for his theories of relativity. These theories explain events that cannot be explained by using Newton's discoveries. A sensitive man, Einstein very much regretted that his ideas were put to use in the atomic bomb.

Today

During the twentieth century the information provided by physics has grown enormously. There is so much knowledge that scientists who study physics tend to concentrate mainly on only one portion of it. Practical physics helps improve such things as communication systems and the streamlining of cars. Research physics helps us understand matter and energy in everything from the smallest known particles to systems of stars.

1.6 Review Assignment

1. Water drops were used to perform the toothpick experiment. What do you think would happen if you were to use milk? (1.2)
2. In the toothpick experiment the wood absorbs some water. This causes the wood to swell. In ancient times the Egyptians made use of this effect when they tried to make uniform stone blocks from huge rocks. Describe how they may have used wood to help them obtain a stone block about one metre on each side. (Assume they were able to drill holes part way into the rock.) (1.2)
3. In the corn-kernel experiment the bubbles of gas that formed on the kernels were composed of carbon dioxide. This is the same gas that provides bubbles in bottled or canned soft drinks. Predict what would happen if the corn kernels were placed in a glass containing ginger ale. (If possible, verify your prediction.) (1.3)
4. Listed below are eight historical ages. Place them in the order in which they occurred. (1.5)

Greek Civilization	Dark Ages	Old Stone Age
New Stone Age	Bronze Age	Renaissance
Twentieth Century	Roman Empire	

5. (a) Name five famous Greek scientists.
   (b) What was the weakness in the method used by the Greeks to solve science problems? (1.5)
6. In what ways did the Moslems influence the development of science? (1.5)
7. List the steps of the scientific method. (1.5)
8. Name three renowned European scientists. (1.5)

2

Measurement

GOALS: After completing this chapter you should be able to:
1. State why we need measurements that have clear meanings.
2. List some advantages of the metric system.
3. Use the metric units of length, area, volume, mass, density and time.
4. Convert metric measurements involving the prefixes milli, centi and kilo.
5. Determine experimentally length, area, volume, mass and density.
6. Estimate lengths, areas, volumes and densities using metric units.
7. Describe useful applications of density.

Knowing the information in this chapter will be especially useful if you plan a career in:
• surveying
• drafting
• map making
• chemistry

2.1 The Need for Measurement

A quick glance through the pages of this book will show that there are many measured quantities in the subject of physics. These quantities must have an exact meaning. See why this is true by comparing some general statements with measured quantities.

Consider the following general statements:

a **roomful** of people

a **bunch** of bananas

a **pocketful** of pennies

a **sack** of potatoes

For each expression there are several possible meanings. Imagine the confusion if such statements were used as a form of measurement.

Now consider examples of measurements that have clear meanings:

180 cm (length of a pair of skis)

5.0 kg (mass of a bag of sugar)

80 W (power for each channel of a stereo)

2 L/(100 km) (gasoline consumption of a motorcycle)

Each measurement listed should have the same meaning for everyone who uses it.

We need measurements with clear meanings in order to compare and discuss quantities in science, as well as in international trade and our daily activities. For instance, assume there are two motorcycles that are identical in every way but one. The first motorcycle has a fuel consumption of 2.2 L/(100 km) and the second 2.0 L/(100 km). In terms of gasoline consumption, we learn that the second motorcycle is a better buy. It uses less fuel for each 100 km.

PRACTICE
1. If one grocery store offered a bunch of bananas for $1.00 and another store offered a bunch of bananas for $2.00, which would be the better buy? Explain your reasoning.
2. State a measurement that has a clear meaning for each of the following:
    (a) speed limit on a highway
    (b) length of a high-school football field
    (c) temperature used to cook a roast in an oven

2.2 Systems of Measurement

Several centuries ago the only measurements used were length, mass and time. Even for those quantities, no standard system of measurement was used. Length was measured relative to the

Figure 2–1 The fathom depends on the size of the person who defines it.

human body using such units as the foot, span, digit, hand, cubit and fathom. Figure 2-1 shows the problem with such units.

Mass was measured with a simple balance, but there was no standard mass. Sometimes a stone was used, and stones varied from one place to another.

Time was measured using the motion of objects in the sky such as the sun or moon. A sundial, for example, was somewhat useful in clear, sunny weather, but it was not practical on cloudy days or at night.

We can appreciate that these measurements of length, mass and time were a good start; but they did not provide uniform or reliable values.

The British (Imperial) system of measurement, which uses inches, feet, miles, ounces, pounds and so on, is obsolete for science and almost obsolete for world trade. The British system is awkward for conversion purposes because it is not based on the number 10. For example, there are 12 inches per foot, 3 feet per yard, 1760 yards per mile, 4840 square yards per acre, etc. Furthermore, some sizes vary from one country to another. A Canadian gallon is larger than an American gallon.

The metric system is an easier system of measurement to use. It is based on multiples of 10; it does not vary from one place to another; and it is used throughout most of the world.

Internationally, the metric system is called the *Système International d'Unités* or simply the SI. Using the SI, the units for length, mass and time are the metre (m), kilogram (kg) and second (s). Whenever you hear the expression "SI", think of "m", "kg", and "s". Other units in SI will be studied later in the text.

PRACTICE
3. List three advantages of the metric system of measurement.

2.3 Using Metric Prefixes

The fact that the metric system is based on multiples of 10 allows the use of prefixes. The prefixes used most often in this book are kilo, centi, and milli. For example, all we have to do is add prefixes to the word metre to obtain kilometre, centimetre, and millimetre. Table 2-1 lists important facts about the three prefixes. (See the inside front cover of the text for a summary of the metric system.)

Table 2–1: Three Common Metric Prefixes

Prefix	Symbol	Meaning	Conversions using length
kilo	k	1000	1 km = 1000 m
centi	c	1/100	1 m = 100 cm
milli	m	1/1000	1 m = 1000 mm (or 1 cm = 10 mm)

Sample problem 1: Convert 7.6 km to metres.
Solution: 1 km = 1000 m
$$7.6 \text{ km} = 7.6 \text{ km} \times \frac{1000 \text{ m}}{1 \text{ km}}$$
$$= 7600 \text{ m}$$

Sample problem 2: Convert 2.45 m to centimetres.
Solution: 1 m = 100 cm
$$2.45 \text{ m} = 2.45 \text{ m} \times \frac{100 \text{ cm}}{1 \text{ m}}$$
$$= 245 \text{ cm}$$

Sample problem 3: Convert 485 mm to metres.
Solution: 1 m = 1000 mm or 1 mm = 0.001 m
$$485 \text{ mm} = 485 \text{ mm} \times \frac{0.001 \text{ m}}{1 \text{ mm}}$$
$$= 0.485 \text{ m}$$

PRACTICE
4. Convert the following measurements to metres:
   (a) 3.1 km   (d) 64 cm
   (b) 8.65 km   (e) 790 mm
   (c) 12.3 km   (f) 2340 mm
5. Convert to centimetres:
   (a) 12 m   (b) 6.8 m   (c) 25.7 mm

2.4 Experiment 3: Measuring and Estimating Lengths

INTRODUCTION
In this experiment you will practise using a centimetre ruler and a metre stick, two simple yet important instruments. As each measurement is made, your ability to estimate lengths should improve. When you record each measurement, use a reasonable number of digits, usually two or three.

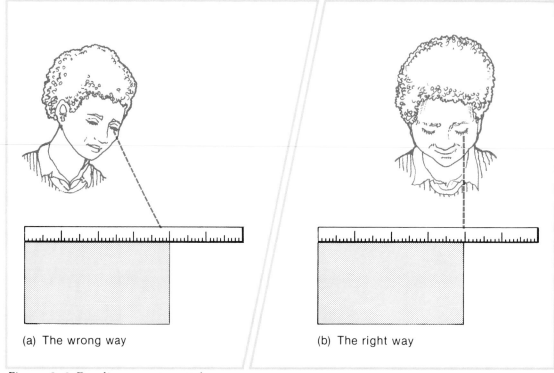

(a) The wrong way

(b) The right way

Figure 2–2 Reading a metre stick

When reading a metre stick or ruler, place your one eye that does the reading directly in line with the object to be measured. This eliminates the effect of **parallax**, which is the apparent shift in the object's position when the observer moves. Refer to Figure 2–2.

PURPOSE: To practise measuring and estimating lengths in metric units.

APPARATUS: metre stick; centimetre ruler; a bicycle (if #5 is performed)

PROCEDURE
1. Measure the length and width of a loose-leaf page in millimetres, centimetres and metres.
2. Measure each of the following lengths in a convenient unit:
   (a) the length and width of your desk top
   (b) your hand span
   (c) your height without shoes

(d) your arm span

(e) the sole of your shoe

3. Estimate the length of your own natural pace. Check your estimate by taking a forward step and having your partner measure the distance from one heel to the other.

4. Without using a metre stick, estimate the length, width and height of the classroom. Describe your method.

5. If time and conditions permit, estimate a large distance outside your school from some point A to another point B. Check your estimate using a bicycle in the following way: With one pedal of the bicycle in its lowest position, make a mark on the road at the place where the front tire touches it. Now pedal the bicycle forward exactly one complete turn of the pedals and again mark the ground by the front tire. Use a metre stick to measure the distance from one mark to the other so the bicycle can be calibrated. Now cycle from A to B, counting the number of turns of the pedals as you go. Finally, calculate the distance from A to B. (For example, if one turn of the pedals = 6.0 m, then 40 turns would yield a distance of 40 × 6.0 m = 240 m.)

QUESTION

1. Which of the lengths measured in the experiment would it be useful for you to memorize?

2.5 Calculating Surface Area

The measurement of length allows us to calculate other quantities such as surface area. The area of a rectangular shape can be determined by the product of length and width. Thus,

$$A = lw$$

Sample problem 4: Calculate the surface area of a rug that measures 3.8 m by 2.6 m.

Solution: $A = lw$

$= 3.8 \text{ m} \times 2.6 \text{ m}$

$= 9.88 \text{ m}^2$

This answer may be rounded off* to 9.9 m² to indicate that the product of measured quantities is not an exact value.

* Rules for rounding off numbers are summarized in Appendix C at the end of the book.

An application of surface area that should be mentioned here is land area. A unit used to measure land area is the hectare (ha), which is equivalent to the area of a square that is one hectometre (1.0 hm or 100 m) on each side. To imagine the size of a hectare, remember that a high-school football field is 100 m long. A field that is as long as a football field and equally wide has a surface area of 1.0 ha.

PRACTICE

6. Calculate the surface area of a rectangle that measures:
   (a) 2.3 cm by 3.6 cm
   (b) 8.7 m by 4.6 m
   (c) 32 mm by 18 mm
   (d) 100 m by 100 m (This is equivalent to 1.0 ha.)

2.6 Experiment 4: Calculating and Estimating Surface Area

INTRODUCTION

As you are making measurements in this experiment, try to estimate areas before you calculate them. Remember that calculated answers are not exact, so you should round off your answers to two or three digits.

PURPOSE: To practise calculating and estimating surface area.

APPARATUS: metric ruler; rectangular objects provided by your teacher

PROCEDURE

1. Measure the dimensions of a loose-leaf page and calculate its surface area in square centimetres.
2. Estimate, then calculate the surface area of one side of each object suggested by your teacher.
3. Estimate the surface area of the irregular shape in Figure 2–3 in square centimetres. Check your estimate, using some logical method. Describe that method.

QUESTION

1. A building contractor wishes to purchase bundles of material to

insulate an attic that measures 12.4 m by 8.5 m. Each bundle covers 10 m². Determine the:

(a) surface area of the attic

(b) number of bundles the contractor needs

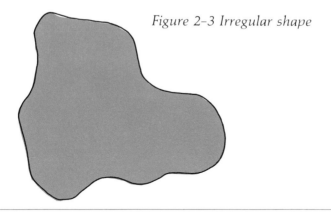

Figure 2–3 Irregular shape

2.7 Calculating Volume

Length is the fundamental measurement from which area and volume can be derived. Length is a one-dimensional measurement having units such as metres (m). Area is a two-dimensional measurement having units such as metres squared (m²). Volume is an extension to three dimensions and, for solid objects, can be stated in metres cubed (m³). The volume of liquids and gases can also be expressed in cubic measures but may be shown in litres (L) and millilitres (mL).

Volume of a regular solid object

The volume of a regular solid object, such as a box, can be found using the product of length and width and depth. Thus,

$$V = lwd$$

> **Sample problem 5:** Calculate the volume of a box that measures 6.1 cm by 4.8 cm by 3.2 cm.
>
> *Solution:* $V = lwd$
> $$= 6.1 \text{ cm} \times 4.8 \text{ cm} \times 3.2 \text{ cm}$$
> $$= 93.696 \text{ cm}^3$$
>
> This answer should be rounded off because the product of measured quantities is not so exact. Rounding it off to two digits, we have 94 cm³ as the volume.

Volume of a liquid

The volume of a liquid can be measured using a graduated cylinder. When liquid is placed in a glass graduated cylinder, a curved shape, called a **meniscus**, may be formed. Readings are most accurate if taken from the bottom of the meniscus, assuming the curve is lower in the middle, as shown in Figure 2–4.

Volume of a small, irregular, solid object

If the volume of a solid irregular object, such as a small stone, is required, a technique called the **displacement of water** can be used. For example, assume a graduated cylinder contains 20 mL of water. A stone is lowered into the water, and the water rises to the 23 mL level. The volume of the stone is 3 mL, which is 3 cm³.

Volume of a large, irregular, solid object

The volume of a large, irregular object can be found by the displacement of water using an overflow can. Refer to Figure 2–5. The can should be filled with water, and the excess water allowed to drip out of the spout. The object, when lowered gently into the water, will force some water out of the can's spout. The water is then caught and measured in a graduated cylinder.

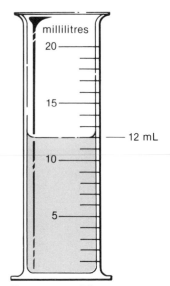

Figure 2–4 The volume of a liquid

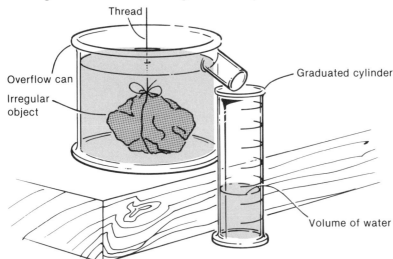

Figure 2–5 Measuring volume by the displacement of water

PRACTICE

7. Calculate the volume of:
   (a) a box that measures 15 cm by 12 cm by 8.0 cm
   (b) a room that measures 8.2 m by 6.1 m by 2.0 m
   (c) a freezer that measures 1.8 m by 0.8 m by 0.9 m

2.8 Experiment 5: Measuring and Estimating Volume

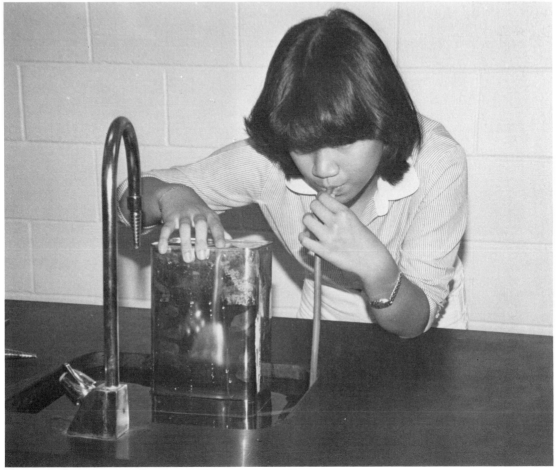

Figure 2-6 Estimating lung capacity

INTRODUCTION

To express volume in appropriate units, the following facts are needed in this experiment:

$$1.0 \text{ mL} = 1.0 \text{ cm}^3$$
$$1000 \text{ mL} = 1.0 \text{ L}$$

PURPOSE: To practise measuring and estimating the volume of solids, liquids and gases.

APPARATUS: metric ruler; regular solid block; graduated cylinder; overflow can; irregular solids (e.g., metal cylinder, large rubber stopper, nuts, bolts); mystery objects and containers; rubber tubing and ditto-fluid can

PROCEDURE

1. Measure the dimensions of a regular solid and calculate its volume in cubic centimetres. Remember to round off your answer.
2. Use the displacement of water in a graduated cylinder to measure the volumes of certain small, irregular objects.
3. Estimate the volume of a large rubber stopper (or similar object). Use the displacement of water in an overflow can to measure the volume of the stopper.
4. Estimate the volumes of some mystery objects and containers set up by your teacher.
5. Estimate to the closest half-litre the volume of air you can exhale from your lungs. Check your estimate by blowing into an inverted water-filled ditto-fluid can that is submerged in water. (See Figure 2-6.) Place your hand tightly over the mouth of the can. Invert the can and measure the volume of water required to refill it.

2.9 Mass and Density

From your past experience, you should have a general idea of the meanings of mass and density. **Mass** is a measure of the quantity of matter contained in some object. In the SI, mass is stated in kilograms (kg). In some situations it may be stated in other metric units such as milligrams (mg), grams (g) and metric tons (t).

Density is a little more difficult to define. You know that cork floats on water, but iron sinks in water. This indicates that cork has a lower density than water, and iron has a higher density.

These general thoughts on density can be made scientific by using a definition. **Density** is the quantity of substance per unit volume. The "quantity of substance" is simply the mass. The "per unit volume" part of the definition means that we must compare the masses of equal volumes of substances. For example, 1.0 m³ of iron has a mass of about 8000 kg and 1.0 m³ of water has a mass of 1000 kg. So for the same volume (1.0 m³), iron has eight times as much mass as water. Thus, the density of iron is eight times greater than the density of water.

Since density is defined as the quantity of matter (mass) per unit volume, an equation for density is written in terms of mass and volume. The equation is:

$$D = \frac{m}{V}$$

Sample problem 6: A 2.5 m³ sample of ore has a mass of 10 000 kg. Calculate the density of the ore.

Solution: $D = \frac{m}{V}$

$$= \frac{10\ 000\ \text{kg}}{2.5\ \text{m}^3}$$

$$= 4000\ \frac{\text{kg}}{\text{m}^3}$$

Sample problem 7: Assume that 2.0 L of gasoline has a mass of 1400 g. Calculate the density of the gasoline.

Solution: $D = \frac{m}{V}$

$$= \frac{1400\ \text{g}}{2.0\ \text{L}}$$

$$= 700\ \frac{\text{g}}{\text{L}}\ (\text{grams per litre})$$

The preferred unit for density in the SI is kilograms per metre cubed (kg/m³). In sample problem 7, the density of 700 g/L can be expressed as 700 kg/m³ because

$$\frac{1.0\ \text{g}}{1.0\ \text{L}} = \frac{1000\ \text{g}}{1000\ \text{L}} = \frac{1.0\ \text{kg}}{1.0\ \text{m}^3}$$

In other words, 1000 g = 1.0 kg and 1000 L = 1.0 m³. Thus, density expressed in grams per litre is equivalent to density expressed in kilograms per metre cubed.

PRACTICE
  8. Calculate the density of a 4.0 m³ wooden log that has a mass of 2000 kg.
  9. Find the density of a liquid if 1.5 L of it has a mass of 2100 g.
  10. Change the answer in #9 to the preferred SI unit.

2.10 Experiment 6: Density

INTRODUCTION

Mass must be measured in this experiment. If you don't remember how to use a triple-beam balance, ask your teacher to review the correct procedure.

The volumes in this experiment are fairly small, so they may be measured in millilitres. To have the density calculations expressed in grams per litre, the millilitres must be changed to litres. For example, assume that a 50 g sample of a substance has a volume of 20 mL. First,

$$1 \text{ L} = 1000 \text{ mL}$$
$$1 \text{ mL} = 0.001 \text{ L}$$
$$\therefore \ 20 \text{ mL} = 20 \text{ mL} \times \frac{0.001 \text{ L}}{1 \text{ mL}}$$
$$= 0.02 \text{ L}$$

Then $D = \dfrac{m}{V} = \dfrac{50 \text{ g}}{0.02 \text{ L}}$

$$= 2500 \frac{\text{g}}{\text{L}} \text{ or } 2500 \frac{\text{kg}}{\text{m}^3}$$

PURPOSE: To determine the densities of various substances.

APPARATUS: graduated cylinder; overflow can; triple-beam balance; samples of various substances (e.g., iron, copper, aluminum, rubber, cork)

PROCEDURE

1. Using the graduated cylinder and the triple-beam balance, determine the density of water in grams per litre. Describe your method.
2. Determine the mass and volume of a sample of some metal. Then calculate its density in grams per litre.
3. Estimate the densities of other substances available. Check your estimates experimentally. (If you use cork, be sure to measure the mass before the volume. Also, you may need a pin to help you submerge the cork in water.)

QUESTIONS

1. Compare your calculated densities with the accepted densities given in the next section (Table 2–2).

2. Assume you are holding a soft, dry sponge in your hand. Without adding any other substance, how could you increase the sponge's density?

2.11 Applications of Density

One important use of density is to determine the purity of a substance. For example, the density of pure lead is 11 340 kg/m³. If we find a sample of a substance that has a density of 10 900 kg/m³, we know the sample is not pure lead.

Every substance has a known density. Table 2–2 lists the densities of several common substances.

Table 2–2 Densities of Common Substances

Substance	Density (kg/m³)	Substance	Density (kg/m³)
Hydrogen	0.1	Aluminum	2 700
Air	1.3	Iron	7 860
Cork	240	Copper	8 950
Ice	820	Mercury	13 600
Water	1000	Gold	19 300

Notice in Table 2–2 that the density of air is much lower than the density of water or any metals. A submarine takes advantage of this fact to sink or float. Whether a submarine floats or sinks is determined by its density relative to the water. A submarine (like the one illustrated in Figure 2–7) has a storage tank (A) for water and air, and another tank (B) for compressed air.

If tank A has enough air, the submarine will float. Water can be allowed into A to cause the submarine's mass to increase until it is

Figure 2–7 A submarine

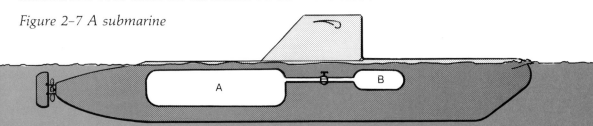

more dense than water. Then the submarine will sink. When the submarine is below the surface of the water, air from tank B can be forced into A to eliminate some water. This causes the submarine to lose mass, become less dense than water and rise again.

PRACTICE

11. Compare the action of a submarine to the action of the corn kernels in Experiment 2.

2.12 Review Assignment

1. Why do we need measurements with clear meanings? (2.1)
2. State the meanings of the metric prefixes kilo, centi and milli. (2.3)
3. Write these lengths in metres:
   (a) A girl is 170 cm tall.
   (b) A boy's hand span is 220 mm.
   (c) An Olympic track is 0.4 km long.
   (d) A mountain top is 4.52 km above sea level. (2.3)
4. What is the surface area of a window that measures 2.2 m by 1.6 m? (2.5, 2.6)
5. Assume that 1.0 L of paint can cover 10 m² of surface area. How much paint would be required to cover a wall that is 8.1 m long and 2.4 m high? (2.5, 2.6)
6. A certain soccer field measures 100 m by 25 m. How many such fields could fit into 1.0 ha? (2.5)
7. The inside dimensions of a refrigerator are 0.5 m by 0.5 m by 1.5 m. What is the capacity (volume) of the refrigerator? (2.7)
8. What is your own mass in kilograms? (This quantity is worth knowing, especially when you have to fill in a passport application.)
9. A steel beam has a mass of 3200 kg. Its dimensions are 10 m by 0.2 m by 0.2 m. Calculate the beam's:
   (a) volume
   (b) density   (2.9)
10. The mass of a chunk of metal is 48 g. The chunk is placed in a graduated cylinder as shown in the diagram. Calculate the metal's:
    (a) volume
    (b) density (State the final answer in the preferred SI units.) (2.9, 2.10)

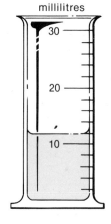

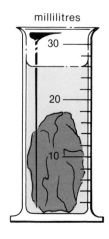

11. A student performed an experiment to find the density of glycerin (a liquid). The measurements are shown below. Use them to calculate the density of glycerin in grams per litre.

mass of graduated cylinder	12.4 g
volume of glycerin added	82 mL
mass of cylinder and glycerin	117.4 g (2.9, 2.10)

12. Rewrite the equation $D = \dfrac{m}{V}$ to express:
    (a) m by itself
    (b) V by itself   (2.9)

13. Calculate the unknown quantities: (2.9)

	Mass (kg)	Volume (m³)	Density (kg/m³)
(a)	22 000	4.0	?
(b)	?	0.2	2500
(c)	200	?	500

14. A piece of wood, having a density of 800 kg/m³, has a volume of 0.5 m³. What is the mass of the wood?   (2.9)

15. The density of a certain grade of petroleum is 750 g/L. What volume would be occupied by 150 g of the liquid?   (2.9)

16. Do research necessary to determine the **cost per litre** of each household product listed below. Give details of the products chosen.
    (a) toothpaste   (d) honey
    (b) milk   (e) soft drink
    (c) shampoo

17. Repeat #16, finding the **cost per kilogram** of these products:
    (a) sugar   (d) coffee
    (b) salt   (e) tea
    (c) breakfast cereal

2.13 Answers to Selected Problems

PRACTICE QUESTIONS
4. (a) 3100 m   (d) 0.64 m
   (b) 8650 m   (e) 0.79 m
   (c) 12 300 m   (f) 2.34 m

5. (a) 1200 cm
   (b) 680 cm
   (c) 2.57 cm
6. (a) 8.3 cm^2
   (b) 40 m^2
   (c) 580 mm^2
   (d) 10 000 m^2
   (Answers are rounded off.)

7. (a) 1440 cm^3 or 1400 cm^3
   (b) 100 m^3
   (c) 1.296 m^3 or 1.3 m^3
8. 500 kg/m^3
9. 1400 g/L
10. 1400 kg/m^3

REVIEW ASSIGNMENT

3. (a) 1.7 m
   (b) 0.22 m
   (c) 400 m
   (d) 4520 m
4. 3.52 m^2 or 3.5 m^2
5. 2.0 L of paint
6. 4
7. 0.375 m^3 or 0.38 m^3
9. (a) 0.4 m^3
   (b) 8000 kg/m^3
10. (a) 14 cm^3
    (b) 3400 kg/m^3

11. 1300 g/L (rounded off)
12. (a) $m = DV$ (b) $V = \dfrac{m}{D}$
13. (a) 5500 kg/m^3
    (b) 500 kg
    (c) 0.4 m^3
14. 400 kg
15. 0.2 L
16. One example: 100 mL of toothpaste cost
    $0.87 ∴ 1000 mL or 1.0 L cost $8.70
17. One example: 5.0 kg of sugar cost $3.50
    ∴ 1.0 kg cost $0.70 or 70¢

II. Fluids

3

Fluids at Rest

GOALS: After completing this chapter you should be able to:
1. Define the words fluid, force and pressure.
2. Use the metric units of force and pressure.
3. Write and apply the equation for pressure in terms of force and area $(p = \frac{F}{A})$.
4. Explain the cause and evidence of atmospheric pressure.
5. State the standard value of atmospheric pressure.
6. Describe how atmospheric pressure is applied in our everyday lives.
7. Describe methods whereby instruments measure pressure.
8. State the difference between gauge pressure and absolute pressure.
9. State what the pressure beneath the surface of a liquid depends on.
10. Write Pascal's law and describe its application to the hydraulic press.
11. Describe the cause of buoyancy in fluids.
12. State Archimedes' principle for objects that either sink in a liquid or float in a liquid.
13. Describe applications of buoyancy in fluids.
14. Describe the forces of adhesion, cohesion and surface tension acting on particles of liquids.

Knowing the information in this chapter will be especially useful if you plan a career in:
• mechanics (hydraulic systems, air-pressure determination, dam building, tunnel construction)
• medicine (for example, determining blood pressure)
• the wine-making industry
• environmental control (for example, weather forecasting)
• boat building
• the dry-cleaning industry

3.1 Developments in the Study of Fluids

A **fluid** is any substance that can flow. Both liquids and gases are fluids.

Throughout the ages, fluids have been used extensively. Two common fluids, water and air, have always been present and were always readily available. For instance, more than 4500 a ago Egyptians used water in an interesting way to solve the problem of constructing level bases for their pyramids. First they marked off the outside dimensions of the pyramid, a size as large as 250 m on each side. They dug a set of trenches parallel to each other and deep enough to hold water, and then they dug another set of trenches at right angles to the first set. The ground likely resembled a huge checkerboard, as illustrated in Figure 3–1. Water was added to the trenches until they were nearly full. Because water is always level, it was used as a guide to make the ground level. Finally, the water was drained away, and construction of the pyramid could begin.

Figure 3–1 The base of a pyramid with workers.

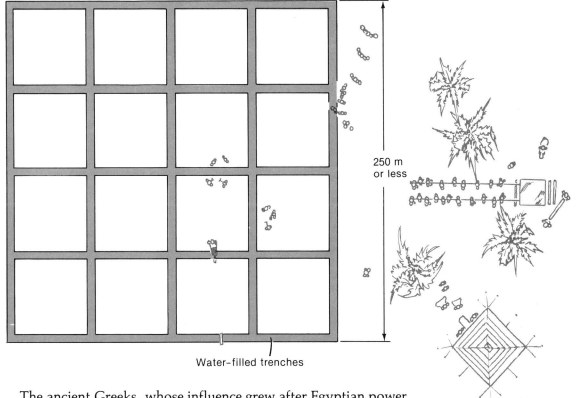

250 m
or less

Water-filled trenches

The ancient Greeks, whose influence grew after Egyptian power decreased, studied the use of fluids. One famous Greek scientist was Archimedes, who lived from about 287 to 212 B.C. A well-known story tells of how Archimedes discovered the answer to a problem put to him by King Hiero. The king wanted to know if his crown was made out of the amount of gold he had paid for, or if some lead or silver had been dishonestly added. Melting the crown would provide the answer, but that was not a sensible solution. The problem bothered Archimedes until one day, in a public bath near his home, he stepped into a tub filled with water. As the water displaced by his body spilled onto the floor, he realized how to solve the problem. Using a piece of pure gold equal in mass to the king's crown, he could find out if the pure gold displaced the same amount of water as the crown. He was so excited by this discovery that he jumped out of the tub, and ran naked through the streets to his home yelling "Eureka!" ("I've found it!") Later he proved that the king had been cheated.

Many other scientists have contributed to our understanding of fluids. Some of these scientists will be mentioned later in this chapter.

PRACTICE

1. Water (H_2O) can be found as ice, water or steam. Which of these can be called a fluid?

Archimedes
(about 287–212 B.C.)

3.2 Force and Pressure

A **force** is a push or a pull. There are many forces in nature. The force of gravity prevents us from falling off the earth as it rotates in space. The force of magnetism allows us to use a compass. We notice an electrical force when we draw a comb through our hair, especially during the dry winter months. A stretched elastic can exert a pulling force, and the force of friction will slow a bicycle down after the cyclist has stopped pedalling.

The SI unit of force is the newton, symbol N. It is named after Sir Isaac Newton, the English scientist who lived from 1642 to 1727.

Before you proceed to the next paragraph, be sure you understand what newtons of force feel like. For example, the force needed to lift an ordinary-sized tumbler filled with water is about 5 N. The force required to lift you off the ground may be somewhere between 400 N and 1000 N, depending on your mass. A direct way to feel 1.0 N is to hold a 100 g mass in your hand. Another way is to obtain a spring scale in newtons and pull on it or lift some objects with it. (Refer to Figure 3-2.)

It is also important to learn the difference between force and pressure. In a general way you can discover the difference by trying the following simple experiment. Place your left forearm on a desk or table. With the palm of your right hand, press down as hard as you can on your left forearm near the elbow. Now try to exert the same push downwards using only the tip of the index finger of your right hand. In both cases you are exerting a force. Why does the second force hurt your left forearm more than the first force? The reason is that the first force is applied over a large

Figure 3-2 The photograph shows that a force of almost 5 N is needed to hold up a mass of 500 g.

surface area (the size of your hand) while the second force is applied over a small surface area (the tip of your finger). We say that the pressure is larger when the same force is applied over a smaller area.

Scientifically defined, **pressure** is the force applied to a given area. In other words, pressure is the ratio of force to area.

$$p = \frac{F}{A}$$

In the SI force is measured in newtons and area in metres squared. Thus, a force of 1.0 N applied over a surface area of 1.0 m^2 gives a pressure of:

$$p = \frac{F}{A}$$
$$= \frac{1.0 \text{ N}}{1.0 \text{ m}^2}$$
$$= 1.0 \frac{\text{N}}{\text{m}^2}$$

The unit newtons per metre squared (N/m^2) is called a pascal, symbol Pa. It is named after a French scientist, Blaise Pascal (1623-1662), who contributed greatly to our knowledge of fluids. (Pascal's law is discussed in Section 3.6.)

A pressure of 1.0 Pa is very small. It is approximately the pressure exerted by a dollar bill resting flat on a table. Scientists prefer to use kilopascals (1.0 kPa = 1000 Pa). The pressure a chair exerts against your seat when you are sitting is approximately 3 or 4 kPa.

Sample problem 1: A large wooden box measures 2.0 m × 3.0 m × 1.0 m high. The force of gravity pulling down on it is 1.2×10^4 N (12 000 N). Calculate the pressure the box exerts on the floor when it is resting upright.
Solution: First we must find the area of the box in contact with the floor.

$$A = lw$$
$$= 2.0 \text{ m} \times 3.0 \text{ m}$$
$$= 6.0 \text{ m}^2$$

Now $p = \dfrac{F}{A}$

$$= \dfrac{12\ 000 \text{ N}}{6.0 \text{ m}^2}$$
$$= 2000 \text{ Pa or } 2.0 \text{ kPa}$$

Sample problem 2: A circus elephant can easily balance on two of its feet. Calculate the pressure exerted by the elephant on a wooden floor when it is standing on two feet. Assume that the force of gravity pulling down on the elephant is 5×10^4 N (50 000 N) and that the area of each foot is 0.01 m².
Solution: The total area is $2 \times 0.01 \text{ m}^2 = 0.02 \text{ m}^2$

Then $p = \dfrac{F}{A}$

$$= \dfrac{50\ 000 \text{ N}}{0.02 \text{ m}^2}$$
$$= 2\ 500\ 000 \text{ Pa}$$
$$= 2\ 500 \text{ kPa (or } 2.5 \times 10^3 \text{ kPa)}$$

Another example can be used to show that pressure depends on area. Imagine if the sharp end of a nail were protruding out of a board that is lying on a floor. You would not wish to stand on that nail in your bare feet because the pressure would be so great on such a small surface that the nail would easily pierce your skin. Now consider lying on a "bed" of nails, like the one shown in

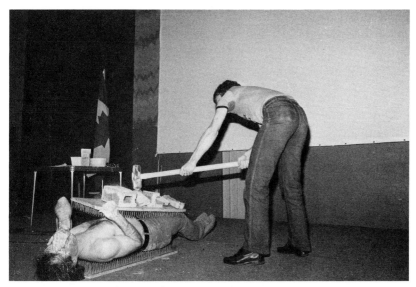

Figure 3–3 While sandwiched between two beds of nails, the man in the photograph chooses not to watch as the concrete block is shattered.

Figure 3–3. If the nails are spaced appropriately, the force will be spread out over the surface area of hundreds of nails, so that it is unlikely that any one nail would break through the skin.

PRACTICE

2. Why are an opponent's gloves an advantage to a boxer being punched in the face? In your answer consider force and pressure.
3. Calculate the pressure in each case:
   (a) $F = 20$ N, $A = 4.0$ m^2
   (b) $F = 400$ N, $A = 0.5$ m^2
   (c) $F = 600$ N, $A = 0.03$ m^2
4. Calculate the pressure applied by the toe of a ballet dancer's shoe when she balances, briefly, on that toe. Assume that the force of gravity on the dancer is 500 N and the surface area of her toe is only 2×10^{-4} m^2 (0.0002 m^2).
5. Compare your answer in #4 above to the answer to sample problem 2 about the circus elephant.
6. Rewrite the equation $p = \dfrac{F}{A}$ to express:
   (a) F by itself
   (b) A by itself

7. Calculate the unknown quantities:

	Pressure (Pa)	Force (N)	Area (m²)
(a)	?	720	4.5
(b)	2.5	?	0.22
(c)	800	640	?

8. Assume that the pressure in a bicycle tire is 400 kPa (or 400 000 Pa), and that the pressure is spread over an area of 0.2 m². What total force acts on the tire?

9. Suppose that ground in a playground can withstand a pressure of 1.1×10^4 Pa (11 000 Pa), and you are asked to design a sandbox that could rest on the ground without sinking into it. The force of gravity on the sand and sandbox is 2.2×10^4 N (22 000 N). What should the surface area of the bottom of the box be? [Hint: Refer to #6(b) above.]

3.3 Atmospheric Pressure

One major reason our earth can support life as we know it is that it has an atmosphere. Our atmosphere (or air) is piled up layer upon layer, each layer pressing down on the one below. The result is a pressure that is called **atmospheric pressure**. It is greatest near the surface of the earth because it must support all the air above that level. The pressure becomes less at higher elevations.

Our ears are sensitive to changes in atmospheric pressure. No doubt you have experienced a "pop" in your ears when your elevation above ground level changed rapidly. This may happen when riding in an elevator in a tall building, in an airplane when it is taking off or in a car on a mountain highway.

Atmospheric pressure, like any other pressure, is measured in pascals or kilopascals. The lowest atmospheric pressure on record is 87.7 kPa on the island of Guam in the South Pacific. The highest pressure on record is 108.38 kPa, measured in Siberia, U.S.S.R. The standard atmospheric pressure, used by scientists for comparison purposes, is 101.3 kPa. This is the average atmospheric pressure at sea level.

If atmospheric pressure conditions are known, forecasting the weather can be fairly accurate. If the pressure is constant, the weather will remain about as it is. Decreasing atmospheric pressure indicates stormier weather, while increasing pressure generally means fair weather.

Besides forecasting weather, there are other everyday uses of atmospheric pressure. Every time we use a straw to drink liquid from a container, we take advantage of atmospheric pressure. Sucking air out of the straw reduces the pressure inside the straw. The pressure on the surface of the liquid is greater than the pressure in the straw, and it forces the liquid up the straw. See Figure 3–4. A syringe and medicine dropper (eye dropper) work on the same principle as the straw.

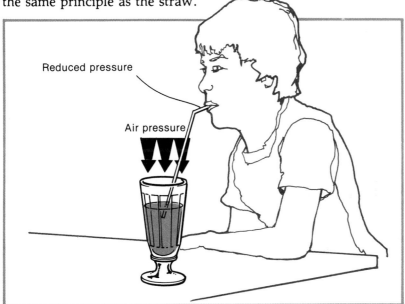

Figure 3–4 The liquid rises in the straw when the atmospheric pressure on the liquid's surface is greater than in the straw.

PRACTICE

10. The elevator in Toronto's CN Tower takes only 60 s to rise to the main observation deck, 342 m above the ground. If a person's ears pop during the ride up the tower, would they pop inwards or outwards? Explain your reasoning.
11. A certain weather report indicates that the atmospheric pressure has gone from 100.8 kPa to 100.1 kPa in 6 h. What prediction could be made about the weather?
12. Suppose that at sea level a student is able to suck water up a certain straw to a height of 210 cm. How would this value compare to the height if the same student tried the experiment at the top of a high mountain? Explain your answer.

3.4 Instruments That Measure Pressure

The type of instrument that measures atmospheric pressure directly is called a **barometer**. The barometer was invented in Italy in 1640 by Evangelista Torricelli (1608-1647). Torricelli, whose teacher was Galileo Galilei, made the first barometer by filling a long glass tube with water so no air could get into the tube. Refer to Figure 3–5. The atmospheric pressure was great enough to hold about 10.3 m of water in the tube. A glass tube over 10 m long was impractical for a science laboratory, so Torricelli tried the same experiment using liquid mercury. He found that the tube needed to be only about 76 cm (0.76 m) long.

Figure 3–5 This diagram shows Torricelli experimenting with a water-filled barometer. He discovered that atmospheric pressure could hold water to a height of over 10 m in the vertical tube.

Torricelli's type of barometer, especially the kind using mercury, is still used today. However, it is not as convenient as the **aneroid barometer**. (The word aneroid means without liquid.) An aneroid barometer consists of an enclosed container having thin metal walls that are sensitive to pressure changes. A needle attached to the container indicates the pressure. (See Figure 3–6.) One use of an aneroid barometer is as an **altimeter**, a device that measures the altitude of an airplane above sea level.

The type of instrument we will use to measure pressure other than atmospheric pressure is called a **manometer**. You will make a simple student manometer when you perform Experiment 7. Such a manometer consists of two pieces of glass tubing (each about 30 cm long) connected by a rubber tube, as shown in Figure 3–7(a). Water is added to a depth of about 15 cm from the bottom of the instrument.

Figure 3–6 The aneroid barometer

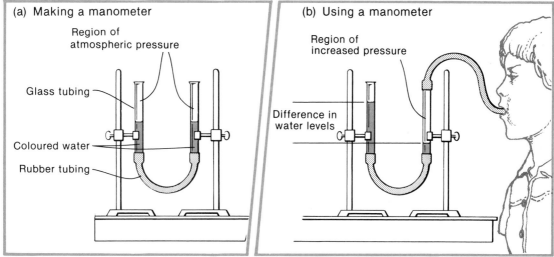

Figure 3–7 A student manometer

In Figure 3–7(b) air is being blown into one side of the manometer causing an increase in pressure there. This causes the water to fall in one side of the tube and rise in the other. The difference in pressure between oné side of the manometer and the other can be calculated using the fact that:

for every centimetre difference between the water levels there is a pressure difference of 100 Pa or 0.1 kPa
(Reference: Appendix F)

The same statement applies if air is sucked out of one side of the manometer.

Sample problem 3: A student blows into one side of a manometer, as illustrated in Figure 3–7(b). The distance between water levels is 12 cm. What is the difference in pressure between the levels?
Solution: For every centimetre the difference in pressure is 0.1 kPa,

so the answer is 12 cm $\times$ 0.1 $\dfrac{\text{kPa}}{\text{cm}}$ = 1.2 kPa.

When an instrument is used to measure pressure other than atmospheric pressure, it is necessary to understand the difference between gauge pressure and absolute pressure. **Gauge pressure** is the reading obtained using some measuring device. For instance, you usually add air to a bicycle tire to a gauge pressure of about

400 kPa. But the **absolute pressure** inside the tire is greater than the gauge pressure because it must include the atmospheric pressure. In fact, the absolute pressure is the sum of the atmospheric and gauge pressures.

absolute pressure = atmospheric pressure + gauge pressure

In the example of the bicycle tire, the absolute pressure would be about 500 kPa (100 kPa + 400 kPa).

PRACTICE

13. A mountain climber takes a barometer reading and discovers an atmospheric pressure of 74 kPa. Later in the same day the climber reads the barometer at 83 kPa. Is the climber ascending or descending the mountain? How can you tell?
14. Calculate the difference in pressure between one side of a manometer and the other if the water levels differ by:
    (a) 10 cm   (b) 18 cm   (c) 34 cm
15. The gauge pressure in a tire is 203 kPa and the atmospheric pressure is 99.8 kPa. What is the absolute pressure in the tire?

3.5 Experiment 7: Pressure in Liquids

INTRODUCTION

In Section 3.4 you learned how a student manometer can be made. In this experiment you will use such a manometer to measure the pressure beneath the surface of liquids.

Remember that every centimetre difference between water levels in a manometer represents a pressure difference of 0.1 kPa.

PURPOSE: To discover how the pressure beneath the surface of a liquid depends on the:
    (a) direction
    (b) depth
    (c) size of the container holding the liquid
    (d) density of the liquid

APPARATUS: fish tank; manometer; thistle tube with rubber diaphragm; glass cylinder (the same height as the fish tank); ruler; water; alcohol; glycerin (See Figure 3–8.)

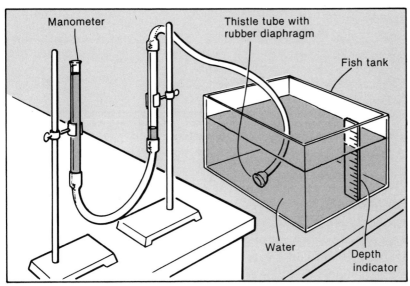

Figure 3–8

PROCEDURE

1. With the apparatus shown in Figure 3–8, hold the thistle tube so the diaphragm, facing sideways, is at a depth of 10 cm. Read the manometer and calculate the pressure. Now aim the diaphragm in other possible directions, including up and down. Be sure the depth remains constant. Record all manometer readings.
2. With the diaphragm on the thistle tube always facing downwards, determine how pressure changes with depth by finding the pressure at depths of 2 cm, 4 cm, . . . 20 cm or so.
3. Place water in the glass cylinder and measure the pressure at a depth of 10 cm. Compare this value to the one found at the same depth using the fish tank.
4. Replace the water in the cylinder with alcohol. Determine the pressure at the same depths used in #2. Compare the values found using alcohol with those found using water. (Remember that alcohol is less dense than water.)
5. Repeat #4 using glycerin, a liquid that is more dense than water.

QUESTION

1. Assuming the pressure under water changes as you discovered in #2, determine the pressure at a depth of:
   (a) 100 cm
   (b) 170 cm (This could be enough pressure to damage the eardrums of a diver.)

3.6 Pascal's Law

An important property of liquid pressure was discovered by the French scientist, Blaise Pascal, mentioned in Section 3.2. His discovery, called **Pascal's law**, states that:

> **pressure applied to an enclosed liquid is transmitted equally to every part of the liquid and to the walls of the container**

Pascal applied his law in the design of the hydraulic press. (The word hydraulic means operating by the force of a liquid.) Figure 3–9 illustrates how such a press works. A small downward force applied to the small movable cylinder can create a large upward force on the large movable cylinder. The way this relates to Pascal's law will become clear when you study sample problems 4 and 5.

Sample problem 4: Assume that for the hydraulic press in Figure 3–9 a force is applied to the small cylinder so that the pressure everywhere in the liquid is 2000 Pa. The surface area of the small cylinder is 0.1 m² and that of the large cylinder is 1.0 m². Calculate the force:
(a) applied on the small cylinder
(b) exerted by the liquid on the large cylinder

Solution: If we take the equation $p = \dfrac{F}{A}$ and rearrange it to find F by itself, we get $F = pA$. According to Pascal's law the pressure (p) remains the same everywhere in the closed container, so we can find the required forces.
(a) $F = pA$
 $= 2000 \text{ Pa} \times 0.1 \text{ m}^2$
 $= 200 \text{ N}$
(b) $F = pA$
 $= 2000 \text{ Pa} \times 1.0 \text{ m}^2$
 $= 2000 \text{ N}$
Notice that if the area is 10 times as large, the force is also 10 times as large.

An important use of the hydraulic press is made in automobile service stations where force is applied to a small cylinder to hoist a large cylinder on which a car is perched.

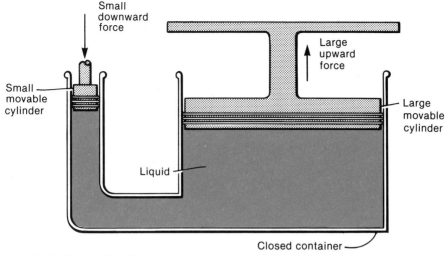

Figure 3–9 The hydraulic press

Sample problem 5: In an automobile service centre the hydraulic hoist can exert a maximum force on the small cylinder of 2000 N. If the surface area of the small cylinder is 0.1 m² and the surface area of the large cylinder is 2.0 m², then what is the maximum force that can be used to lift a car?
Solution: The area is 20 times as large (2.0 m² ÷ 0.1 m²), so the force is 20 times as large or
    20 × 2000 N = 40 000 N (4.0 × 10⁴ N)

Another important application of the hydraulic press is the use of hydraulic brakes on automobiles. Figure 3–10 explains the basic functioning of hydraulic brakes.

Figure 3–10 The hydraulic-brake system

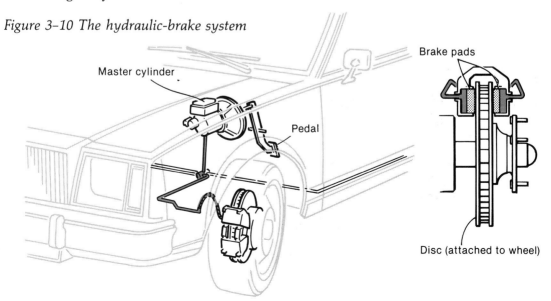

PRACTICE
16. Find the force applied to the cylinder of a hydraulic press if:
    (a) the pressure is 500 Pa and the area is 1.5 m²
    (b) the pressure is 6000 Pa and the area is 3.2 m²
    (c) the pressure is 8000 Pa and the area is 0.6 m²
17. The large cylinder in a hydraulic press has 3 times the surface area of the small cylinder. What force should be applied to the small cylinder to create a lifting force of 7200 N?
18. In a hydraulic-brake system, a force of 25 N can be applied to a surface area of 5.0 cm². What force can thus be exerted on each brake cylinder having an area of 100 cm²?

3.7 Buoyancy in Fluids

The force that pushes upwards on objects in fluids, causing the objects to seem lighter, is called **buoyancy**. Buoyancy helps hold a swimmer up in water and weather balloons up in air. Because buoyancy is a force, it is measured in newtons (N).

To discover why buoyancy exists, recall the results of Experiment 7. In that experiment you learned that the pressure increases as the depth in the liquid increases. Consider a block of aluminum held in water as illustrated in Figure 3–11. Besides the force of gravity on the block, there are forces exerted by the water on the block. A downward force of the water occurs at surface A and an upward force at surface B. Since B is lower than A, the pressure, and therefore the force, is greater at B. The amount by which the upward force at B exceeds the downward force at A is the force of buoyancy.

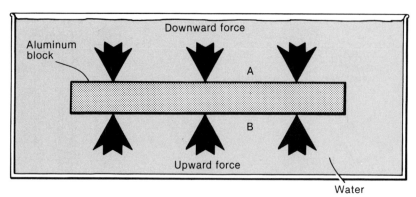

Figure 3–11 Illustrating the force of buoyancy

Sample problem 6: The force of gravity on a man on the ground is 700 N. When the man lies in water, the buoyancy force is 680 N. What force would be required to keep the man from sinking?
Solution: Only 20 N (700 N − 680 N) would be required.

Sample problem 7: The force of gravity on a block of wood in air is 80 N. The block is placed in water where it floats easily. How much is the buoyancy force of the water on the block?
Solution: The buoyancy force must be 80 N, enough to support the block.

PRACTICE

19. Compare the direction of the force of gravity with the force of buoyancy in fluids.
20. A man is trying to remove a rock from the water near the shore of a lake. The buoyancy force on the rock is 200 N and the force required to lift the rock when it is completely submerged is 420 N. How much force will the man have to exert in order to lift the rock once it is in air? Do you think the man should do this task himself?
21. For an object that floats, how does the size of the force of gravity in air compare to the size of the buoyancy force on it when it is floating?

3.8 Experiment 8: Archimedes' Principle

INTRODUCTION

At the beginning of this chapter you learned that Archimedes, who lived in ancient Greece, discovered that he could solve certain problems using his knowledge that an object can displace water. He discovered a principle that applies both to objects that float and objects that sink. **Archimedes' principle**, for water, states that:

the buoyant force on an object in water equals the force of gravity on the water displaced by the object

All forces in this experiment must be measured in newtons. To save time in calculating the force of gravity on water, use the fact that each millilitre of water is pulled toward earth with a force of 0.01 N.

That is, the force on 1.0 mL of water = 0.01 N
the force on 10 mL of water = 0.1 N
the force on 100 mL of water = 1.0 N, etc.
(Reference: Appendix F)

PURPOSE: To prove Archimedes' principle for objects that
(a) sink in water, and
(b) float in water

APPARATUS: force scales (in newtons); overflow can; graduated cylinder; 2 blocks of wood that can fit into the overflow can; 2 hooked metal masses

PROCEDURE

1. Fill the overflow can with water and let the excess water drip away.
2. Use the force scales to measure the force of gravity on one metal object.
3. With the graduated cylinder ready to catch the overflowing water (Figure 3–12), lower the metal object into the water and determine the:
   (a) amount of water displaced
   (b) force of gravity on the submerged object
4. Calculate and compare these forces:
   (a) the buoyancy force of the water on the metal object [This is a subtraction, #2 − #3(b).]
   (b) the force of gravity on the displaced water
5. Repeat #1 to #4 using a second metal object.
6. Measure and record the force of gravity pulling down on one block of wood.
7. Use the overflow can to determine how much water the block displaces when it is gently lowered into the water. (Do not submerge the wood.)
8. Calculate and compare these forces:
   (a) the buoyancy force of the water on the block
   (b) the force of gravity on the water displaced by the block
9. Repeat #6 to #8, using a second block of wood.

QUESTION

1. Would Archimedes' principle apply if a liquid other than water were used? Explain your answer.

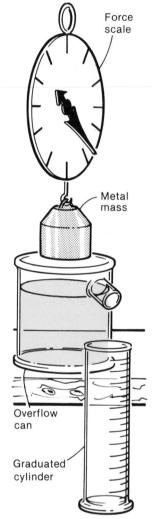

Figure 3–12
Archimedes'
principle for an
object that sinks

3.9 Applications of Buoyancy

Long before Archimedes discovered how the force of buoyancy related to the force of gravity on the displaced water, people were applying the concept of buoyancy in the making of ships. Their wooden ships floated easily despite the fact that they had never heard of Archimedes' ideas. However, such ships were small and weak and would not be able to serve the purposes for which we require ships nowadays.

Today's ships can be made large and strong using metal construction. Since metals are much more dense than water, why does a ship float so easily, even when loaded with cargo? The answer lies in the shape of the ship. The ship must be built so that it contains a large amount of air. The average density of the air, metal and cargo combination is less than the density of water, so the ship floats.

Another common application of buoyancy is in the use of hydrometers. A **hydrometer** is a long, hollow tube weighted at the bottom so that it can float upright in liquids. Each hydrometer has a scale, which indicates the density of the liquid in which it is floating. As you learned in Experiment 7 (Section 3.5), the more dense the liquid the greater is the pressure. So the force of buoyancy becomes greater as the density increases. Thus, a hydrometer will sink lower in alcohol than in a more dense liquid such as water. (See Figure 3–13.)

Figure 3–13 The photograph shows a hydrometer in each of three liquids, glycerin, water and alcohol respectively.

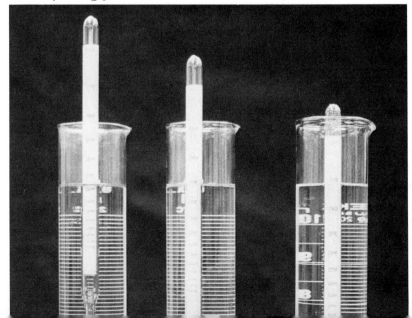

Hydrometers are used to check the densities of battery water and antifreeze in automobiles. They also serve useful functions in the making of many liquid products such as syrup, milk, wine and the by-products of oil.

PRACTICE
22. How could a hydrometer be used to help determine whether or not a sample of water is pure?

3.10 Experiment 9: Forces Acting on Particles of Liquids

INTRODUCTION
You learned in Section 3.2 that a force is a push or a pull. So far in this chapter, force has been studied in connection with either pressure or gravity. Now we will study forces that act on the tiny particles that make up liquids. We cannot see these forces, but we know they exist because of the way liquids act.

In this experiment you will study adhesion, cohesion and surface tension. **Adhesion** is the force of attraction between particles that are not alike. **Cohesion** is the force of attraction between particles that are alike. **Surface tension** is the special name given to the force of cohesion of particles at the surface of a liquid.

PURPOSE: To determine how certain factors affect the forces of adhesion, cohesion and surface tension.

APPARATUS: samples of various materials (e.g., aluminum foil, wax paper, writing paper and paper towel); 250 mL beaker; eye dropper; 2 paper clips; soap solution; 20 cm of copper wire; water; alcohol; pin; liquid soap

PROCEDURE
1. Set the samples of materials (foil and paper) on a table and be sure they are smooth. Use an eye dropper to place a drop of water gently onto each material. Describe each drop and compare the forces of adhesion and cohesion in each case.
2. Add nine more drops to each drop begun in #1 above. Describe the resulting drops and compare adhesion and cohesion. (Diagrams may help.)
3. Repeat #1 and #2, using alcohol instead of water.

4. Shape the copper wire into a loop and tie a thread loosely across the loop, as shown in Figure 3–14(a). Dip the loop into the soap solution in the beaker, then use a pin to break the soap film on one side of the thread. Use a diagram and the concept of cohesion to explain the result.

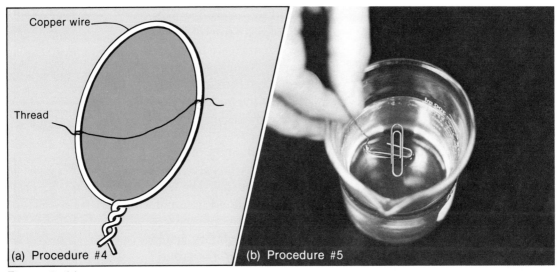

(a) Procedure #4   (b) Procedure #5

Figure 3–14

5. Add water to a **clean** beaker until it is about ¾ full. Use a paper clip made into a handle [Figure 3–14] to lower another paper clip gently onto the surface of the water. Describe the water surface where the clip is resting. Save the setup for the next step.
6. Put a small amount of liquid soap onto the tip of one finger and, as you watch very carefully, touch your finger to the surface of the water (#5 above) as far from the paper clip as possible. Describe what happens.

QUESTIONS

1. Of the products listed below, which would require a large adhesion to water particles? Explain why.
   (a) cloth towel
   (b) floor polish
   (c) facial tissue
   (d) glass
   (e) garden soil
2. Surface tension in water prevents water particles from getting between strands of dirty clothing. Describe how soap could help overcome this problem.

3.11 Applications of Forces on Particles of Liquids

The forces of adhesion, cohesion and surface tension are evident in many instances in the science laboratory and our everyday lives.

A **meniscus**, the curved shape on the top of a column of liquid, occurs in a glass tube or graduated cylinder. Figure 3–15 illustrates two general shapes of curves of a meniscus. Mercury particles have low adhesion for glass but high cohesion for each other. Thus, a mercury meniscus is curved upwards in the middle. Water particles display adhesion to glass, so a water meniscus is curved downwards in the middle.

In Figure 3–15(b) the water appears to be crawling up the walls of the graduated cylinder. This crawling action is even more noticeable if the cylinder or tube is smaller in diameter. The term **capillary action** is used to describe the rising of a liquid up a narrow tube (capillary) due to the adhesion between the particles of the liquid and the particles of the tube. In nature, capillary action is one mechanism responsible for moving water from the ground through the stems of plants or trunks of trees to the leaves.

Surface tension is sometimes a benefit. A water strider (Figure 3–16) is an insect that can walk on water without piercing the surface. The surface tension prevents the strider's thin legs from sinking into the water.

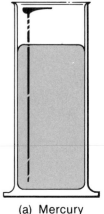

(a) Mercury

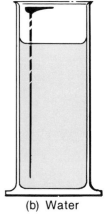

(b) Water

Figure 3–15
The formation
of a meniscus

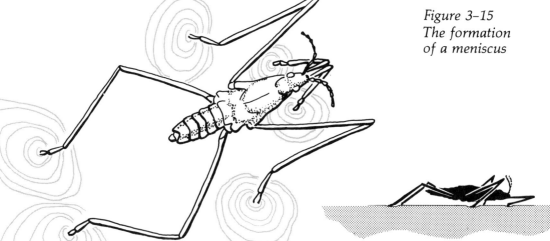

Figure 3–16 A water strider takes advantage of surface tension

Surface tension can also be a problem. When dirty or greasy clothes are being washed in water, the cohesion of the water particles prevents the water from getting between the fibres of the dirty clothes. The surface tension can be reduced by adding soap to the water, as you learned in Experiment 9. Then the water can seep into the places where it is needed.

3.12 Review Assignment

1. Define:
   (a) fluid   (d) pressure
   (b) force   (e) pascal (Pa)
   (c) newton (N)   (3.1, 3.2)

2. Assume that the force of gravity pulling down on a man is 800 N. If the area of the sole of his shoe is 0.02 m², how much pressure does he apply to the floor when he stands on one foot?   (3.2)

3. If the man in #2 above stands on a snowshoe (area = 0.2 m²), how much pressure does he apply to the snow?   (3.2)

4. Use your answers to #2 and #3 above to explain why it is easier to walk on snow using snowshoes or skis than using boots.   (3.2)

5. A pressure of 200 kPa (200 000 Pa) inside a car tire is exerted over a surface area of 1.2 m². Calculate the total force on the inside of the tire.   (3.2)

6. A piece of styrofoam can withstand 30 kPa (30 000 Pa) of pressure without being crushed. A 6000 N box is to be placed on the styrofoam.
   (a) Calculate the area of the bottom of the box needed to prevent crushing the styrofoam.
   (b) If the box is placed on its side (area = 0.1 m²), will the styrofoam be crushed? Explain your answer.   (3.2)

7. Explain what causes atmospheric pressure.   (3.3)

8. Explain how an eye dropper is able to suck a liquid up its tube.   (3.3)

9. A vacuum cleaner is sucking air from a hose connected to one side of a student manometer. The resulting difference in height of the water columns is 72 cm. Find the difference in pressure.   (3.4)

10. The gauge pressure in a truck tire is 263 kPa, and the atmospheric pressure is 102 kPa. What is the absolute pressure in the tire?   (3.4)

11. The human heart exerts pressure on the blood to cause it to circulate through the body. This results in a blood pressure of between 110 kPa and 120 kPa, absolute values. If a doctor measures someone's blood pressure using a gauge, what is the approximate range of gauge pressures the doctor could expect? (Assume an atmospheric pressure of 101 kPa.)   (3.4)

12. State the effect that each of the following has on the pressure beneath the surface of a liquid:
    (a) decreasing the depth beneath the surface
    (b) increasing the density of the liquid
    (c) changing the direction from facing downward to facing upward
    (d) going from a large lake to a swimming pool (at the same depth)   (3.5)

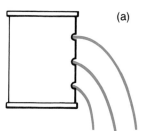
(a)

13. A swimmer may experience a popping sensation in the ears when diving to a depth of more than about 1 m into the water. Explain why this happens.   (3.5)

14. A can with three holes of equal size is filled with water so that the water can run out of all three holes. Which of the diagrams best illustrates the way the water would flow from the holes? Explain your choice. (You may wish to set up an experiment to check your answer.)   (3.5)

15. Assume that every metre increase in depth beneath the surface of water causes a pressure increase of 10 kPa. Calculate the gauge pressure at a depth of (a) 10 m   (b) 25 m.   (3.5)

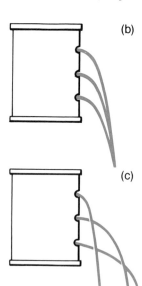
(b)

(c)

16. (a) State Pascal's law.
    (b) Describe how Pascal's law is applied to the operation of a hydraulic press.   (3.6)

17. In a hydraulic press, the pressure on the liquid is 5000 Pa. The small cylinder has an area of 0.5 m² and the large cylinder has an area of 7.0 m². Calculate the force on each cylinder. (3.6)

18. A force of 150 N is applied to a 6 cm² cylinder in a hydraulic press. How much force will be exerted on the 24 cm² larger cylinder?   (3.6)

19. What is the cause of buoyancy in a fluid?   (3.7)

20. The force of gravity on a human brain is about 15 N. The fluid around the brain exerts a buoyancy force of 14.5 N. What is the force exerted by the skull on the brain to keep the brain in its place?   (3.7)

21. The measurements listed below were taken by a student during an experiment involving Archimedes' principle:

    Force of gravity on metal block = 16 N

    Volume of water displaced by the block = 400 mL

    Calculate the:

    (a) buoyancy force acting on the block

    (b) force pulling down on the block when it is submerged in water   (3.8)

22. The force of gravity on a certain object in air is 3.0 N. The object is placed into water in an overflow can, and 300 mL of water overflow.

    (a) Calculate the force of gravity on the displaced water.

    (b) What is the buoyancy force acting on the object?

    (c) Does the object float in water?   (3.8)

23. What is the difference between cohesion and adhesion? (3.10)

24. The diagram shows a soap film in the shape of a rectangle. Draw the approximate shape of the soap film that remains after the film has been broken on the side of the thread indicated.   (3.10)

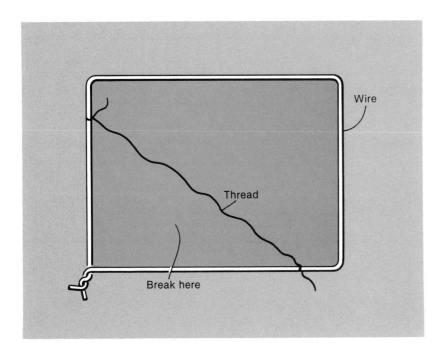

25. On what factors does the action of liquid in a capillary tube depend?   (3.11)

3.13 Answers to Selected Problems

PRACTICE QUESTIONS

3. (a) 5 Pa
   (b) 800 Pa
   (c) 20 000 Pa or 20 kPa
4. 2 500 000 Pa or 2500 kPa
6. (a) $F = pA$

   (b) $A = \dfrac{F}{p}$
7. (a) 160 Pa
   (b) 0.55 N
   (c) 0.8 m²
8. 80 000 N or 8.0 × 10⁴ N

9. 2.0 m²
14. (a) 1.0 kPa
    (b) 1.8 kPa
    (c) 3.4 kPa
15. 302.8 kPa or 303 kPa
16. (a) 750 N
    (b) 19 200 N or 1.9 × 10⁴ N
    (c) 4800 N
17. 2400 N
18. 500 N
20. 620 N

REVIEW ASSIGNMENT

2. 40 000 N or 4.0 × 10⁴ N
3. 4000 N
5. 240 000 N or 2.4 × 10⁵ N
6. (a) 0.2 m²
9. 7.2 kPa
10. 365 kPa
11. 9 kPa to 19 kPa
15. (a) 100 kPa
    (b) 250 kPa

17. 2500 N and 35 000 N
    (or 3.5 × 10⁴ N)
18. 600 N
20. 0.5 N
21. (a) 4.0 N
    (b) 12 N
22. (a) 3.0 N
    (b) 3.0 N

4

Fluids in Motion

GOALS: After completing this chapter you should be able to:
1. State the causes of and corrections for turbulence in moving fluids.
2. Describe applications of streamlining in liquids and gases.
3. State how the speed of a fluid in motion and the pressure are related (Bernoulli's principle).
4. Describe applications of Bernoulli's principle in liquids and gases.

Knowing the information in this chapter will be especially useful if you plan a career in:

• vehicle design (streamlining)
• sports (for example, golf, baseball, sailing, flying)

4.1 Introduction

Both liquids and gases are fluids.

The phrase fluids in motion applies to two situations—a fluid moving through an object (e.g., water through a pipe), and an object moving through a fluid (e.g., a baseball through air).

In this chapter, you will study three concepts involving fluids in motion—turbulence, streamlining and the effects of changes in a fluid's speed. These concepts will be explained with the aid of demonstrations.

4.2 Turbulence and Streamlining

Turbulence is a disturbance that results when fluids cannot move smoothly around or through objects. To observe an example of turbulence, hold a 6 cm by 6 cm piece of cardboard 4 or 5 cm from a lit candle and blow toward the cardboard, as shown in Figure 4–1(a). The candle's flame bends toward you instead of away from you. If we could see air, its motion in this demonstration might appear like that shown in Figure 4–1(b).

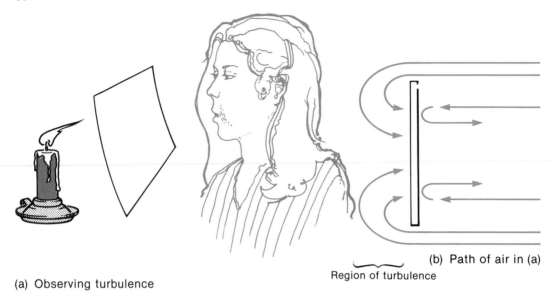

(b) Path of air in (a)

Region of turbulence

(a) Observing turbulence

Figure 4-1 Turbulence

Streamlining is the process of reducing the turbulence of fluids moving around or through objects. It is accomplished by making the surfaces of the objects smooth and curved. To observe a simple example of streamlining, hold a piece of paper, folded into a curve, in front of a lit candle and blow toward the flame. (See Figure 4-2.) This shape reduces turbulence greatly.

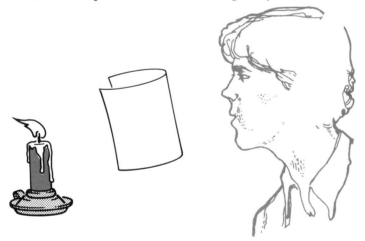

Figure 4-2 Simple streamlining

PRACTICE
1. What shape would provide streamlining that is even better than that shown in Figure 4-2?

4.3 Applications of Streamlining

Streamlining helps overcome turbulence around any object that must move rapidly in a fluid. Nature has provided streamlining to animals, such as many birds and fish, that must move quickly in air or water. The transportation industry continually tries to improve streamlining of cars, trucks, motorcycles, trains, boats, airplanes, spacecraft and other vehicles. Streamlining may improve the appearance of a vehicle, but more importantly, it improves safety and reduces fuel consumption.

Trailer-hauling trucks are good examples of the need for streamlining. Most trailers are box shaped and create much air turbulence. (Recall the turbulence illustrated in Figure 4-1.) The air resistance causes a waste of fuel, and the air patterns can be dangerous to small nearby vehicles. Figure 4-3(a) shows the pattern of turbulence around a truck. Such patterns are studied using models in water tanks or wind tunnels. Experiments using various designs of models are performed to try to improve streamlining. One streamlining design is shown in Figure 4-3(b). Refer also to Figure 4-3(c).

Figure 4-3 Streamlining

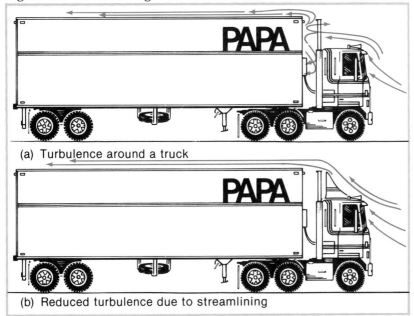

(a) Turbulence around a truck

(b) Reduced turbulence due to streamlining

(c) *The photograph shows a modern wind-generating fan which is over 13 m in diameter. It is used to create wind in an aerodynamic wind tunnel designed to test automotive streamlining.*

PRACTICE

2. Name four animals that are streamlined in shape.

4.4 Bernoulli's Principle

The speed of a moving fluid has an effect on the pressure exerted by the fluid.

Consider water flowing through a pipe that is shaped like the one in Figure 4–4(a). As the water flows from the wide section to the narrow section, its speed increases. This makes sense if you think of a river. A river flows slowly when it is wide but speeds up when it passes through a narrow gorge.

Figure 4–4 Water flowing in a pipe

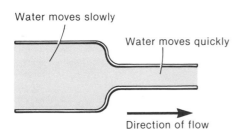

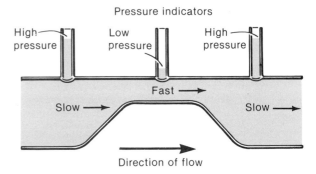

(a) The speed of the water depends on the pipe's diameter.

(b) The pressure of the water depends on the speed.

Now consider Figure 4–4(b). It illustrates an apparatus that shows what happens to the pressure of the water as the speed changes. The pressure is highest where the speed is lowest, and the pressure is lowest where the speed of the water is highest.

This concept was analysed in detail by Daniel Bernoulli (1700-1782), a Swiss scientist. His conclusion, called **Bernoulli's principle**, states that:

> **where the speed of a fluid is low, the pressure is high, and where the speed of a fluid is high, the pressure is low**

Figure 4–5 A spinning ball in the path of moving air

The situation of water flowing in a pipe is a simple example of Bernoulli's principle. Now consider an example in which a ball is spinning as air is blowing by it. Refer to Figure 4–5. As the ball spins, it exerts a dragging force on the air near its surface. This causes the speed of the air above the ball (at A) to be greater than the speed below the ball (at C). Where the speed of the air is greater, the pressure is less (Bernoulli's principle). Thus, the pressure upwards at C is greater than the pressure downwards at A. The result is an upwards pressure on the ball.

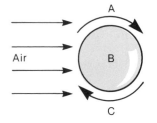

Figure 4-6 A ball in moving air

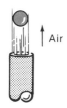

(a) Vertical path of air   (b) Air path at an angle to the vertical

The concept of a spinning ball in moving air can be demonstrated as shown in Figure 4–6. The air is coming from a hose connected to the exhaust of a vacuum cleaner. The ball chosen may be a ping-pong ball, styrofoam ball, or tennis ball, depending on the strength of the air flow. As the air's path is slowly changed to a small angle from the vertical, the ball begins to spin in such a way that an upward pressure prevents it from falling.

PRACTICE

3. As a convertible car with its top up cruises along a highway, the top bulges outward. Explain why.
4. **Demonstration experiment** (which you can perform yourself): Hold the short edge of a 10 cm × 20 cm piece of paper above your mouth and blow air under the paper to try to lift it. [See Figure 4–7(a).] Now place the paper below your mouth [diagram (b)], and blow again. Explain your observations.

Figure 4-7

(a) Blow air under the paper.   (b) Blow air above the paper.

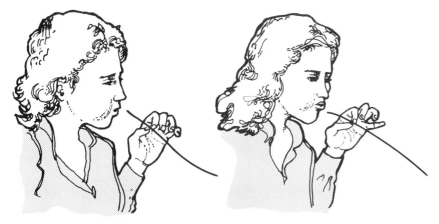

4.5 Applications of Bernoulli's Principle

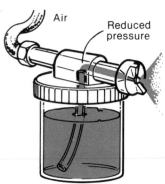

Figure 4–8
A paint sprayer

(1) A **paint sprayer**, shown in Figure 4–8, applies Bernoulli's principle. Air from a pump moves rapidly across the top end of a tube, reducing the pressure in the tube. Atmospheric pressure forces the paint up the tube to be mixed with the flowing air and create a spray.

(2) **Airplane wings**, as illustrated in Figure 4–9, are designed to direct the air a larger distance above the wing than below. This causes the air above the wing to have an increased speed and thus a reduced pressure. The pressure below the wing is greater, exerting an upward force, or lift, on the wing. (Lift is also provided by air particles that bounce off the lower surface of the wings.)

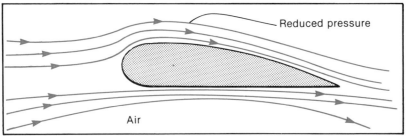

Figure 4–9 An airplane wing

(3) Some modern **skis** used in ski-jumping competitions have flexible rubber extensions at the rear part of the skis. As the skier glides through the air, the air above the skis travels faster than the air below. The result is a higher pressure below the skis than above, creating an upward lift on the skis. See Figure 4–10.

Figure 4–10 Ski jumping

(4) A **carburetor** in a car has a barrel in which air flow controls the amount of gasoline sent to the engine. Figure 4–11 shows air flowing by the gasoline intake. The fast-moving air has a reduced pressure, so the gasoline, under atmospheric pressure, is forced into the carburetor. There it mixes with the air and goes to the engine.

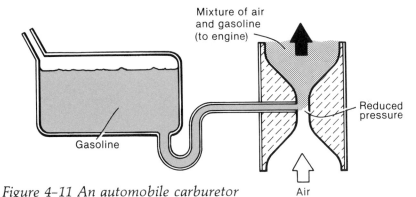

Figure 4–11 An automobile carburetor

(5) **Throwing a curve** in baseball is the last application of Bernoulli's principle discussed here. In Figure 4–12 a ball is thrown to the right which means that, at least for discussion purposes, we can assume that the air is moving to the left. The ball is thrown with a clockwise spin which causes air to be dragged along with the ball. Above the ball the speed of the air is slow, so the pressure is high. This forces the ball to curve downwards, following the path indicated in diagram (c).

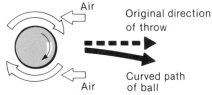

(a) No spin on a thrown ball; motion is in a straight line.

(b) Spinning ball; air is dragged around near the surface of the ball.

(c) Ball thrown with a spin; the pressure above the ball is greater than the pressure below.

Figure 4–12 Throwing a curve

PRACTICE

5. A baseball is thrown as indicated by the broken arrow in the diagram. The ball is spinning counter-clockwise. Determine the approximate direction of the path of the ball. Use diagrams to help explain your answer.

4.6 Review Assignment

1. (a) What causes turbulence around objects that move in fluids?
   (b) How can this turbulence be reduced?   (4.2)
2. In the sport of car racing, one car might be seen following the car ahead very closely. This reduces the effort needed by the car that is following. Explain why this is possible. A sketch of the turbulence may help.   (4.2)
3. For a transportation vehicle of your choice, describe design features that have helped improve streamlining of the vehicle.   (4.3)
4. State Bernoulli's principle.   (4.4)
5. In a windstorm it is possible for the roof of a house to be lifted off while the walls remain undamaged. Use Bernoulli's principle to explain how this is possible.   (4.4, 4.5)

4.7 Answers to Selected Problems

PRACTICE QUESTIONS
1. A pear shape would provide excellent streamlining.
5. The ball will curve downward to the left.

III. Motion, Mechanical Energy and Machines

5

Uniform Motion

GOALS: After completing this chapter you should be able to:
1. Define uniform motion and non-uniform motion.
2. Apply the equation for speed in terms of distance and time ($v = \frac{d}{t}$).
3. Draw line graphs of distance versus time, given a chart of distance-time data.
4. Measure time accurately in experiments.
5. Examine lines on a distance-time graph to determine whether the motion plotted is uniform and whether one motion is faster than another.
6. Calculate the slope of a straight line on a graph, using the equation for slope ($m = \frac{\Delta y}{\Delta x}$).
7. Know that the slope of a line on a distance-time graph represents speed.
8. Plot a speed-time graph, given a distance-time graph.

Knowing the information in this chapter will be especially useful if you plan a career in:

• the auto industry (especially research and design)
• environmental work (for example, studying movement of air patterns)
• aviation, flight, parachuting

5.1 Uniform and Non-Uniform Motion

Uniform motion is movement in a straight line at a constant speed. One example of uniform motion is a cyclist travelling along a straight path at a speed of 10 m/s.

Any movement that is not in a straight line or not at a constant speed is called **non-uniform motion**. An example of such motion is a car travelling around a corner.

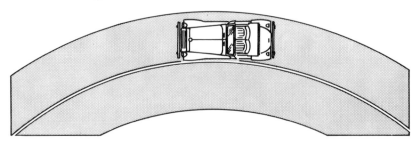

Sample problem 1: Decide whether each motion listed below is uniform or non-uniform:
(a) A steel ball is dropped from your raised hand to the floor.
(b) A car is travelling north at a constant speed of 80 km/h for two hours.
(c) A jogger is running around a circular track at a constant speed.

Solution:
(a) non-uniform motion (The speed of the ball increases as it falls.)
(b) uniform motion
(c) non-uniform motion (The jogger's direction is constantly changing.)

PRACTICE
1. Describe each motion below as either uniform or non-uniform. If you think it is non-uniform, explain why.
   (a) A car is travelling a steady 60 km/h due west.
   (b) A leaf flutters toward the ground.
   (c) A rocket is launched from the earth toward the moon.
   (d) A motorcyle rider applies the brakes in order to come to a stop.
   (e) A ball on the end of a rope is whirled around a person's head at a constant speed of 8.0 m/s.

5.2 Speed

In order to analyse motion, we must be able to measure speed. **Speed** is the rate of change of an object's position. In order to determine an object's speed, two quantities must be known—the change of position or **distance travelled** and the **time** for that change. Using those quantities, the speed can be calculated with the equation:

$$\text{speed} = \frac{\text{distance travelled}}{\text{time}} \quad \text{or} \quad v = \frac{d}{t}$$

In the SI, distance is measured in metres and time in seconds, so speed is stated in metres per second (m/s). It may also be expressed in centimetres per second and kilometres per hour. (Most texts use the symbol v for speed from the word velocity, which is speed with a direction.)

Sample problem 2: Calculate the speed of a runner who takes 500 s to run 2000 m.

Solution: $v = \dfrac{d}{t}$

$= \dfrac{2000 \text{ m}}{500 \text{ s}}$

$= 4.0 \text{ m/s}$

PRACTICE

2. Calculate the speed if:
   (a) $d = 8.0$ m and $t = 4.0$ s
   (b) $d = 25$ m and $t = 0.5$ s
   (c) $d = 6.2$ m and $t = 0.1$ s
3. In 3.0 s sound travels 1000 m in air. What is the speed of sound in air?
4. In the human body blood travels in the largest blood vessel, the aorta, faster than in any other blood vessel. Calculate the blood's speed (in centimetres per second) in the aorta if the blood takes 0.2 s to travel 6.0 cm.

5. Rewrite the equation $v = \dfrac{d}{t}$ to express:
   (a) d by itself
   (b) t by itself.
6. Calculate the unknown quantities:

	Speed (m/s)	Distance (m)	Time (s)
(a)	?	35	0.7
(b)	14.5	?	12
(c)	16	256	?

7. How far would a cyclist travel in 120 s at a constant speed of 12 m/s?
8. At a speed of 950 km/h, how many kilometres would an airplane travel in 12 h?
9. Assume the distance around the earth at the equator is 40 000 km (4.0×10^4 km). Calculate how many hours it would take a supersonic jet to travel once around the world at a speed of 1600 km/h.

5.3 Graphing Uniform Motion

A **graph** is a diagram that represents how one set of variables depends on another set. The type of graph most useful in physics is the **line graph**. It is made by plotting points on paper that is marked off with equal spaces and joining the points with a line.

Often the numbers obtained from experiments or equations are placed in a chart of ordered pairs. The ordered pairs are plotted as points on a graph and the points are joined to complete the line graph. For example, the equation $y = 3x$ has these ordered pairs:

x	0	2	4	6	8	10
y	0	6	12	18	24	30

This set is plotted on the graph in Figure 5-1. The x values are always plotted along the horizontal axis and the y values always along the vertical axis.

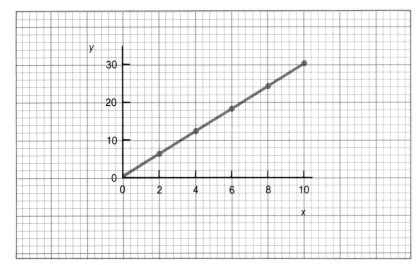

Figure 5-1 The graph of $y = 3x$

In experiments involving motion, the variables usually measured are distance and time. The third variable, speed, is often found by calculation rather than direct measurement.

In uniform motion, the speed is constant, so the distance moved is the same in equal time intervals. For instance, assume that an object moves exactly 3.0 m every second. Then its speed is steady at 3.0 m/s and we can make a chart of ordered pairs of its motion:

$t(s)$	0	1	2	3	4
$d(m)$	0	3	6	9	12

Figure 5–2 shows a graph of this motion. Notice that for uniform motion a graph of distance versus time yields a straight line.

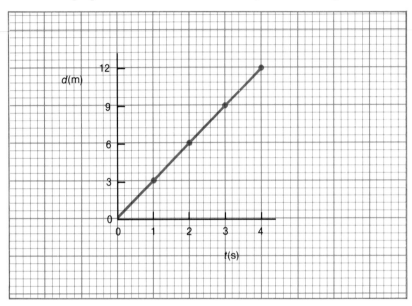

Figure 5–2 A distance-time graph of uniform motion

PRACTICE

10. On one x-y graph, plot the points and draw the line for each set of ordered pairs listed below:

(a)

x	0	2	4	6
y	0	1	2	3

(b)

x	0	1	2	3
y	2	3	4	5

(c)

x	0	2	4	6
y	6	6	6	6

11. On one distance-time graph, draw the line for each set of ordered pairs listed below. Remember to label each axis of the graph.

(a)

$t(s)$	0	10	20	30
$d(m)$	0	2	4	6

(b)

$t(s)$	0	10	20	30
$d(m)$	0	5	10	15

(c)

$t(s)$	0	5	10	15
$d(m)$	0	5	10	15

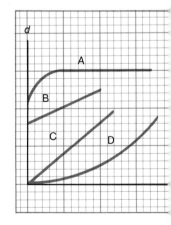

12. Which lines on the distance-time graph shown represent uniform motion? How can you tell?

5.4 Measuring Time

Time is an important quantity in the study of motion. (Remember that speed $= \dfrac{\text{distance}}{\text{time}}$.) You will be required to measure time in various experiments involving motion. A simple **stopwatch** gives acceptable values of time intervals, but if accurate results are required, a better method must be used.

Most physics laboratories have instruments that measure time very accurately for demonstration purposes. A **digital timer** is an electronic device that measures time intervals to a fraction of a second. An **electronic stroboscope** has a light that flashes on and off at time intervals controlled by adjusting a dial. A stroboscope is used to light up a moving object in a dark room as a camera records the object's motion on film. The motion can be analysed knowing the time between flashes of the strobe.

Another device, called a **recording timer** or **bell timer**, is excellent for student experimentation. A recording timer, shown in Figure 5.3, has a metal arm that vibrates at constant time intervals. A needle on the arm strikes carbon paper and records dots on white paper that is pulled through the timer. The recorded dots give an exact picture of the motion. The faster the motion, the greater is the space between the dots.

Figure 5–3
The recording timer

Most recording timers make 60 dots each second. We say they have a frequency of 60 Hz or 60 vibrations/s. (The unit hertz, symbol Hz, is named after a German physicist, Heinrich Hertz, 1857-1894.) Figure 5–4 illustrates why an interval of six spaces on a recording tape represents a time of 0.1 s.

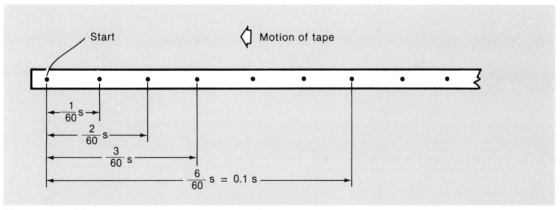

Figure 5–4 Measuring time using a recording timer

Sample problem 3: A student pulls a tape through a 60 Hz recording timer and obtains the tape shown. Plot a graph of distance (from the starting position) versus time.

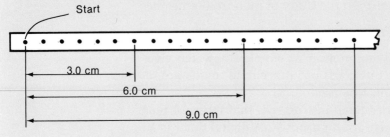

Start

3.0 cm

6.0 cm

9.0 cm

Solution: The ordered pairs are:

t(s)	0	0.1	0.2	0.3
d(cm)	0	3.0	6.0	9.0

The required graph is shown at the right:

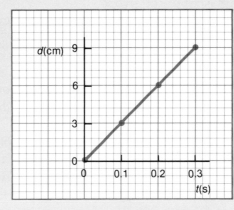

PRACTICE

13. The motion of a tape pulled through a 60 Hz recording timer is shown in the diagram below. Use the results to plot a graph of distance (from the starting position) versus time.

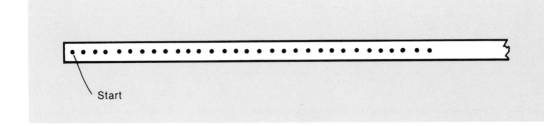

Start

5.5 Experiment 10: Graphing Motion

INTRODUCTION

Motion experiments can be performed using many different techniques. In this experiment three different techniques are described, one in each of Procedures A, B and C. Your choice of procedure will depend on the equipment available in your laboratory for measuring time.

Graphing is an important activity in this experiment. When you draw a graph, use a large portion of the graph paper and place a title at the top of the page. Label each axis, including the units.

For a graph of uniform motion, the ordered pairs should form a straight line. However, due to experimental limitations, this may not happen. Thus, it is acceptable to join ordered pairs with a **line of best fit**, a straight line that, in your judgment, fits the points best.

PURPOSE: To use graphing to determine how well experimenters can exhibit or create uniform motion.

APPARATUS: metre stick; graph paper; and, for Procedure
  A: several stopwatches; bicycle (if desired)
  B: recording timer and related apparatus
  C: stroboscope; instant camera; linear air track and related accessories; pin; overhead projector

PROCEDURE A: STOPWATCH TECHNIQUE

1. Choose an appropriate place to perform the experiment—in the classroom, in the halls or outside on the track. Mark a starting position, then mark positions at equal intervals from the starting position. For example:
    In the classroom use (0), 1, 2, 3, . . . m
    In the hall use (0), 3, 6, 9, . . . m
    On the track use (0), 10, 20, 30, . . . m
2. Locate students with stopwatches at the positions named in #1.
3. Choose a member of the group to be the student trying to create uniform motion. That student will either walk, run or ride a bicycle, depending on the location chosen.
4. Start all the stopwatches at the instant that the student, already moving at a constant speed, crosses the starting position. Stop the watches, one at a time, at the instant the moving student passes each timer.

5. Set up a chart of ordered pairs indicating the distance from the starting position and the time.
6. Plot a distance-time graph of the motion and use it to determine whether the motion was uniform.

PROCEDURE B: THE RECORDING-TIMER TECHNIQUE

1. Obtain a white tape about 1 m long. Pull the tape through the timer at a constant speed.
2. Find a point, A, close to the beginning of the motion where the spaces become fairly uniform. (Refer to Figure 5–5.) From A, count 6 spaces to B. Measure AB, which is the distance travelled during the first time interval.

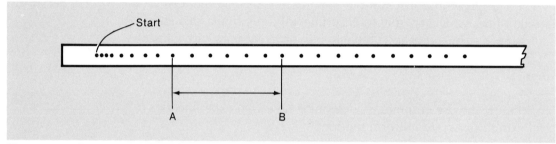

Figure 5–5 *One interval of time*

3. From B count another 6 spaces to C. Measure AC to obtain the distance moved during 2 time intervals. (See Figure 5–6.)

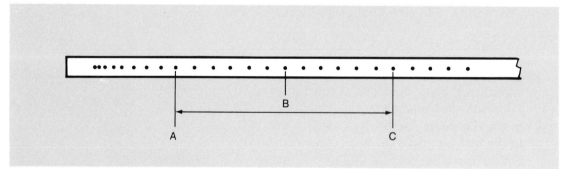

Figure 5–6 *Two intervals of time*

4. Repeat this procedure for a maximum of 10 intervals, always measuring the distance from point A. Record your observations in a chart of ordered pairs, indicating the distance from the starting position and the time.
5. Plot a distance-time graph of the motion and use it to determine whether the motion was uniform.

PROCEDURE C: PHOTOGRAPHIC TECHNIQUE

1. Set up and level the air track and hold a reference metre stick just behind it. Place the camera about 1.5 m from the track and at the same height. Focus the camera on the plastic straw attached to the cart. Position the stroboscope so its light will reflect from the straw to the camera. Adjust the stroboscope scale to have a time interval of 0.2 s between flashes. (This is a frequence of 5 Hz.)

2. With the room lights out, set the cart in motion at a constant speed. Cover the light-exposure meter of the camera, open the shutter and leave it open for the duration of the motion. Develop the photograph.

3. Poke two small holes along the metre stick in the photograph so that a reference distance (e.g., 50 cm) can be projected onto the blackboard. Poke holes in the photograph wherever the straw appears. With the result projected onto the blackboard, measure the distance from the start to each point.

4. Record the observations in a chart of ordered pairs, indicating the distance from the starting position and the time.

5. Plot a distance-time graph of the motion and use it to determine whether the motion was uniform.

5.6 Slopes of Lines on Uniform Motion Graphs

If a graph of distance versus time is drawn for uniform motion at various speeds, a set of straight lines at different slopes will result. In order to calculate the value of the slope of a straight line on a graph, we use the equation from mathematics:

$$\text{slope} = \frac{\text{rise}}{\text{run}} \text{ or } m = \frac{\Delta y}{\Delta x}$$

In this equation, m is the slope of the line, Δy is the change in the value of the y variable, and Δx is the change in the value of the x variable. (Δ is the letter delta from the Greek alphabet.)

On a distance-time graph, distance is the y variable and time is the x variable, so the equation for slope is

$$m = \frac{\Delta d}{\Delta t}$$

Units of measurement must be included when calculating the slope of a line on a graph. This is shown in the example that follows.

Sample problem 4: Calculate the slope of the line on the graph shown.

Solution: $m = \dfrac{\Delta d}{\Delta t}$

$= \dfrac{15 \text{ cm}}{3.0 \text{ s}}$

$= 5.0 \text{ cm/s}$

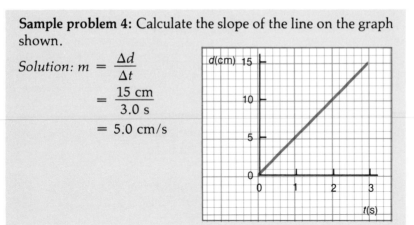

The answer to sample problem 4 has units of centimetres per second, which represent speed. Thus, we can conclude that the slope of a line on a distance-time graph indicates speed.

PRACTICE

14. Calculate the slope of each line on the graphs shown:

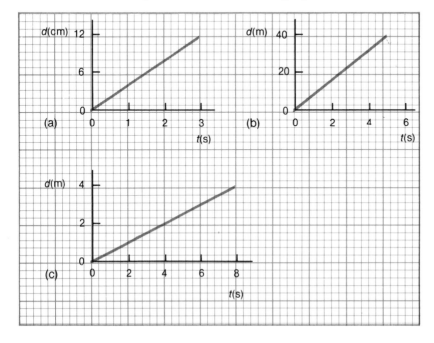

5.7 Graphs of Speed Versus Time for Uniform Motion

Speed can be found by calculating the slope of the line on a distance-time graph. Since the slope of a straight line is constant, this means that the speed is constant. This fact can be used to plot a speed-time graph of the motion. For example, if the slope of a line on a distance-time graph is 5.0 m/s, then the line on the corresponding speed-time graph is horizontal (constant) at a value of 5.0 m/s.

Sample problem 5: Given the distance-time graph shown, determine the speed of the motion and plot the corresponding speed-time graph.

Solution: speed = slope

$$= \frac{\Delta d}{\Delta t}$$

$$= \frac{36 \text{ m}}{4.0 \text{ s}}$$

$$= 9.0 \text{ m/s}$$

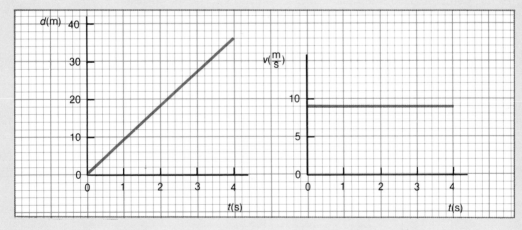

The line on the speed-time graph in sample problem 5 ends at the same time as the line on the distance-time graph. This is because we do not have data after that time.

PRACTICE

15. Plot a speed-time graph that corresponds to each distance-time graph in practice question 14, at the end of the previous section.

5.8 Experiment 11: Uniform Motion at Various Speeds

INTRODUCTION

In this experiment, as in Experiment 10, the technique you use depends on which timing device (stopwatch, recording timer or flashing light) you choose.

PURPOSE: To draw and compare graphs of uniform motion at various speeds.

APPARATUS: as in Experiment 10

PROCEDURE

1. Use a technique suggested by your teacher to obtain observations of motion at three distinct constant speeds—slow, medium and fast.
2. Record all the observations in a chart indicating the distance from the starting position and the time.
3. On one distance-time graph plot lines representing all three motions. Remember to label each axis as well as each line. (If the points plotted do not lie on a straight line, draw the line of best fit so that the slope can be found.)
4. Calculate and compare the slopes of the three lines.
5. On one speed-time graph, plot the speeds of the three motions. Be sure each line ends at the appropriate time. Explain how this graph relates to the experiment.

QUESTION

1. For the graph shown, list the speeds in order of slowest to fastest.

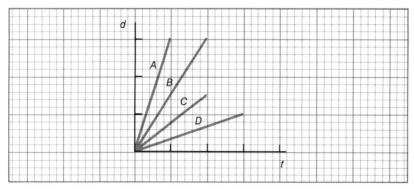

5.9 Review Assignment

1. Define and give an example of uniform motion.　(5.1)
2. Define and give an example of non-uniform motion.　(5.1)
3. What speed is needed by a bionic track star in order to run 100 m in 8.0 s?　(5.2)
4. Light travels from the sun to the earth, a distance of 1.5×10^{11} m, in only 500 s. Calculate the speed of light.　(5.2)
5. The record speed for a motorcycle race is about 82 m/s (239 km/h). If the cyclist took 180 s to make one lap of the track, what is the track's length?　(5.2)
6. The record speed for car racing is about 100 m/s (356 km/h). The record was set on a track that is 4280 m in length. How long did it take the driver to complete one lap?　(5.2)
7. The motion of a tape pulled through a 60 Hz recording timer is shown in the diagram below.

0.2

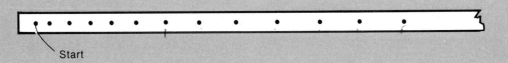

Start

   (a) What is the time interval between the beginning of the tape and the end?
   (b) Is the motion uniform? Explain your answer.　(5.4)
8. What does the slope of a line on a distance-time graph represent?　(5.6)
9. Find the slope of each line on the distance-time graph shown. Then draw the corresponding speed-time graph for the motions.　(5.6, 5.7)

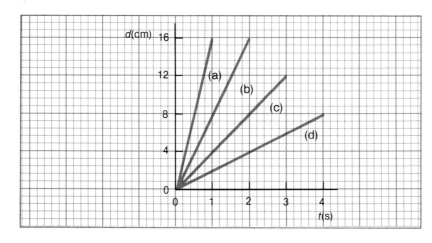

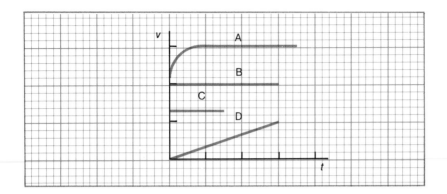

10. Which lines on the speed-time graph shown represent uniform motion? Explain your answer.   (5.7)

11. Explain why the diagram at the right summarizes much of the content of this chapter.

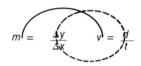

5.10 Answers to Selected Problems

PRACTICE QUESTIONS

2. (a) 2.0 m/s
   (b) 50 m/s
   (c) 62 m/s
3. 333 m/s
4. 30 cm/s
5. (a) $d = vt$
   (b) $t = \dfrac{d}{v}$
6. (a) 50 m/s
   (b) 174 m
   (c) 16 s

7. 1440 m
8. 11 400 km or 1.1×10^4 km
9. 25 h
12. B, C
14. (a) 4.0 cm/s
    (b) 8.0 m/s
    (c) 0.5 m/s

REVIEW ASSIGNMENT

3. 12.5 m/s
4. 3.0×10^8 m/s
5. 14 760 m or 14.76 km
6. 42.8 s
7. (a) 0.2 s

9. (a) 16 cm/s
   (b) 8.0 cm/s
   (c) 4.0 cm/s
   (d) 2.0 cm/s
10. B, C

6

Uniform Acceleration

GOALS: After completing this chapter you should be able to:
1. Define uniform acceleration that is either positive or negative.
2. Calculate the acceleration of an object, given its change in speed and the time during which that change occurred ($a = \dfrac{\Delta v}{t}$).
3. State the SI unit of acceleration and recognize other metric units of acceleration.
4. Determine the slope of a straight line on a speed-time graph and use it to plot an acceleration-time graph.
5. Given a distance-time graph of uniform acceleration, be able to plot a speed-time graph and calculate the acceleration.
6. Determine experimentally the acceleration of an object that, starting from rest, accelerates down a sloping ramp.
7. State the average accleration due to gravity at the surface of the earth.
8. ·Determine experimentally the acceleration due to gravity.
9. State applications of acceleration.

Knowing the information in this chapter will be especially useful if you plan a career in:

• the auto industry (especially research and design)
• environmental work (for example, studying movement of air patterns)
• aviation, flight, parachuting

6.1 Comparing Uniform Motion and Uniform Acceleration

In Chapter 5 you studied uniform motion in which an object travels at a steady speed in a straight line. Figure 6–1 reminds you that a line on a speed-time graph of uniform motion is horizontal.

Most moving objects do not display uniform motion. Any change in an object's speed or direction means that its motion is not uniform. This non-uniform motion is called **acceleration**. An example of acceleration would be a car ride at rush hour in a city when the car has to speed up, slow down and turn corners.

One type of acceleration is called **uniform acceleration**. This is motion of an object that travels in a straight line and has its speed

$t(s)$	0	1	2	3	4
$v(\frac{m}{s})$	15	15	15	15	15

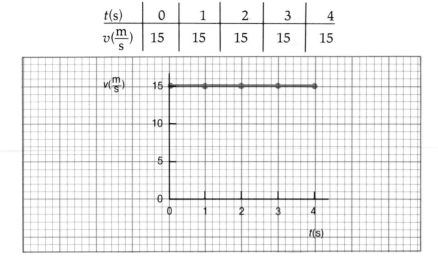

Figure 6-1 Uniform motion

changing steadily with time. Figure 6-2 gives an example of uniform acceleration for an object that starts from rest and increases its speed by 5.0 m/s every second.

$t(s)$	0	1	2	3	4
$v(\frac{m}{s})$	0	5	10	15	20

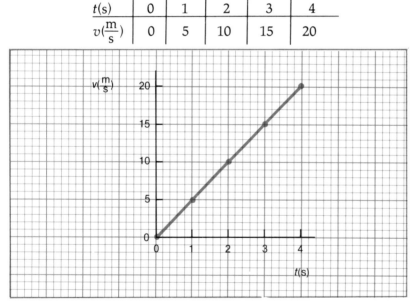

Figure 6-2 Uniform acceleration

Uniform acceleration also occurs when an object, travelling in a straight line, is slowing down steadily. In this case, the object is actually **decelerating**, but mathematically it is said to have **negative**

$t(s)$	0	1	2	3	4
$v(\frac{m}{s})$	20	15	10	5	0

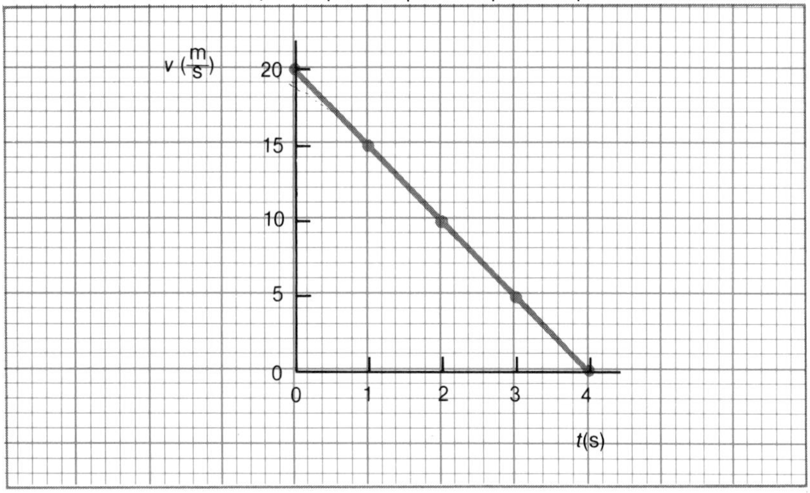

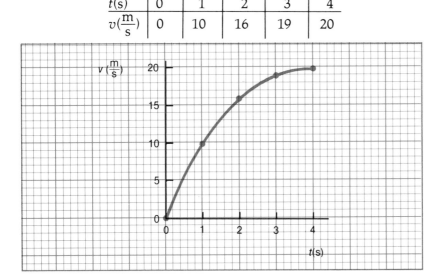

Figure 6–3 Uniform negative acceleration

acceleration. Refer to Figure 6–3, which gives an example of uniform negative acceleration in which a moving object slows down steadily from 20 m/s to 0.0 m/s in 4.0 s.

$t(s)$	0	1	2	3	4
$v(\frac{m}{s})$	0	10	16	19	20

Figure 6–4 Non-uniform acceleration

If an object is changing its speed in an unsteady fashion, its acceleration is non-uniform. Such motion is difficult to analyse, but an example is given in Figure 6–4 for comparison purposes.

PRACTICE

1. The table shows five different sets of speeds at times of 0, 1, 2 and 3 s. Which sets represent uniform positive or negative acceleration?

t (s)	0	1	2	3
(a) v $(\frac{m}{s})$	0	10	20	30
(b) v $(\frac{m}{s})$	0	5	10	10
(c) v $(\frac{m}{s})$	5	5	5	5
(d) v $(\frac{m}{s})$	15	16	17	18
(e) v $(\frac{m}{s})$	15	10	5	0

2. Choose which speed-time graphs represent uniform positive or negative acceleration:

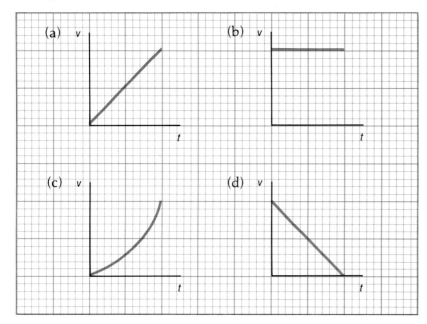

6.2 Calculating Acceleration

For motion in a straight line, **acceleration** is defined mathematically as the rate of change of speed. In order to calculate an object's acceleration, we must know its **change of speed** and the **time** during which that change occurred. In equation form:

$$\text{acceleration} = \frac{\text{change of speed}}{\text{time}} \quad \text{or } a = \frac{\Delta v}{t}$$

The units of acceleration must be units of speed divided by units of time. In the SI, speed is measured in metres per second, so acceleration is stated in metres per second ÷ seconds, or (m/s)/s. SI units will be necessary in Chapters 7 and 8, but for now it is acceptable to use centimetres per second per second, (cm/s)/s, and kilometres per hour per second, (km/h)/s.

Sample problem 1: A motorbike, starting from rest, reaches a speed of 20 m/s in 8.0 s. Find the bike's acceleration.

Solution: $a = \dfrac{\Delta v}{t}$

$$= \frac{20\frac{m}{s}}{8.0 \text{ s}}$$

$$= 2.5 \; \frac{(\frac{m}{s})}{s} \text{ or } 2.5 \text{ (m/s)/s}$$

In sample problem 1, the acceleration of 2.5 (m/s)/s means that the speed of the motorbike increases 2.5 m/s every second. Thus, the bike's speed is 2.5 m/s after 1.0 s, 5.0 m/s after 2.0 s, and so on.

Sample problem 2: An airline flight is behind schedule so the pilot increases the air speed from 135 m/s to 165 m/s in 60 s. What is the aircraft's acceleration?

Solution: $a = \dfrac{\Delta v}{t}$

$$= \frac{30\frac{m}{s}}{60 \text{ s}}$$

$$= 0.5 \; \frac{(\frac{m}{s})}{s} \text{ or } 0.5 \text{ (m/s)/s}$$

If an object is slowing down, its acceleration is negative, as shown in the next example.

> **Sample problem 3**: A cyclist, travelling at a speed of 12 m/s, brakes smoothly and stops in 4.0 s. What is the cyclist's acceleration?
>
> *Solution*: The change in speed (Δv) is negative because the speed decreases from 12 m/s to 0.0 m/s. Thus,
>
> $$a = \frac{\Delta v}{t}$$
>
> $$= \frac{-12\frac{m}{s}}{4.0\ s}$$
>
> $$= -3.0\ (\frac{m}{s})\ \text{or} \ -3.0\ (m/s)/s$$

PRACTICE

3. Calculate the acceleration if:
   (a) Δv = 72 m/s and t = 6.0 s
   (b) Δv = 8.4 m/s and t = 0.5 s
   (c) Δv = -35 m/s and t = 7.0 s
4. The world record for motorcycle acceleration occurred when a cycle took only 6.0 s to go from rest to 78 m/s (281 km/h). Calculate the record acceleration.
5. A car, travelling at 60 km/h, increases its speed to 100 km/h in 10 s. What is its acceleration?
6. Calculate the acceleration needed by a train, travelling at 12 m/s, to stop in 120 s.
7. Rewrite the equation $a = \frac{\Delta v}{t}$ to express:
   (a) Δv by itself
   (b) t by itself
8. Calculate the unknown quantities:

	$a\ [\frac{(\frac{m}{s})}{s}]$	$\Delta v\ (\frac{m}{s})$	t (s)
(a)	?	220	11
(b)	4.2	?	15
(c)	2.1	42	?

9. In the second stage of a rocket launch the rocket's speed increased from 1000 m/s to 10 000 m/s with an average acceleration of 30 (m/s)/s. How long did the acceleration last?

10. A truck driver, travelling at 90 km/h, applies the brakes to prevent hitting a stalled car. In order to prevent a collision, the truck would have to be stopped in 20 s. At an acceleration of –4.0 (km/h)/s, will a collision occur?

11. Assume that when a ball is thrown upwards it accelerates at a rate of –10 (m/s)/s. With what speed must a ball leave a thrower's hand in order to climb for 2.0 s before stopping?

6.3 Using Speed-Time Graphs to Find Acceleration

In Chapter 5 you learned that the slope of a line on a distance-time graph indicates the speed. Let us now use the equation $m = \frac{\Delta y}{\Delta x}$ to see what the slope of a line on a speed-time graph gives.

Consider the graph in Figure 6–5. The slope of the line is

$$m = \frac{\Delta y}{\Delta x}$$

$$= \frac{\Delta v}{\Delta t}$$

$$= \frac{30 \frac{m}{s}}{10 \text{ s}}$$

$$= \frac{3.0 \left(\frac{m}{s}\right)}{s}$$

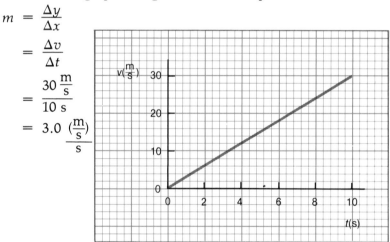

Figure 6–5 Speed-time graph

The units for this slope represent acceleration. Thus, we can conclude that the slope of a line on a speed-time graph equals the acceleration. In equation form:

acceleration = slope on a *v–t* graph or $a = \frac{\Delta v}{\Delta t}$

The equation for acceleration in Section 6.2, $a = \frac{\Delta v}{t}$, is actually an easier way of writing the equation $a = \frac{\Delta v}{\Delta t}$. Both equations can be used to solve acceleration problems.

Sample problem 4: Find the acceleration of the motion shown in the graph.

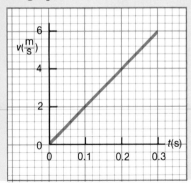

Solution: $a = \dfrac{\Delta v}{\Delta t}$

$= \dfrac{6\,\frac{m}{s}}{0.3\ s}$

$= 20\dfrac{\left(\frac{m}{s}\right)}{s}$

PRACTICE

12. Calculate the acceleration for the motion in each of the three graphs below.

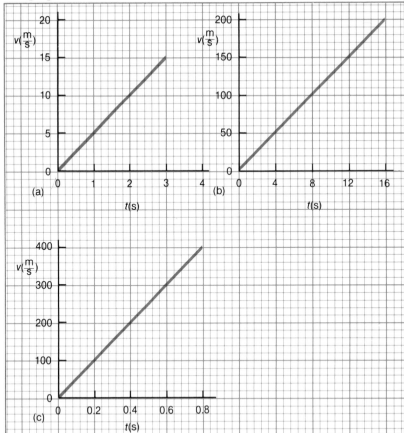

13. For the motions shown in the graph at the right, which line represents the greatest acceleration? Explain why.

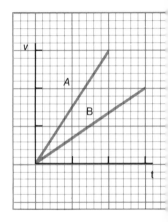

6.4 Investigating Acceleration in the Laboratory

Assume you are asked to calculate the acceleration of a car as it goes from 0.0 km/h to 100 km/h. The car has a speedometer that can be read directly, so the only instrument you need is a watch. The average acceleration can be calculated knowing the time it takes to reach 100 km/h. For instance, if the time taken is 10 s, the average acceleration is

$$a = \frac{\Delta v}{t}$$

$$= \frac{100\,\frac{km}{h}}{10\ s}$$

$$= \frac{10\,(\frac{km}{h})}{s}$$

In a science laboratory, however, an acceleration experiment is not so simple. Objects that move (e.g., a cart, ball or metal mass) do not have speedometers, so the speed cannot be calculated directly.

This problem can be solved by measuring distance travelled and time, rather than change of speed and time. Then a distance-time graph of the motion can be plotted and a mathematical procedure followed to calculate the average acceleration.

An example of how to calculate acceleration from a distance-time graph is described below. The calculations are based on the assumption that the acceleration is uniform, which is the only type of acceleration we wish to study. Similar calculations will be done in the next two experiments.

To begin, consider Figure 6–6, which shows a typical distance-time graph of uniform acceleration for a skier starting from rest and accelerating downhill. Notice that the line is curved, not straight as for uniform motion. This is because the skier travelled 1.0 m in the first 1.0 s, 3.0 m in the next 1.0 s, 5.0 m in the next 1.0 s, and 7.0 m in the last 1.0 s.

Because the line is curved, we cannot find its slope easily, the way we did for uniform motion in Chapter 5. In fact, the slope keeps changing. Thus, we will find the slope of the curved line at one instant. The instant we will choose is 2.0 s, at "half-time" in the motion.

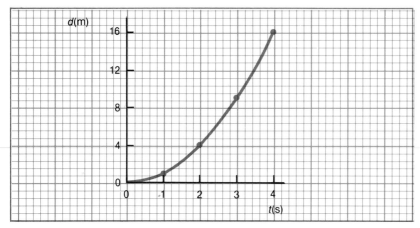

Figure 6–6 Distance-time graph of uniform acceleration

To find the slope of the curved line at the "half-time", we draw a straight line from the final point to the origin of the graph. Refer to Figure 6–7(a). Then we find the slope of that straight line, which has the same slope as the curved line at 2.0 s. Now we know the "half-time" speed.

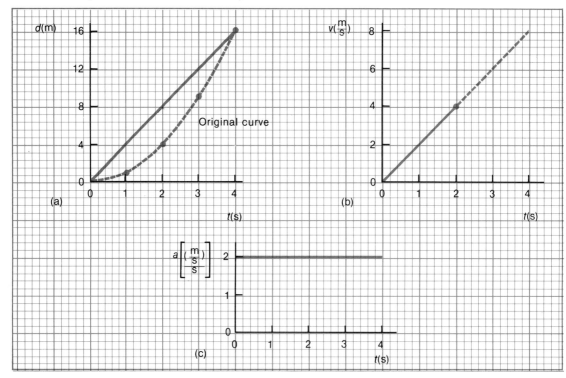

Figure 6–7 Graphing uniform acceleration

Next we plot a speed-time graph of the motion, using the fact that at "half-time" (2.0 s) the skier's speed is 4.0 m/s. In order to complete the graph, as shown in Figure 6–7(b), we assume that the acceleration is constant so we extend the straight line to 4.0 s.

Finally, we calculate the slope of the line on the speed-time graph. This tells us that the average acceleration is 2.0 (m/s)/s, which is constant. The resulting acceleration-time graph is shown in Figure 6–7(c).

PRACTICE
14. In the charts below are three sets of distance-time data for uniform acceleration. In each case,
   (a) plot a distance-time graph
   (b) find the "half-time" speed of each motion
   (c) plot a speed-time graph
   (d) plot an acceleration-time graph

(i)
t (s)	0	2	4	6	8
d (m)	0	8	32	72	128

(ii)
t (s)	0	1	2	3	4
d (m)	0	5	20	45	80

(iii)
t (s)	0	0.1	0.2	0.3	0.4
d (m)	0	1	4	9	16

6.5 Experiment 12: Uniform Acceleration

INTRODUCTION
This experiment, like Experiments 10 and 11, may be performed using stopwatches (Procedure A), a recording timer (Procedure B), or a flashing light (Procedure C). In each case acceleration is achieved by allowing an object, starting from rest, to travel down a ramp elevated at one end.

PURPOSE: To observe and analyse uniform acceleration.

APPARATUS: metre stick; graph paper; and for Procedure:
   A: 5 or 6 stopwatches; steel ball; 2.5 m board; 2.5 m curtain rod

B: recording timer and related apparatus; cart; masking tape; 2.5 m board

C: stroboscope; instant camera; linear air track and related apparatus; pin; overhead projector

PROCEDURE A: STOPWATCH TECHNIQUE

1. Elevate one end of the board about 4.0 cm. Attach the curtain rod securely to the board to act as a track for the steel ball. See Figure 6–8. Mark a starting position near the top of the track. Also mark positions at intervals of 50 cm along the track from the start.

Figure 6–8

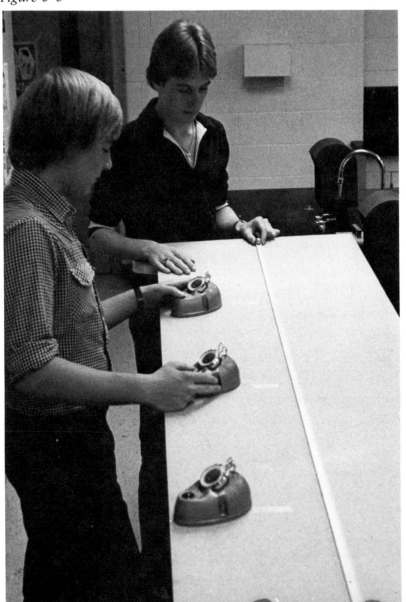

2. Assign each student with a stopwatch to a particular position. Place the ball at the starting position. At a signal allow the ball to begin rolling and start the stopwatches. Measure the time it takes the ball to reach each position. Take the average of several trials to improve results.

3. Record the observed results in a chart of ordered pairs of distance from the starting position and time; for example:

d (cm)	0	50	100	
t (s)	0			

. . .

4. Plot the ordered pairs on a d-t graph. Try to determine from the smoothness of the graph whether or not the acceleration was uniform.

5. Use the d-t graph to find the "half-time" speed of the motion. Use that speed to plot a v-t graph.

6. From the v-t graph calculate the acceleration and plot an a-t graph of the motion.

PROCEDURE B: RECORDING-TIMER TECHNIQUE

1. Elevate one end of the board about 30 cm. Use masking tape to attach a 2.0 m length of white tape to the cart. Feed the tape through the recording timer, as shown in Figure 6–9. Hold the cart absolutely still, turn on the timer, then release the cart.

Figure 6–9

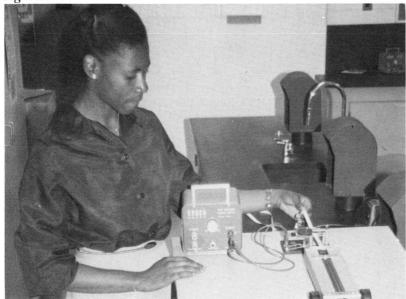

2. Choose as a starting position the first distinct point, A. Count 6 spaces to the next position, B, and measure AB. See Figure 6–10.

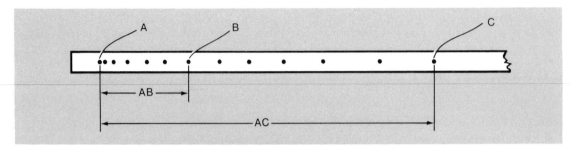

Figure 6–10 Measuring the distance from the starting position

3. From B count another 6 spaces to C and measure AC, again shown in Figure 6–10. Continue this procedure for as many sets of 6 spaces as possible, always measuring distance from the starting position, A. Tabulate the observations in a chart of ordered pairs; for example:

t (s)	0	0.1	0.2
d (cm)	0		

4. Follow steps 4, 5 and 6 in Procedure A.

PROCEDURE C: PHOTOGRAPHIC TECHNIQUE

1. Elevate one end of the linear air track about 10 to 15 cm above the other end. Hold the cart with the straw near the elevated end of the track. Position the camera so the cart is at one edge of the field of view. Hold a reference metre stick along the track. Focus the camera. Adjust the stroboscope to a time interval of 0.2 s, which corresponds to a frequency of 5.0 Hz.

2. With the camera's light-exposure meter covered and the room lights out, take a time-exposure photograph of the cart as it accelerates down the track. Develop the photograph.

3. Poke two small holes along the metre stick in the photograph so that a reference distance (e.g., 50 cm) can be projected onto the blackboard. Poke holes in the photograph wherever the straw appears. With the result projected onto the blackboard, measure the distance from the start to each point. Tabulate the results in a chart of ordered pairs; for example:

t (s)	0	0.2	0.4
d (cm)	0		

4. Follow steps 4, 5 and 6 in Procedure A.

6.6 Acceleration Near the Surface of the Earth

If two solid metal objects of different mass (e.g., 50 g and 100 g) are dropped from the same height above the floor, they should land at the same time. This means that the acceleration of falling objects near the surface of the earth does not depend on mass.

The acceleration of falling objects does depend, however, on air resistance. Try the following demonstration. Fold a loose-leaf piece of paper once. Hold the paper and a textbook horizontally about 50 cm above a table top. Release them at the same instant. Which lands first? Why? Now place the paper on the top of the book, hold the book horizontally and drop it. The book eliminates the effect of air resistance.

It was Galileo (1564-1642) who first proved that, if we do not consider air resistance, the acceleration of falling objects is constant. He proved this by measuring the acceleration of steel balls rolling down a ramp. He found that, for a constant slope of the ramp, the acceleration was constant, no matter what the mass of the ball. The reason he could not measure vertical acceleration was that he had no way of measuring short periods of time accurately. The best clocks available in his day were dripping-water clocks, and sometimes Galileo even used his pulse to measure time. You might appreciate the difficulty of measuring time when you perform Experiment 13.

Had Galileo been able to evaluate the acceleration of objects falling near the surface of the earth, he would have found it to be 9.8 (m/s)/s. This value does not apply to objects influenced by air resistance. It is an average value that changes slightly, depending on factors to be discussed in Chapter 7. It is an acceleration caused by the force of gravity.

The quantity of 9.8 (m/s)/s is so common that from now on we will give it the symbol g, the **acceleration due to gravity**. (Do not confuse this g with the g used as a symbol for gram.)

PRACTICE
15. A steel ball falls freely for 3.0 s at an acceleration of 9.8 (m/s)/s. Calculate the speed of the ball after:
    (a) 1.0 s (b) 2.0 s (c) 3.0 s

6.7 Experiment 13: The Acceleration Due to Gravity

INTRODUCTION
In this experiment, as in the previous three motion experiments, you may choose one or more of Procedures A, B and C. If Procedure A is performed, the object must be dropped at least 5.0 m, but preferably about 10 m. If Procedure B is chosen, the results may be affected slightly by friction between the tape and the recording timer. Procedure C may be done either in the laboratory or by using the illustration of a falling ball shown in Figure 6–12.

PURPOSE: To determine the acceleration due to gravity and to compare the result to the known value of 9.8 (m/s)/s or 980 (cm/s)/s.

APPARATUS: metre stick; and for Procedure:
  A: stopwatches; steel ball
  B: recording timer and related apparatus; metal mass (200 g); masking tape; retort stand and clamps
  C: stroboscope; camera; pin; overhead projector; steel ball

PROCEDURE A: STOPWATCH TECHNIQUE
1. Choose a location where a steel ball can be dropped 5.0 m or more. Measure the distance. Choose one student to drop the ball and announce the exact instant of the drop. Have several students measure the time it takes the ball to fall to the floor or ground. Average the times obtained. (For greater accuracy take the average of three trials.)
2. Calculate the speed of the ball at the "half-time" point of the fall. Use that speed to plot a v-t graph.
3. Use the v-t graph to calculate g.
4. Explain any weaknesses in this experiment.

PROCEDURE B: RECORDING-TIMER TECHNIQUE
1. Set up the apparatus so you can hold the tape in a vertical position, as shown in Figure 6–11. Allow at least 130 cm between the timer and the floor.
2. Attach a 150 cm white tape to the 200 g mass, using masking tape. Position the tape and the mass so that when it is later released, the falling motion of the mass will be recorded on the tape.

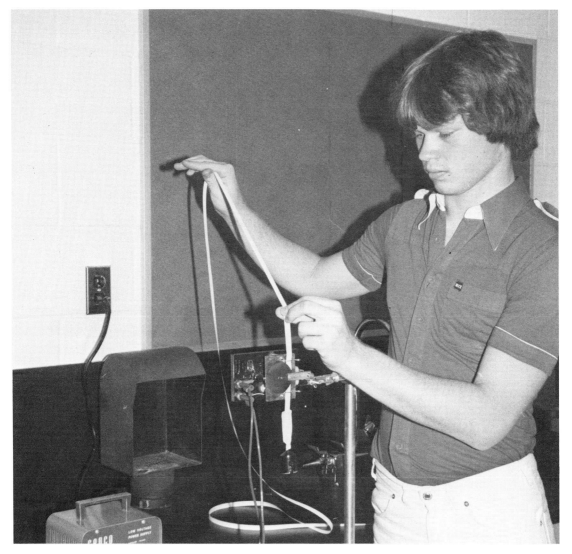

Figure 6–11

3. Place cardboard on the floor where the mass will land. Hold the mass still, turn on the recording timer and release the mass, guiding the tape through the timer.

4. Choose as a starting position the first distinct point. From that position, mark off every 6 spaces as you did in Procedure B of Experiment 12. Measure and record the distances from the starting position.

5. Plot a *d-t* graph of the motion. Find the "half-time" speed and plot a *v-t* graph. Then calculate *g*.

6. Explain any weaknesses in the experiment.

PROCEDURE C: PHOTOGRAPHIC TECHNIQUE

1. Select one student to drop a shiny steel ball alongside a set of two reference metre sticks held vertically. Select another student to aim the stroboscope at the falling ball. Adjust the strobe scale to 0.05 s between flashes (a frequency of 20 Hz). Position the camera to obtain a full view of the 2.0 m drop.

2. With the camera's light-exposure meter covered and the room lights out, take a time-exposure photograph of the falling ball. Develop the photograph and analyse it in the familiar way (as in Procedure C of Experiment 12). Set up a chart of ordered pairs; for example:

t (s)	0	0.05	0.10	0.15	. . .
d (cm)	0				

3. Plot a d-t graph of the motion. Find the "half-time" speed and plot a v-t graph. Then calculate g.

4. Explain any weaknesses in the experiment.

5. If your experiment was unsuccessful, repeat the calculations, using the information for the same experiment shown in Figure 6-12.

6.8 Applications of Acceleration

Galileo Galilei began the mathematical analysis of acceleration, and the topic has been studied by physicists ever since. However, only during the twentieth century has acceleration become a topic that relates closely to our everyday lives.

The study of acceleration is important in the field of transportation. Humans undergo acceleration in automobiles, airplanes, rockets and other vehicles. The positive acceleration in cars and airplanes is small, but in a rocket it can be great enough to cause damage to the human body. Modern experiments have shown that the maximum acceleration a human can withstand is 30 g [294 (m/s)/s]. Astronauts experience up to 10 g [98 (m/s)/s] when a rocket is launched. At that acceleration, if the astronauts were

Figure 6–12 The diagram represents a photograph of a ball falling freely in a dark room. Assume the stroboscope flashed on every 0.05 s.

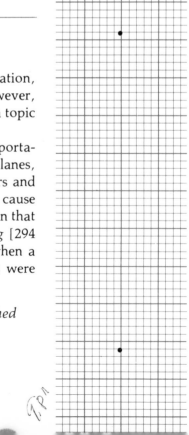

standing, they would faint due to loss of blood to the head. To prevent this problem, the astronauts must be lying down during blast-off.

In our day-to-day lives we are more concerned with the negative acceleration in cars and other vehicles than with the positive acceleration in rockets. Studies are continually being made to determine the effect on the human body when a car has a collision or must stop quickly. Seat belts, headrests and air bags help prevent many injuries caused by a large negative acceleration.

Another application is the study of the effect of acceleration rates on gasoline consumption of cars. It is logical that a driver with a "heavy foot" wastes valuable resources by accelerating at excessive rates. Some cars are now equipped with a light that indicates when the acceleration is greater than a certain energy-saving value.

In the exciting sport of sky diving, the diver jumps from an airplane and accelerates toward the ground, experiencing "free fall". The parachute is not opened until the diver is at a predetermined distance from the ground. During free fall, the diver's speed will increase to a maximum amount called **terminal speed**. Air resistance prevents a higher speed. At terminal speed the acceleration becomes zero; in other words, the speed remains constant. For humans, terminal speed in air is about 53 m/s or 190 km/h.

One other application is the acceleration due to gravity on heavenly bodies other than the earth. If an astronaut standing on the moon dropped a ball, it would accelerate at about one-sixth of that on the earth, or about 1.6 (m/s)/s. Table 6–1 lists the acceleration caused by gravity on the nine planets in our solar system.

Table 6–1

Planet	Acceleration due to gravity [(m/s)/s]
Mercury	3.8
Venus	8.9
Earth	9.8
Mars	3.7
Jupiter	25.8
Saturn	11.1
Uranus	10.5
Neptune	13.8
Pluto	uncertain

PRACTICE

16. How do you think the terminal speed of a stone and a mouse compare to that of a human?
17. Sketch a speed-time graph for a sky diver who accelerates, then reaches terminal speed, then opens the parachute.

6.9 Review of Motion

Two types of motion, uniform motion and uniform acceleration, have been studied in Chapters 5 and 6. Both types deal with motion in a straight line. Uniform motion is motion at a constant speed. Uniform acceleration is motion with a steady change in speed. Refer to Figures 6–13 and 6–14.

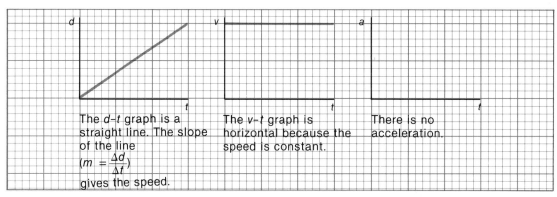

The d–t graph is a straight line. The slope of the line $(m = \frac{\Delta d}{\Delta t})$ gives the speed.

The v–t graph is horizontal because the speed is constant.

There is no acceleration.

Figure 6–13 Summary of uniform motion

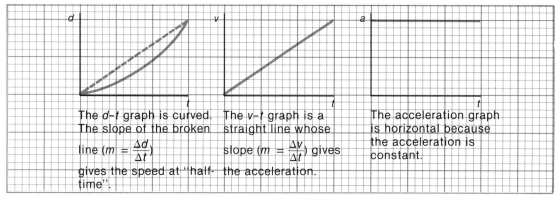

The d–t graph is curved. The slope of the broken line $(m = \frac{\Delta d}{\Delta t})$ gives the speed at "half-time".

The v–t graph is a straight line whose slope $(m = \frac{\Delta v}{\Delta t})$ gives the acceleration.

The acceleration graph is horizontal because the acceleration is constant.

Figure 6–14 Summary of uniform acceleration

The acceleration due to gravity near the surface of the earth is 9.8 (m/s)/s, assuming air resistance is not considered.

6.10 Review Assignment

1. Define uniform acceleration.   (6.1)
2. Sketch a speed-time graph that represents
   (a) uniform acceleration that is positive
   (b) uniform acceleration that is negative   (6.2)
3. In drag racing a car accelerates from rest and must reach a maximum speed in about 400 m. Calculate the world-record accelerations of the cars in each situation described below:
   (a) A piston-engine car reached 112 m/s (403 km/h) in 5.6 s.
   (b) A rocket-powered car reached 170 m/s (608 km/h) in 4.6 s.
   (6.2)
4. Give reasons we should or should not waste gasoline on sports activities such as the races described in #3 above.
5. Find the acceleration of a truck that increases its speed from 22 m/s to 28 m/s in 30 s.   (6.2)
6. What acceleration is needed by a car, travelling at 100 km/h, to come to a stop in 5.0 s?   (6.2)
7. Determine the change of speed of a bullet that accelerates at 400 (m/s)/s for 0.1 s.   (6.2)
8. At an acceleration of 6 (m/s)/s, how long would it take a car to change its speed from rest to 21 m/s?   (6.2)
9. What does the slope of a line on a speed-time graph indicate? (6.3)
10. Calculate the acceleration for each line on the speed-time graph shown.   (6.3)

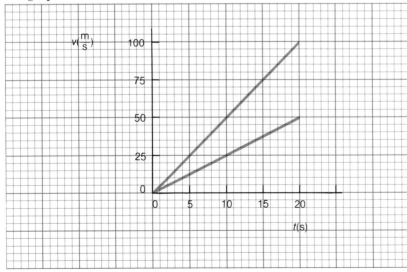

11. The table shows the results of a uniform acceleration experiment.

t (s)	0	2	4	6	8
d (m)	0	16	64	144	256

(a) Plot the results on a d-t graph. Is the curve smooth?
(b) Calculate the "half-time" speed of the motion. $\sqrt{} \frac{16m}{25} = 8.6 \frac{m}{s}$
(c) Draw a v-t graph of the motion.
(d) Calculate the acceleration.   (6.4)

12. A 60 Hz recording timer is used in an experiment to measure the acceleration of a cart travelling down a ramp. Use the results, shown in the table, to calculate the acceleration.

t (s)	0	0.1	0.2	0.3	0.4
d (cm)	0	0.8	3.2	7.2	12.8

     (6.4, 6.5)

13. What is the average value of the acceleration due to gravity at the surface of the earth?   (6.6)

14. Calculate the acceleration experienced by the driver of a rocket-powered dragster rated at 3.5 g.   (6.6)

15. Describe an experiment that could be set up to show that, in the absence of air, a feather and a steel ball fall at the same rate under the influence of gravity.   (6.6)

16. The maximum acceleration a human can survive is 294 (m/s)/s. In a head-on collision with a solid wall, a car stops in about 0.05 s. Determine whether or not a person, wearing a safety belt, can possibly survive such a collision in a car travelling:
(a) 28 m/s (This is about 100 km/h.)
(b) 14 m/s   (6.2, 6.8)

17. The acceleration due to gravity on a planet increases as the mass of the planet increases. Which planet listed in Table 6–1 has the greatest mass?   (6.8)

6.11 Answers to Selected Problems

PRACTICE QUESTIONS

1. (a), (d), (e)   (b) 16.8 (m/s)/s
2. (a), (d)   (c) –5.0 (m/s)/s
3. (a) 12 (m/s)/s   4. 13 (m/s)/s

5. 4.0 (km/h)/s
6. −0.1 (m/s)/s
7. (a) $\Delta v = at$
   (b) $t = \dfrac{\Delta v}{a}$
8. (a) 20 (m/s)/s
   (b) 63 m/s
   (c) 20 s
9. 300 s (or 5.0 min)
10. A collision will occur because it would take 22.5 s to stop.
11. 20 m/s

12. (a) 5.0 (m/s)/s
    (b) 12.5 (m/s)/s
    (c) 500 (m/s)/s
13. A
14. The final accelerations are:
    (i) 4.0 (m/s)/s
    (ii) 10 (m/s)/s
    (iii) 200 (m/s)/s
15. (a) 9.8 m/s
    (b) 19.6 m/s
    (c) 29.4 m/s

REVIEW ASSIGNMENT

3. (a) 20 (m/s)/s
   (b) 37 (m/s)/s
5. 0.2 (m/s)/s
6. −20 (km/h)/s
7. 40 m/s
8. 3.5 s

10. (a) 2.5 (m/s)/s
    (b) 5.0 (m/s)/s
11. (d) 8.0 (m/s)/s
12. 160 (cm/s)/s
14. 34.3 (m/s)/s
16. (a) no
    (b) yes

7

Force and Newton's Laws of Motion

Knowing the infor-
mation in this
chapter will be
especially useful if
you plan a career in:

• the transportation
industry (for exam-
ple, roads, tires, fric-
tion reduction)
• engineering (for ex-
ample, bridge design
and building design
in civil engineering)

7.1 Forces

A **force** is a push or a pull. Forces act on everything from the smallest particles imaginable to the largest objects in the universe. The invisible force of gravity prevents tiny air particles from escaping to outer space and keeps the earth revolving around the sun. Electric forces help objects maintain their shape, and both electric and magnetic forces are needed to operate telephones. Nuclear forces hold invisible particles close to each other, and the force of friction helps cars stay on the road. In fact, forces play a major role in the study of physics because they determine how matter interacts with matter. (Force was mentioned in connection with pressure in Chapter 3.)

7.2 Measuring Force

The title of this chapter referred to one of the greatest scientists in history, Sir Isaac Newton (1642-1727). Newton was born the same year Galileo Galilei died. Many of Newton's ideas resulted from Galileo's discoveries. Since Newton developed important ideas about force, it is fitting that the unit of force is called the **newton** (symbol N).

Sir Isaac Newton (1642–1727)

The newton is an SI unit, so it is necessary to measure distance in metres, mass in kilograms and time in seconds.

If you have not had experience measuring force in newtons, you should arrange to do so now. One way to do this is to hang a 100 g mass on the end of a force scale available in the laboratory. The force needed to hold up the mass is about 1.0 N. You could also try pulling on the force scale. (See Figure 7–1.) Each force scale has a spring that extends when a pulling force is applied. The spring is attached to a needle that indicates the force.

Figure 7–1 Force is measured in newtons using a spring scale.

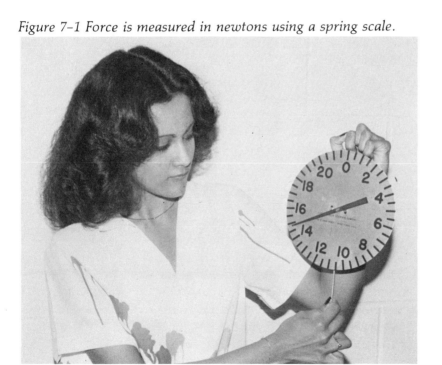

PRACTICE
1. What force would be required to hold up a mass of:
   (a) 200 g?   (b) 400 g?   (c) 1000 g?

7.3 The Force of Friction

Friction is a force that resists motion. It occurs at the surfaces of two objects in contact. No one would put on a pair of ice skates to try to glide along a cement sidewalk! It is the friction between the sidewalk and the blades that would prevent any skating.

One type of friction, called **static friction**, is the force that prevents a motionless object from starting to move. (Static means at rest.) Assume, for example, that the force of static friction between an exercise mat and a gymnasium floor is 100 N. In order to get the mat to start sliding across the floor, a force of at least 100 N would have to be exerted on it. See Figure 7-2.

Figure 7-2 Static friction must be overcome before an object begins moving.

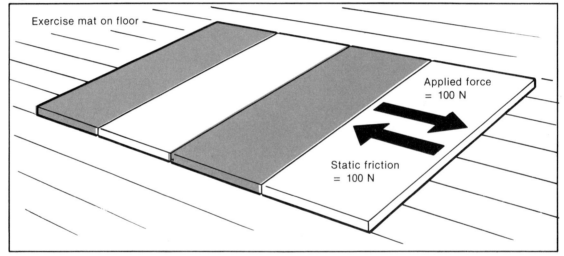

Exercise mat on floor

Applied force = 100 N

Static friction = 100 N

Sometimes static friction is helpful, and sometimes it is not. A carpenter can stand safely on a sloping roof with the help of the friction between his boots and the roof. However, a person trying to move a heavy filing cabinet across a room does not appreciate static friction.

Once the force applied to an object overcomes static friction, the object begins moving. Then **moving friction** replaces static friction. Moving friction is the force that acts against an object's motion in a direction opposite to the direction of motion.

Different types of moving friction have different names, depending on the situation. Sliding friction would affect a toboggan, rolling friction would affect a bicycle and air resistance would affect a sky diver.

PRACTICE
2. State one example of each of the following types of friction.
   (Choose examples different from those given in this section.)
   (a) static friction that is useful
   (b) static friction that is unwanted
   (c) sliding friction
   (d) rolling friction
   (e) air resistance

7.4 Experiment 14: Static and Moving Friction

INTRODUCTION
This experiment will help you gain experience with force measured in newtons. Each measurement should be made several times to obtain a reasonable average. When applying a force (in this experiment), it is important to keep it horizontal.

PURPOSE: To determine what factors affect the force of friction.

APPARATUS: flat wooden block with a hook at one end; 2 masses (0.5 kg and 1.0 kg); force scale (to 10 N); 3 pieces of sandpaper or emery cloth of varying degrees of roughness

PROCEDURE
1. Place the wooden block on the lab bench and the 0.5 kg mass on the block near the end with the hook. (Refer to Figure 7.3.) Attach the force scale to the block and **very gently** pull with a horizontal force while watching the scale. The maximum value you will observe on the scale before the block just starts moving is the force needed to overcome static friction. Repeat the measurement until you are sure of the force, then record it.

Figure 7–3

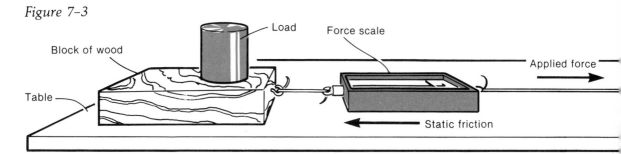

2. Determine the force needed to overcome moving friction (in this case sliding friction) by finding the force required to keep the block, with the 0.5 kg mass on it, moving at a **constant** speed. Again repeat the measurement until you are sure of the value and record it.
3. Repeat #1 and #2, using the 1.0 kg mass. (Be sure to place the 1.0 kg mass in the same position as the 0.5 kg mass.)
4. Repeat #1 and #2, using the three different surfaces of sandpaper or emery cloth and the 0.5 kg mass. Then try the 1.0 kg mass.
5. Set up a chart of observations like the one shown.

Surface	Load (kg)	Static friction (N)	Moving friction (N)
Lab bench	0.5		
	1.0		

QUESTIONS
1. For two surfaces in contact, which is larger, static or moving friction?
2. What factors, besides those found in the experiment, do you think might affect friction?

7.5 Friction and Technology

More than 4500 a ago the Egyptians built enormous pyramids using huge stone blocks that were difficult to move by sliding. To move the blocks they placed logs underneath them and pushed. By doing this, they were taking advantage of the fact that rolling friction is much less than sliding friction. (See Figure 7–4.)

Modern technology uses the same principles as the Egyptians, though in a more sophisticated way. We try to reduce unwanted friction for many reasons. For instance, all machines have moving parts that rub together during operation causing friction. We know that friction can wear out the machines, reduce efficiency and cause unwanted heat. (If you rub your hands hard together you can feel the heat produced by friction.) Excess friction in machines can be overcome by making smooth surfaces, lubricating with grease or oil and using bearings.

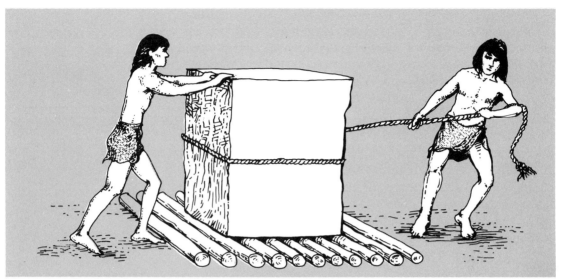

Figure 7–4 Rolling logs reduced friction for the ancient Egyptians.

Bearings function on the principle of the rolling logs used by the Egyptians to move stones. A **bearing** is a device containing many rollers or balls that reduce friction while supporting a load. Bearings change sliding friction into rolling friction, reducing friction by up to 100 times. Figure 7–5 illustrates the application of bearings to the wheel-and-axle assembly of a common device.

Figure 7–5 Ball bearings reduce friction in the wheel of a skate board.

Although friction is often unwanted, it can also be useful. Consider the technology involved in designing roads, bridges and automobile tires. Without friction, driving on highways would not only be dangerous, it would be impossible. Friction between tires and roads aids all types of acceleration—speeding up, slowing down and changing direction. Engineers consider friction when they design treads for tires and surfaces of roads and bridges. Assume that the moving friction between rubber and ice is 1. Then the amount of moving friction between rubber and wet asphalt is 40; in other words, there is 40 times as much friction to help stop a car on wet asphalt as there is on a patch of ice! Table 7–1 compares the friction between rubber and common surfaces used on roads. All automobile drivers should be aware of the consequences of the reduction of helpful friction.

Table 7–1   Friction between Rubber and Other Surfaces

Materials	Friction (compared to rubber on ice)
Rubber on ice	1
Rubber on wet asphalt	40
Rubber on dry asphalt	80
Rubber on wet concrete	60
Rubber on dry concrete	140

PRACTICE

3. Explain why it is necessary to streamline modern aircraft and rockets.
4. Students in a certain physics class refused to believe that friction helps us walk. In an effort to show the students that friction is needed, the teacher threw thousands of hard, plastic beads onto the floor and told the students to walk out of the room as fast as possible. What point would be proved by such a demonstration? (Do not include safety concerns in your answer.)
5. Sliding friction occurs in gasoline engines as the pistons move back and forth in the cylinders. What two methods would help reduce that sliding friction?
6. Explain why friction between your hand and a doorknob helps you open the door.
7. Give an explanation involving friction for a sign that reads "Reduce speed on wet highway".

7.6 Balanced and Unbalanced Forces

To study the effects of forces acting on objects, we must distinguish between forces that are balanced and forces that are unbalanced. Imagine that you are holding a book as shown in Figure 7-6. Two forces affect it. One is the force of gravity, which acts to pull the book downward. The other is the force exerted by your hand holding up the book. Those forces are equal and act in opposite directions. When such forces act on a single object, they are called **balanced forces**.

Now consider forces that are not balanced. Imagine that the hand holding the book in Figure 7-6 is suddenly removed. The only force acting on the book now is the downward pull of gravity. This force will cause the book to accelerate downward. In this case, the force of gravity is an example of an **unbalanced force**, one that is greater in one direction than any other.

Figure 7-6 An example of balanced forces

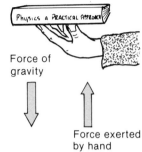

Force of gravity

Force exerted by hand

The two forces on the book in the diagram are balanced.

PRACTICE

8. A brick is at rest on a table. Describe the forces acting on the brick.
9. Name the type of force that causes acceleration.

7.7 Newton's First Law of Motion

An object is either at rest or in motion. One that is at rest requires an unbalanced force to get it to move. One that is in motion requires an unbalanced force to get it to stop. In both cases, static and moving, the object tends to resist change. For instance, a train is hard to get moving, and once it is in motion, it is hard to stop.

These facts are summarized by Newton's **first law of motion**. It states that:

an object maintains its state of rest or uniform motion unless it is acted upon by an unbalanced force

First, think about objects that are at rest. They will remain at rest unless an unbalanced force acts on them. Here are examples of this:

(1) It is hard to push a stalled car to the side of the road.

(2) Some "magicians" can jerk a smooth tablecloth from beneath a table setting of glasses and silverware.

(3) A passenger standing on a bus or subway train tends to fall backward as the vehicle accelerates forward from rest.

Now consider an object in uniform motion. It will continue in a straight line at a constant speed unless an unbalanced force acts on it. A car in uniform motion is subject to balanced forces, as shown in Figure 7–7. The downward force of gravity is balanced by the upward force of the road on the car. The backward force of moving friction is balanced by the forward force produced by the engine. As long as the forces remain balanced, the car does not accelerate.

Figure 7–7 Forces acting on an object in uniform motion are balanced.

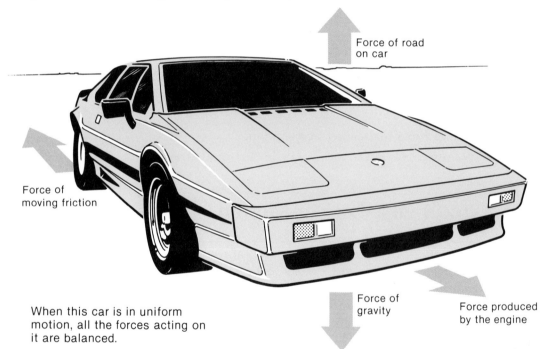

Force of road on car

Force of moving friction

Force of gravity

Force produced by the engine

When this car is in uniform motion, all the forces acting on it are balanced.

Here are examples of objects in motion tending to remain in motion:

(1) A spacecraft travelling toward Mars does not need its engine on most of the time. Outer space is basically a vacuum, so there is no moving friction on the craft. The craft tends to maintain a constant speed in a straight line (uniform motion).

(2) A speeding car approaching a curve on an icy highway has the tendency to continue in a straight line, thus failing to follow the curve. That is why care must be taken when driving on slippery roads.

(3) People in cars and other vehicles have the tendency to maintain uniform motion. This can be dangerous for them if the vehicle comes to a sudden stop, and the passengers keep on going. The danger can be greatly reduced by the proper wearing of safety belts.

PRACTICE

10. **Class demonstrations:** Your teacher will perform demonstrations of Newton's first law of motion. Describe those demonstrations.
11. Explain why astronauts are placed horizontally in the space capsule during blast-off from the launching pad.
12. Some thrill rides at exhibitions or fairs create sensations that may be explained using Newton's first law of motion. Describe two such rides.
13. Some people claim to have seen unidentified flying objects (UFOs) stop and start instantaneously or make right-angle turns at high speeds. What arguments can be given for or against such motion?

7.8 Newton's Second Law of Motion

Newton's first law deals with situations in which the forces on an object are balanced. No acceleration occurs. His second law deals with situations in which the forces on an object are unbalanced (greater in one direction than any other). The object accelerates in the direction of the unbalanced force.

Newton's **second law of motion** states that:

the acceleration of an object increases as the unbalanced force increases, and it decreases as the mass increases

A simple example will illustrate this second law. In Figure 7-8(a), an unbalanced force of 5.0 N is accelerating a 1.0 kg cart. (The applied force may be 5.1 N with 0.1 N overcoming moving friction. Thus, the unbalanced force is 5.1 N – 0.1 N = 5.0 N.)

This unbalanced force causes a certain acceleration of the cart.

In diagram (b), a 10 N unbalanced force acts on the same cart, and the acceleration is twice as great. In other words, for a constant mass, if the force is doubled, the acceleration is doubled.

In diagram (c), the original unbalanced force of 5.0 N acts on two carts. The acceleration is only half that in diagram (a). In other words, for a constant force, if the mass is doubled, the acceleration is cut in half.

Figure 7-8 Illustrating Newton's second law of motion

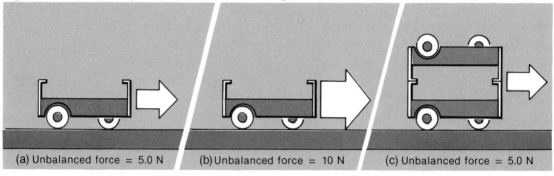

(a) Unbalanced force = 5.0 N   (b) Unbalanced force = 10 N   (c) Unbalanced force = 5.0 N

These ideas of Newton's second law of motion can be summarized in mathematical form.

$$\text{acceleration} = \frac{\text{unbalanced force}}{\text{mass}} \text{ or } a = \frac{F}{m}$$

This equation is often written in the rearranged form

$$\textbf{unbalanced force} = \textbf{mass} \times \textbf{acceleration} \text{ or } F = ma$$

In these equations, the units used must be from the SI. Force must be measured in newtons, mass in kilograms and acceleration in metres per second per second.

The equation $F = ma$ can be used to define the newton in terms of metres, kilograms and seconds.

$$F = ma$$

$$N = (kg)\ \frac{\left(\frac{m}{s}\right)}{s}$$

$$\text{or } N = \frac{kg\left(\frac{m}{s}\right)}{s}$$

Thus, **one newton** is the force required to give a 1.0 kg object an acceleration of 1.0 (m/s)/s.

Sample problem 1: Calculate the acceleration of a 5.0 kg cart being pushed by a 40 N unbalanced force.

Solution: $a = \dfrac{F}{m}$

$$= \dfrac{40 \text{ N}}{5.0 \text{ kg}} \qquad \dfrac{\cancel{kg}\; \dfrac{(\frac{m}{s})}{s}}{\cancel{kg}}$$

$$= 8.0 \; \dfrac{(\frac{m}{s})}{s}$$

Sample problem 2: Calculate the unbalanced force required to give a 1500 kg car an acceleration of 2.0 (m/s)/s.

Solution: $F = ma$

$$= 1500 \text{ kg} \times 2.0 \; \dfrac{(\frac{m}{s})}{s}$$

$$= 3000 \text{ N}$$

PRACTICE

14. Compare the direction of the acceleration of an object with the direction of the unbalanced force causing that acceleration.
15. Calculate the acceleration in each situation:
    (a) A cyclist exerts an unbalanced force of 15 N to a total mass (cyclist + bicycle) of 60 kg.
    (b) A bowler exerts an unbalanced force of 17.5 N on a 7.0 kg bowling ball.
    (c) An unbalanced force of 8.0 N is applied to a 0.1 kg model rocket.
16. Find the unbalanced force if:
    (a) A cannon gives a 5.0 kg shell an acceleration of 5000 (m/s)/s before it leaves the muzzle.
    (b) A 0.05 kg arrow is given an acceleration of 2500 (m/s)/s.
    (c) A 500-passenger Boeing 747 jet (with a mass of 1.64×10^5 kg) undergoes an acceleration of 1.0 (m/s)/s.
17. Rearrange the equation $F = ma$ to express m by itself.
18. In parts of outer space where the gravitational force is practically zero, mass cannot be measured using a balance. However, it can be measured by experiment using Newton's second law of motion. Calculate the mass of an object in such an experiment if an unbalanced force of 8.0 N gives a measured acceleration of 5.0 (m/s)/s.

7.9 The Force of Gravity

We can apply Newton's second law of motion, in the form $a = \dfrac{F}{m}$, to calculate the acceleration due to gravity at the surface of the earth. For example, if a 1.0 kg object is held up by a force scale, the scale reads 9.8 N. This means that the force of gravity pulling down on a 1.0 kg object is 9.8 N. If this object is allowed to fall freely, it will accelerate at a value

$$a = \frac{F}{m}$$

$$= \frac{9.8 \text{ N}}{1.0 \text{ kg}}$$

$$= 9.8 \frac{\left(\frac{m}{s}\right)}{s}$$

This value is, of course, the acceleration due to gravity (g) discussed in Section 6.6.

Now we can use the value $g = 9.8$ (m/s)/s to calculate the force required to hold up an object, even though the object is not accelerating. The equation we use is:

$$F = mg \text{ where } g = 9.8 \text{ (m/s)/s}$$

The F in this equation may be called the force required to hold up an object, the force of gravity acting on an object or the force required to lift an object without acceleration.

> **Sample problem 3:** What is the force of gravity acting on a 0.5 kg book?
> Solution: $F = mg$
> $\qquad = 0.5 \text{ kg} \times 9.8 \left(\frac{\frac{m}{s}}{s}\right)$
> $\qquad = 4.9 \text{ N}$

(Some texts refer to the force of gravity as "weight". What is important to remember is that mass is measured in kilograms and the force of gravity, or weight, is measured in newtons.)

From the equation $F = mg$, it is evident that the force of gravity on a known mass changes if g changes. Although the average value of g on earth is 9.8 (m/s)/s, that value may change slightly, depending on location. In general, the further you are from the centre of the earth, the less the value of g. Thus g would be less at the top of a mountain than at the bottom of a valley. Also, g at the equator is somewhat less than at the North or South Pole because the earth is slightly flattened at the Poles. Table 7–2 lists the values of g at various locations on the earth. From that list it is apparent that the force of gravity changes with location.

Table 7–2   Acceleration Due to Gravity at Various Locations on Earth

Location	Latitude	Altitude (m)	g [(m/s)/s]
Equator	0°	0	9.781
North Pole	90°	0	9.832
Toronto	44°	162	9.805
Brussels	51°	102	9.811
Denver	40°	1638	9.796

If you were to travel to the moon, you would find that the force of gravity there is only 1/6 of that on the earth. That is why the acceleration caused by gravity on the moon is only 1.6 (m/s)/s. Accelerations caused by the force of gravity on different planets were shown in Table 6–1, Section 6.8.

Sample problem 4: Calculate the force of gravity on a fully outfitted astronaut (mass 110 kg) (a) on the earth   (b) on the moon
Solution:
(a) $F = mg$
$= 110 \text{ kg} \times 9.8 \; \dfrac{\left(\frac{m}{s}\right)}{s}$
$= 1078 \text{ N}$

(b) $F = ma$
$= 110 \text{ kg} \times 1.6 \; \dfrac{\left(\frac{m}{s}\right)}{s}$
$= 176 \text{ N}$

Notice, of course, that the mass does not change when the astronaut is on the moon.

PRACTICE
19. Calculate the force of gravity acting on:
   (a) a 1500 kg car   (b) yourself

20. (a) Determine the force of gravity pulling down on a 15 kg
     curling stone (on the earth).
    (b) What force would you have to exert to lift the curling stone
     in (a) off the ice (without any acceleration)?
21. The 1968 Summer Olympics were held in Mexico City, which
    has an elevation of 2200 m above sea level and a latitude of
    about 20°. Several jumping records were broken at those
    Olympics. Explain what conditions helped contribute to this.
22. Calculate the force of gravity on a 10 kg object on the surface
    of (a) Venus   (b) Jupiter (Refer to Table 6-1, Section 6.8.)
23. Some science-fiction movies and television programs show
    people walking around spacecraft the way we walk around a
    room. How scientifically accurate is this representation?

7.10 Newton's Third Law of Motion

Newton's first law of motion is descriptive and his second law
mathematical. In both cases we consider the forces acting on only
one object. When a force is applied to one object, however, it must
be applied by a second object. This brings us to the third law,
which considers forces acting as pairs on two objects.

Newton's **third law of motion**, often called the action-reaction
law, states that:

> **for every action force on an object there is an equal reac-
> tion force in the opposite direction on a second object**

To understand the third law, imagine a boy jumping off the rear
end of a rowboat in a calm lake, as shown in Figure 7-9. The boy
exerts a force against the boat, and this force is noticed because the
boat darts forward. We call this the **action** force. The boy moves
in a direction opposite to the boat because of the **reaction** force of
the boat on him. The action and reaction forces are equal in size,
but opposite in direction, and act on different objects.

As you read the examples of the third law that follow,
remember there are always two objects to consider. One object exerts the **action** force, and the other exerts the **reaction** force. In certain cases, one of the "objects" may be a gas, such as air.

(1) In swimming the **action** force is exerted by the hands moving
    backward against the water. The water exerts a **reaction** force
    forward against the hands, pushing the body forward.

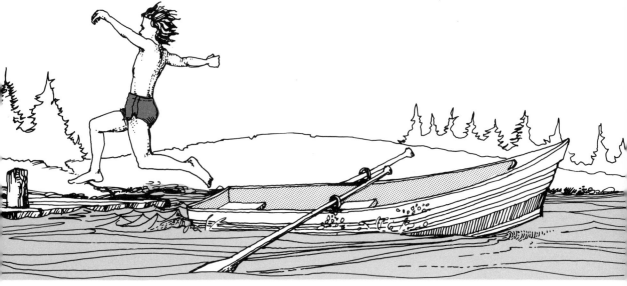

Figure 7–9 A swimmer exerts an action force on the boat; the boat exerts a reaction force on the swimmer.

(2) The propeller blades on a helicopter are designed to force air in one direction as the propeller spins rapidly. Thus, the **action** force is exerted downward by the blades against the air. The **reaction** force is exerted by the air upward against the blades, sending the helicopter in a direction opposite to the motion of air.

(3) A jet engine on an aircraft allows air to enter a large opening at the front of the engine. The engine heats the air and expels it rapidly out the rear. The **action** force is exerted by the engine backward on the expelled air. The **reaction** force is exerted by the expelled air forward on the engine, forcing the engine, and thus the entire airplane, in the opposite direction.

(4) A squid is a marine animal with a body size ranging from about 3 cm to 6 m. The squid propels itself by taking in water and expelling it in sudden spurts. The **action** force is applied by the discharged water backward on the surrounding water. The **reaction** force of the sea water forward against the discharged water sends the squid in the opposite direction.

PRACTICE

24. **Class demonstrations:** Your teacher will perform demonstrations of Newton's third law of motion. Describe those demonstrations.

25. Explain each event described below in terms of Newton's third law of motion. In each case state what exerts the **action** force and what exerts the **reaction** force.
    (a) A rifle recoils when it is fired.
    (b) When a toy balloon is blown up and released, it flies violently around the room.
    (c) A paddle is used to propel a canoe.
    (d) A person with ordinary shoes is able to walk on a sidewalk.

7.11 More Applications of Forces

The force we call gravity has interesting applications besides those already discussed. To all of us who inhabit the earth, gravity is taken for granted. It pulls us toward the centre of the earth, so that no matter where we are we always know which way is "up". To an observer outside our world, however, things may appear somewhat different. Figure 7–10 illustrates that, to an outside observer, only someone at the "top" of the world can really be upright. However, people in Australia do not think they are up-side down!

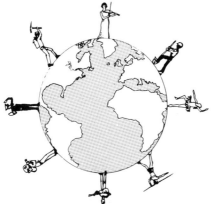

Figure 7–10 Which way is up?

In Section 7.1 it was mentioned that gravity keeps the earth revolving around the sun. The same kind of force keeps our moon revolving around the earth. Not only does the earth exert a force on the moon, but the moon also exerts a force on the earth. That force is evident in the creation of ocean tides on the earth.

Under certain conditions, molecules that make up substances can exert large forces. For example, in the cylinder of a gasoline engine a mixture of gasoline and air molecules is ignited and the resulting explosion forces the piston to move rapidly in the cylinder.

Forces are also exerted on very tiny particles. All matter is made up of atoms, which in turn are made up of protons and neutrons (in the central nucleus) with high-speed electrons travelling in regions around the nucleus. The force of electrical attraction between the electrons and the protons keeps the fast-moving electrons from escaping.

Other applications of forces will be introduced later in this book.

7.12 Review Assignment

1. Define force.   (7.1)
2. State the SI unit used to measure force.   (7.2)
3. In general, which is greater in size, static friction or moving friction?   (7.3, 7.4)
4. List factors that affect the force of friction.   (7.4)
5. A meteor or "shooting star" is a chunk of rocky material from outer space that burns and gives off light as it enters the earth's atmosphere. What is the likely cause of the burning?   (7.5)
6. A surface that appears smooth to the human eye may appear irregular under the magnification of a microscope. For example, the surface of a "smooth" block of wood may look like this: ⌇⌇⌇⌇⌇⌇⌇⌇⌇⌇⌇⌇⌇⌇⌇⌇⌇⌇⌇⌇⌇⌇⌇⌇
   Use a diagram to explain why friction occurs between such a surface and another surface that is even smoother.   (7.3 to 7.5)
7. A parachutist is falling at a constant speed straight toward the ground. Describe the forces acting on the parachute. Are those forces balanced?   (7.6)
8. State what happens (if anything) to an object that is:
   (a) at rest and has balanced forces acting on it
   (b) in motion and has balanced forces acting on it
   (c) at rest and has an unbalanced force applied to it   (7.6, 7.7)
9. Explain why it is wise to wear a safety belt when riding in a car.   (7.7)

10. An unbalanced force of 5.0 N is applied to a toy electric train. The mass of the train is 2.5 kg. Calculate the acceleration of the train.   (7.8)
11. What unbalanced force is needed to give a 120 kg boat an acceleration of 2.5 (m/s)/s?   (7.8)
12. An unbalanced force of 29.4 N causes a certain object to accelerate at 9.8 (m/s)/s. Calculate the mass of the object. (7.8)
13. Calculate the force of gravity pulling down on a 60 kg person. [Use $g = 9.8$ (m/s)/s.]   (7.9)
14. Describe how the force of gravity on an object depends on location.   (7.9)
15. A rocket in outer space, where there is no air, can change its direction by firing a small engine on the side of the rocket. Explain how this is possible.   (7.10)
16. Imagine you are stranded in the middle of a **frictionless** pond of ice the size of a football field. You have in your hand a basketful of hockey pucks, which you don't mind losing. Describe how you would get to one side of the pond. (Try to use all three of Newton's laws of motion to explain your answer.)

7.13 Answers to Selected Problems

PRACTICE QUESTIONS

1. (a) 2.0 N
   (b) 4.0 N
   (c) 10 N
15. (a) 0.25 (m/s)/s
    (b) 2.5 (m/s)/s
    (c) 80 (m/s)/s
16. (a) 25 000 N (2.5 × 10⁴ N)
    (b) 125 N
    (c) 1.64 × 10⁵ N

17. $m = \dfrac{F}{a}$
18. 1.6 kg
19. (a) 14 700 N or 1.5 × 10⁴ N
20. (a) 147 N
    (b) 147 N
22. (a) 89 N
    (b) 258 N

REVIEW ASSIGNMENT
10. 2.0 (m/s)/s
11. 300 N
12. 3.0 kg
13. 588 N

8

Mechanical Energy and Power

GOALS: After completing this chapter you should be able to:
1. Describe forms of energy, in particular the two forms of mechanical energy called gravitational potential energy and kinetic energy.
2. Calculate the work done on an object that moves a given distance under a known force ($E = Fd$).
3. Use the SI unit of energy.
4. Calculate the gravitational potential energy of an object given its mass and height above a reference level ($E_P = mgh$).
5. Describe applications of gravitational potential energy.
6. Calculate the kinetic energy of an object given its mass and speed ($E_K = \dfrac{mv^2}{2}$).
7. State the law of conservation of energy and give examples that illustrate it.
8. Define power and use the SI unit to measure it.
9. Calculate the power of an object consuming energy for a given amount of time ($P = \dfrac{E}{t}$).
10. Determine experimentally the power of a person participating in an activity.

Knowing the information in this chapter will be especially useful if you plan a career in:
• energy conservation (power systems, mechanics)
• natural resources (mining, oil, etc.)

8.1 The Importance of Energy

Energy is important both in physics and in our everyday lives. Earlier peoples consumed only the energy stored in the food they ate. That energy helped them perform work with their hands. As civilization progressed, more energy was used. The greater the progress, the greater was the consumption of energy.

Nowadays the nations in the world with the highest technology consume the most energy. That energy is used to heat homes in winter, cool homes in summer, manufacture necessities and luxuries, cook food, transport people and goods—the list is endless.

People living in our society should learn about the advantages and disadvantages of consuming energy for such purposes.

Energy exists in many forms. That is why we will not try to give a simple definition of it. Table 8–1 gives a list of forms of energy and Figure 8–1 shows some examples of those forms.

Table 8–1 Forms of Energy

Form of energy	Comment
Radiant	examples include visible light and X rays
Heat	results from the motion of particles
Electrical	results from the force of electron repulsion
Nuclear	stored in the central parts of atoms
Sound	allows us to hear vibrations
Chemical potential	stored in substances such as fuel and food
Gravitational potential	energy of position
Kinetic	energy of motion
Elastic potential	stored in stretched or compressed objects

The forms of energy listed in Table 8–1 can change from one to another. For example, in an electric oven electrical energy changes into heat energy. Other examples of energy changes will be discussed in various parts of the text.

In this chapter the forms of energy studied are gravitational potential energy and kinetic energy, which together are called mechanical energy.

PRACTICE
1. Name at least one form of energy associated with each object in italics:
   (a) A *bonfire* roasts a marshmallow.
   (b) A *baseball* smashes a window.
   (c) A *solar collector* heats water for a swimming pool.
   (d) The *siren* of an ambulance warns of an emergency.
   (e) A watch's *spring* is fully wound.

(a) Chemical potential energy is released when fireworks explode over Ontario Place in Toronto, Ontario.

(b) Elastic potential energy is stored in the archer's stretched bow.

Figure 8–1 Two examples of potential energy

8.2 Using Force to Transfer Energy

Consider a situation in which some groceries must be lifted from the floor to a higher level. An example of this is shown in Figure 8–2. In order to lift the groceries up, a force must be exerted over a certain distance. When a force is exerted to move an object some distance, energy is transferred to the object.

Figure 8–2 Energy is transferred to the box to lift it from the floor to the rollers.

When energy is transferred to an object, we say that **work** has been done on the object. The amount of work done depends on the force exerted and the distance moved, as shown in the relation:

$$E = Fd$$

The symbol E is used to remind you that work done on an object is a measure of the energy transferred to that object.

Since force is measured in newtons and distance in metres, work is measured in newton metres, N·m. The newton metre is called a joule, J, in honour of James Joule, an English physicist (1818-1889) who studied heat and electrical energy.

The equation $E = Fd$ applies when the force and distance moved are in the same direction. This is true whether the force pushes an object horizontally, lifts it vertically or anything in between.

When calculating the work done in lifting an object vertically, you may have to find the force needed to lift the object. If the object's mass is known, use the equation $F = mg$ to find the force. (Refer to Section 7.9.)

James Joule (1818–1889)

Sample problem 1: The mass of the box with groceries in Figure 8–2 is 20 kg, and the height of the rollers above the floor is 1.0 m. Calculate the:
(a) force required to lift the box
(b) work done on the box in lifting it to the rollers
Solution:
(a) $F = mg$
$= 20 \text{ kg} \times 9.8 \text{ (m/s)/s}$
$= 196 \text{ N}$

(b) $E = Fd$
$= 196 \text{ N} \times 1.0 \text{ m}$
$= 196 \text{ J}$

Sample problem 2: A 200 N force, parallel to the ramp, is needed to push a loaded wheelbarrow up the ramp, as shown in the diagram. If the ramp is 6.0 m long, calculate the work done on the wheelbarrow.
Solution: $E = Fd$
$= 200 \text{ N} \times 6.0 \text{ m}$
$= 1200 \text{ J}$

Sample problem 3: A dog team is pulling a loaded sled across the snow with a total force of 150 N. The team and sled move with uniform motion on a level surface.
(a) What is the function of the 150 N force?
(b) Calculate the work done after the team and sled have travelled 1.0 km (1000 m).

Solution:
(a) The force of 150 N is needed to overcome the sliding friction of the sled in the snow.
(b)
$$E = Fd$$
$$= 150 \text{ N} \times 1000 \text{ m}$$
$$= 150\ 000 \text{ J or } 1.5 \times 10^5 \text{ J}$$

In sample problem 3 the force exerted by the dogs balances the force of friction. This allows uniform motion. In such a situation the work that is done simply overcomes friction.

PRACTICE

2. A girl pushes a box 5.0 m with a force of 25 N. How much work has she done on the box?
3. A boy pushes against a large maple tree with a force of 250 N. How much work has he done on the tree?
4. A 0.1 kg book is lifted from the floor to a shelf 2.0 m above. Calculate the:
   (a) force needed to lift the book
   (b) work done on the book in lifting it to the shelf
5. Calculate the amount of work you would have to do in climbing 3.0 m up a ladder.
6. A cyclist exerts an average force of 40 N in a uniform motion trip and covers a distance of 2.0 km (2000 m).
   (a) What is the function of the 40 N force?
   (b) How much work does the cyclist do?
7. Express joules in terms of metres, kilograms and seconds. (Hint: Refer to the definition of a newton in Section 7.8.)
8. Rearrange the equation $E = Fd$ to express:
   (a) F by itself   (b) d by itself
9. Calculate the unknown quantities:

	E (J)	F (N)	d (m)
(a)	?	200	4.0
(b)	620	?	31
(c)	40	80	?

8.3 Experiment 15: Work

INTRODUCTION
Be sure that the force scale needed for this experiment is properly zeroed. If it isn't, ask your teacher to either adjust it or show you how to take the error into account.

PURPOSE: To compare measured work with calculated work.

APPARATUS: cart; board (up to 2.5 m long); 2 masses (each 1.0 kg); force scale (to 10 N); metre stick

PROCEDURE
1. Elevate and support one end of the board so the angle between the board and the table is about 4° or 5°. (See Figure 8–3.) Measure or calculate each of the following:
   (a) distance cart travels up the ramp (d)
   (b) height of the ramp (h)
   (c) force of gravity on the cart
   (d) force of gravity on each 1.0 kg mass

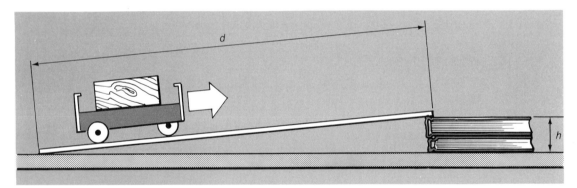

Figure 8–3

2. Determine the force required to pull the cart, with a 1.0 kg mass on it, up the ramp at a constant speed. (Try to keep the force parallel to the ramp.) Repeat this measurement for accuracy.
   Now find the work done on the cart in moving it a distance d up the ramp (E = Fd).
3. Calculate the work required to lift the cart and its load vertically a distance equal to the height (h) of the ramp (E = Fh). (Note that the force in this case is different from the force in #2.)

4. Compare the measured work (#2) with the calculated work (#3). Explain any differences.
5. Repeat the experiment, adding the second 1.0 kg mass to the cart.

QUESTIONS
1. What is the advantage of moving an object up a ramp instead of lifting it vertically?
2. How could you determine whether friction influenced the results of this experiment?

8.4 Gravitational Potential Energy

Suppose that you must pound a peg part of the way into the ground using only a rock. You could lift the rock above the peg, as shown in Figure 8–4. At the raised position, the rock will have the potential to help you do some work on the peg. This potential is due to the fact that the force of gravity is pulling down on the rock.

The type of energy possessed by an object because of its position is called **gravitational potential energy**, E_p. This potential energy can be used to do work on some object.

Figure 8–4 Raising a rock gives it gravitational potential energy.

In order to lift the rock in Figure 8–4 a height h, you would have to transfer energy $(E = Fh)$ to it. That transferred energy, or work, equals the change in the rock's potential energy; that is, $E_p = Fh$, where E_p is the potential energy of the object that is a height h above some level. Since the force F is the force due to gravity, we can now write the common equation for potential energy.

$$E_p = (F)h \text{ and } F = mg$$
$$\therefore E_p = (mg)h, \text{ or:}$$

$$E_p = mgh \text{ where } g = 9.8 \text{ (m/s)/s}$$

In the SI energy is measured in joules, mass in kilograms and height (or distance) in metres.

When performing calculations of potential energy, be careful about your choice of **reference level**, the level to which an object may fall.

128

Sample problem 4: The 20 kg box of groceries (Figure 8-2 and Sample problem 2) is 0.5 m above the level of the loading platform outside the store. Calculate the potential energy of the box relative to that level.

Solution: $E_p = mgh$

$\qquad = 20 \text{ kg} \times 9.8 \text{ (m/s)/s} \times 0.5 \text{ m}$

$\qquad = 98 \text{ J}$

Sample problem 4 illustrates a useful application of potential energy. An object is lifted to a higher position, then gravity helps it accelerate down a slope. The object's speed helps it move a required distance. Ski lifts provide an example of this application. Another example is shown in Figure 8-5 in which a type of roller coaster is at a high position where its potential energy is greatest. Then gravity causes the coaster to accelerate downward. The rest is fun.

Figure 8-5 At the instant shown the cars in this thrill ride have a large amount of gravitational potential energy.

Sample problem 5: Assume a loaded roller coaster has a mass including passengers of 5000 kg. From the loading platform it is raised 12 m to its highest level. What is the potential energy of the coaster relative to the platform?

Solution: $E_p = mgh$

$\qquad = 5000 \text{ kg} \times 9.8 \text{ (m/s)/s} \times 12 \text{ m}$

$\qquad = 588\ 000 \text{ J or } 5.88 \times 10^5 \text{ J}$

Figure 8–6 shows two more applications of potential energy. In diagram (a), a pile driver is about to be lifted by motor high above the pile. Then it will have the potential energy to do the work of driving the pile into the ground.

Diagram (b) shows a source of potential energy in nature. A waterfall's potential energy can be used to drive turbines to create hydro-electricity.

(a) The pile driver

(b) Waterfalls at Niagara, Ontario

Figure 8–6 Applications of gravitational potential energy

PRACTICE

10. A 0.5 kg book is resting on a desk that is 0.6 m high. Calculate the potential energy of the book relative to the:
   (a) desk
   (b) floor
11. Calculate the potential energy of a skier (total mass = 70 kg, including equipment) at the top of a 100 m hill.
12. Rearrange the equation $E_p = mgh$ to express:
   (a) m by itself
   (b) h by itself
13. Calculate the unknown quantities:

	E_p (J)	m (kg)	h (m)
(a)	?	2.0	5.0
(b)	19.6	?	4.0
(c)	29.4	2.0	?

8.5 Kinetic Energy

A bowling ball resting on the floor has no energy of motion. One that is rolling along a bowling alley has energy of motion. Energy due to the motion of an object is called **kinetic energy**, E_K. (Kinetic comes from the Greek word **kinema**, which means motion. A cinema is a place to watch motion pictures.)

Two factors determine the amount of kinetic energy possessed by a moving object. They are the mass and speed of the object. If either quantity increases, the kinetic energy increases. This is evident in the equation used to calculate kinetic energy:

$$E_K = \frac{mv^2}{2}$$

Again, the energy is measured in joules, mass in kilograms and speed in metres per second.

Although the derivation of $E_K = \frac{mv^2}{2}$ is beyond the goals of this chapter, you should be aware that the equation stems from the fact that energy must be transferred to an object to increase its speed. For example, if 100 J of work is done on an object to increase its speed, its kinetic energy increases by 100 J. (The derivation of the equation is shown in Appendix F.)

Sample problem 6: Find the kinetic energy of a 6.0 kg bowling ball rolling at 5.0 m/s.

Solution: $E_K = \dfrac{mv^2}{2}$

$$= \dfrac{(6.0 \text{ kg}) (5.0 \,\tfrac{m}{s})^2}{2}$$

$$= 3 \text{ kg} \times 25 \, \dfrac{m^2}{s^2}$$

$$= 75 \text{ J}$$

PRACTICE

14. Calculate the kinetic energy of each object:
    (a) A 4.0 kg shot leaves an athlete's hand during the shot-put at a speed of 10 m/s.
    (b) A 2000 kg car is travelling at 20 m/s.
    (c) A 0.15 kg hockey puck has a speed of 40 m/s.

15. Rearrange the equation $E_K = \dfrac{mv^2}{2}$ to express:
    (a) m by itself
    (b) v by itself

16. Calculate the unknown quantities:

	E_K (J)	m (kg)	$v \, (\tfrac{m}{s})$
(a)	?	4.0	8.0
(b)	50	?	5.0
(c)	25	2.0	?

8.6 The Law of Conservation of Energy

The quantities of energy transferred (or work), potential energy and kinetic energy are all measured in joules, and one can change to another. Scientists say that, when any such change occurs, energy is "conserved". In other words:

when energy changes from one form to another, no energy is lost

This statement is called the **law of conservation of energy**. (This law applies to other forms of energy. It does not, however, apply to nuclear reactions. Such reactions obey the law of conservation of mass-energy, mentioned in Chapter 25.)

As an example of the law of conservation of energy, consider a situation in which a man has the job of breaking a large rock into smaller chunks. The rock is too heavy for him to lift. The man has no tools, only a ramp and some smaller nearby stones. He can use physics to solve his unusual problem.

Figure 8-7 illustrates a logical solution that uses the resources available. First the man builds a ramp with the smaller stones as a base. Then he does work on the rock, rolling it up the ramp, as shown in diagram (a). Assume the amount of work he does is 1000 J. At the top of the ramp, in diagram (b), the rock has 1000 J of potential energy relative to the rocky ground below. (Recall from Experiment 15, Section 8-3, that the work done moving an object up a ramp equals the work done lifting the object vertically to the same height.)

In diagram (c), the potential energy changes to kinetic energy as the rock accelerates downward. Just before striking the ground the rock has maximum speed and 1000 J of kinetic energy.

Then, in diagram (d), the kinetic energy changes into other types of energy, and the rock shatters into smaller chunks. Again, the amount of energy equals 1000 J.

The law of conservation of energy can also be applied to several situations already mentioned in this chapter:
  —sending groceries to a pick-up area (Figure 8-2)
  —using a rock to pound a peg (Figure 8-4)
  —operating a thrill ride (Figure 8-5)
  —using a pile driver (Figure 8-6)
  —creating hydro-electric power at waterfalls (Figure 8-6)

PRACTICE
17. An object, moving with a kinetic energy of 20 J, has 30 J of energy transferred to it.
    (a) What is the new kinetic energy?
    (b) What law is this based on?
18. Use the law of conservation of energy to describe the energy changes that occur in the operation of a pile driver, Figure 8-6(a).
19. A ball is dropped vertically from a height of 1.5 m and bounces back to a height of 1.3 m. Does this violate the law of conservation of energy? Explain.

Figure 8–7 The law of conservation of energy

(a) $E = Fd$ (b) $E_p = mgh$

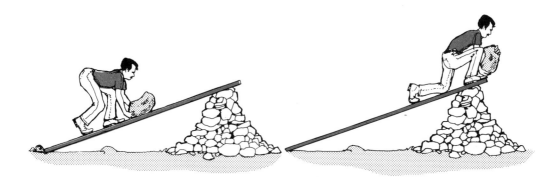

(c) $E_k = \dfrac{mv^2}{2}$ (d) The energy is conserved.

8.7 Power

It takes approximately 2000 J of work for an average person to climb one flight of stairs. The amount of work remains the same whether the person climbs the stairs in 5 s or 30 s. If the work is the same but the time changes, something else must change. That something else is called power.

Power is the rate of consuming energy.

$$P = \frac{E}{t}$$

Since energy is measured in joules and time in seconds, power is measured in joules per second. This SI unit has the special name watt (W), in honor of James Watt, a Scottish physicist (1736-1819) who was first to invent a practical steam engine. Watts and kilowatts are commonly used to indicate the power of electrical appliances.

Sample problem 7: What is the power of a cyclist who transfers 20 000 J (2.0×10^4 J) of energy in 100 s?

Solution: $P = \dfrac{E}{t}$

$\qquad = \dfrac{20\ 000\ J}{100\ s}$

$\qquad = 200\ W$

Sample problem 8: A 60 kg student takes 6.0 s to climb a 3.0 m ladder. Calculate the:
(a) potential energy of the student at the top of the ladder
(b) student's power for the climb

Solution: (a) $E_p = mgh$

$\qquad\qquad = 60\ kg \times 9.8\ (m/s)/s \times 3.0\ m$

$\qquad\qquad = 1764\ J$

$\qquad$ (b) $P = \dfrac{E}{t}$

$\qquad\qquad = \dfrac{1764\ J}{6.0\ s}$

$\qquad\qquad = 294\ W$

PRACTICE

20. If 150 J of work are done in 30 s, what is the power?
21. An electric clock uses 150 J of electrical energy every 60 s. What is the clock's power rating?
22. A certain stereo uses 1.44×10^5 J of energy in 1 h (3600 s). Calculate its power rating.
23. Rearrange the equation $P = \dfrac{E}{t}$ to express:
    (a) E by itself   (b) t by itself

24. Calculate the unknown quantities:

	P (W)	E (J)	t (s)
(a)	?	64	8.0
(b)	120	?	60
(c)	15	750	?

8.8 Experiment 16: Student Power

INTRODUCTION

How powerful are you? You can discover your maximum power for certain activities in this experiment. (The power is "maximum" because the time is short. You would not be able to exhibit so much power all day long.)

Common sense is an important consideration in this experiment. If you are running up a flight of stairs, be careful not to trip. If you are exerting an effort with your arms, do not overstretch your muscles.

To complete the calculations you will need to know either your mass (in kilograms) or the force of gravity on you (in newtons).

It may be interesting to compare your power with that of an average horse that can display about 750 W of power for an entire working day. (This is an old-fashioned "horsepower".)

PURPOSE: To determine student power and factors that affect it.

APPARATUS: stopwatch; metre stick; 4 or 5 hard-cover texts; bathroom scales (in kilograms or newtons)

PROCEDURE
1. Measure the vertical height of a flight of stairs. (The greater the height, the better!) Determine the time it takes you to run up the stairs as quickly and safely as possible. (A running start is allowed.) Calculate your own power for the run.
2. Obtain four or five texts and measure their total mass. Find the time it takes you to lift the books a known distance for 25 repetitions. Calculate your own power for this activity.
3. Compare your power for each activity with the power of the other students in the class. From the comparison try to determine factors that affect student power.

QUESTION
1. Name other activities that may be interesting in this type of experiment.

8.9 Summary of Mechanics

Mechanics, the branch of physics dealing with motion and forces, involves many fundamental definitions, ideas and equations. At this stage it is wise to review the most important concepts studied in Chapters 5 to 8 to help you get an overall view of the subject.

Motion and forces have been studied for thousands of years. The important names associated with the development of the study of mechanics include Aristotle, Galileo, Newton, Joule and Watt. The last three have SI units named after them.

Following is a list of important concepts in mechanics:

(1) **Uniform motion:** $v = \frac{d}{t}$; $m = \frac{\Delta y}{\Delta x}$

(2) **Uniform acceleration:** $a = \frac{\Delta v}{t}$; $g = 9.8$ (m/s)/s

(3) **Force:** static and moving friction; balanced and unbalanced forces; Newton's three laws of motion; $F = ma$; $F = mg$; newtons

(4) **Energy transferred** (or work): $E = Fd$; joules

(5) **Gravitational potential energy:** $E_P = mgh$; joules

(6) **Kinetic energy:** $E_K = \frac{mv^2}{2}$; joules

(7) **Law of conservation of energy**

(8) **Power:** $P = \frac{E}{t}$; watts

As you progress through other parts of this book you may wish to refer to this summary to refresh your memory regarding specific facts.

8.10 Review Assignment

1. What is the scientific term for:
   (a) energy of position?
   (b) energy of motion?   (8.1)

2. A black bear's greatest enemy is a grizzly bear. To escape a grizzly attack, a black bear does what its enemy cannot do—it climbs a tree. Calculate the work done by a 150 kg black bear in climbing 20 m up a tree.   (8.2)

3. A baseball is given 120 J of energy by a pitcher who exerts a force over a distance of 1.0 m. Calculate the force exerted. (8.2)

4. A force of 200 N is required to keep a motorcycle moving with uniform motion. How far will 6.0×10^5 J of work take that motorcycle?   (8.2)

5. Calculate the gravitational potential energy (relative to the floor) of a 1.5 kg hammer when it is 2.0 m above the floor. (8.4)

6. An object of unknown mass, located 10 m above the ground, has a potential energy of 98 J relative to the ground. Calculate the object's mass.   (8.4)

7. How high above the floor must a 0.1 kg ball be lifted to give it a potential energy of 4.9 J relative to the floor?   (8.4)

8. Find the kinetic energy of a 2.0 kg ball that is travelling:
   (a) 2.0 m/s
   (b) 4.0 m/s
   (c) 6.0 m/s   (8.5)

9. A discus travelling at 20 m/s has 400 J of kinetic energy. Find the mass of the discus.   (8.5)

10. Calculate the speed of a 16 kg curling stone that has a kinetic energy of 72 J.   (8.5)

11. An interesting and practical feature of the Montreal subway system is that, in some cases, the level of the station is higher than the level of the adjoining tunnel, as illustrated in Figure 8-8. Explain the advantages of this design. (Take into consideration such concepts as force, acceleration, work, potential energy and kinetic energy.)

Figure 8–8 The Montreal subway system

12. Calculate the power rating of a light bulb that uses 3600 J of energy each 60 s.   (8.7)

13. Some people perform difficult tasks to raise money for charity. For example, walking up the stairs in Toronto's CN Tower helps charity and personal fitness. If a 70 kg man climbs the 342 m 10 times in 4 h, calculate his:
    (a) work done each trip up the stairs
    (b) total work for the 10 trips
    (c) power   (8.2, 8.7)

14. How much energy is used by a 1200 W electric kettle during 5 min (300 s) of operation?   (8.7)

15. Find the time it takes a 60 W light bulb to consume the energy stored in 1.0 kg of coal (3.0×10^7 J).   (8.7)

8.11 Answers to Selected Problems

PRACTICE QUESTIONS

2. 125 J
3. 0.0 J
4. (a) 0.98 N
   (b) 1.96 J
6. (b) 80 000 J or 8.0×10^4 J
8. (a) $F = \dfrac{E}{d}$
   (b) $d = \dfrac{E}{F}$
9. (a) 800 J
   (b) 20 N
   (c) 0.5 m
10. (a) 0.0 J
    (b) 2.94 J
11. 68 600 J or 6.9×10^4 J
12. (a) $m = \dfrac{E_P}{gh}$
    (b) $h = \dfrac{E_P}{mg}$
13. (a) 98 J
    (b) 0.5 kg
    (c) 1.5 m

14. (a) 200 J
    (b) 400 000 J or 4.0×10^5 J
    (c) 120 J
15. (a) $m = \dfrac{2E_K}{v^2}$
    (b) $v = \sqrt{\dfrac{2E_K}{m}}$
16. (a) 128 J
    (b) 4.0 kg
    (c) 5 m/s
17. (a) 50 J
20. 5.0 W
21. 2.5 W
22. 40 W
23. (a) $E = Pt$
    (b) $t = \dfrac{E}{P}$
24. (a) 8.0 W
    (b) 7200 J
    (c) 50 s

REVIEW ASSIGNMENT

2. 29 400 J or 2.9×10^4 J
3. 120 N
4. 3000 m or 3.0 km
5. 29.4 J
6. 1.0 kg
7. 5.0 m
8. (a) 4.0 J
   (b) 16 J
   (c) 36 J

9. 2.0 kg
10. 3.0 m/s
12. 60 W
13. (a) 2.35×10^5 J
    (b) 2.35×10^6 J
    (c) 163 W
14. 3.6×10^5 J
15. 500 000 s or about 139 h

9

Machines

GOALS: After completing this chapter you should be able to:
1. List the six simple machines.
2. State the main functions of machines.
3. Indicate the positions of the fulcrum, load and effort for all three classes of the lever.
4. Explain why the pulley and wheel-and-axle belong to the lever family of machines.
5. Explain why the wedge and screw belong to the inclined-plane family of machines.
6. Define torque and calculate its value around a fulcrum ($T = F \times d$).
7. Use the SI unit of torque.
8. State the law of the lever and verify it experimentally.
9. Apply the law of the lever to solve problems involving torque
$$(F_E = \frac{F_L \times d_L}{d_E}).$$
10. Define and calculate the mechanical advantage of a simple machine
$$(MA = \frac{F_L}{F_E}).$$
11. State the conditions required for a machine to have a mechanical advantage equal to 1.0, greater than 1.0 or less than 1.0.
12. Describe applications of simple machines in the design of complex machines.

Knowing the information in this chapter will be especially useful if you plan a career in:
• tool-and-die making
• machine operation or design
• mechanics
• construction

9.1 Simple Machines

A **machine** is any device that helps us transfer or transform energy.

Although some modern machines are large and complex, many of them operate on the basic principles of simple machines that have been in use for many centuries. Museums often display ancient tools and weapons whose sharp edges are examples of a

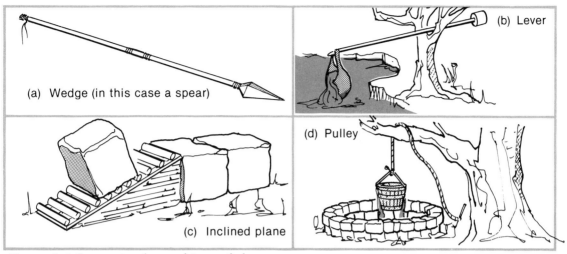

Figure 9–1 Some simple machines of the past

wedge. The Egyptians of early times used an inclined plane (a ramp) to move the blocks used to construct pyramids. They also used pulleys and levers to lift water out of wells and rivers. Figure 9–1 illustrates the use of these machines in ancient times.

Scientists say there are six simple machines. These can be divided into two families—the lever family and the inclined-plane family. Figure 9–2(a) shows that the members of the lever family rotate around some point. Diagram (b) shows that the wedge is a double-inclined plane and the screw is an inclined plane wrapped around a central shaft.

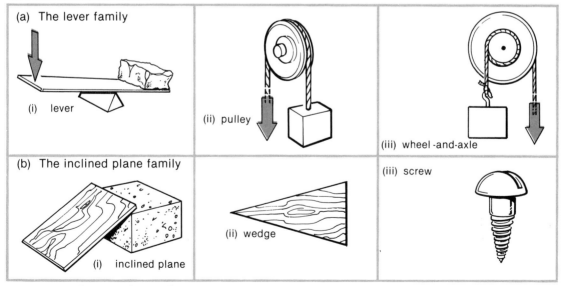

Figure 9–2 Six simple machines

PRACTICE
1. State the kind of simple machine each of the following is:
   (a) a car's steering wheel   (d) a seesaw
   (b) a stairway   (e) a doorknob
   (c) an axe   (f) a car jack

9.2 Functions of Machines

Every machine, whether simple or complex, performs at least one main function. Following is a list of the main functions of machines.

(1) **Changing energy from one form to another** (e.g., a hydro-electric generator changes the kinetic energy of falling water into electrical energy).

(2) **Transferring energy from one place to another** (e.g., the transmission of a car transfers the energy from the motor to the wheels).

(3) **Changing the direction of a force** (e.g., to raise a flag up a pole a person pulls down on a rope attached to a pulley at the top of the pole).

(4) **Increasing or decreasing force** (e.g., a system of pulleys helps a mechanic exert a small force to hoist a heavy engine out of a car).

(5) **Increasing or decreasing distance** (e.g., the rear wheel of a bicycle moves a much greater distance than the sprocket wheel attached to the pedals).

PRACTICE
2. What is the main function of the pulley shown in Figure 9–1(d)?

9.3 The Lever Family of Machines

The lever family of simple machines consists of the lever, the pulley and the wheel-and-axle.

A **lever** is a rigid bar that may rotate freely around a support called a **fulcrum**. (Refer to Figure 9–3.) An **effort** is a force applied to one part of the lever to move a **load** at another part.

Levers are divided into three classes, depending on the positions of the load, effort and fulcrum. A **first-class lever** has the fulcrum between the load and the effort; a **second-class lever** has the load between the fulcrum and the effort; and a **third-class lever** has the effort exerted between the fulcrum and the load. Figure 9–3 illustrates these three classes and an example of each class.

Figure 9–3 Levers

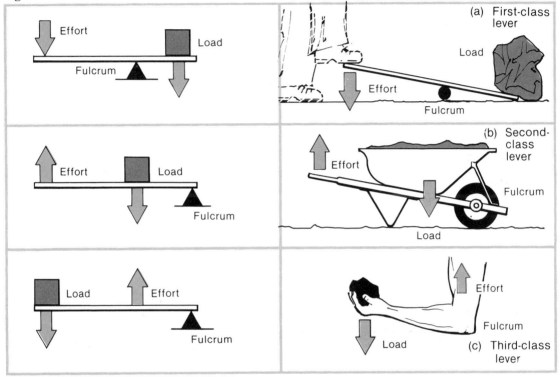

A **pulley** is a wheel with a grooved rim in which a rope runs. The wheel is free to rotate around a fulcrum, which makes the pulley like a lever. Figure 9–4 illustrates two simple pulley arrangements.

Figure 9–4 Two simple pulleys

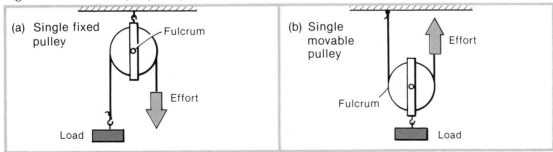

A **wheel-and-axle** is a large-diameter circular disk (the wheel) connected rigidly to a small-diameter rod (the axle). Both the wheel and the axle rotate around a fulcrum, so this machine is also similar to a lever. (See Figure 9–5.)

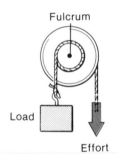

PRACTICE

3. State the class of lever illustrated by each device shown in Figure 9–6 **and** draw a diagram showing the fulcrum, load and effort.

Figure 9–6

Figure 9–5 The wheel-and-axle

(a) Seesaw

(b) Nutcracker

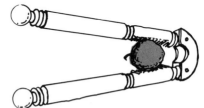

(c) Paddling a canoe (Assume the fulcrum is at the top end of the paddle.)

(d) Rowing a boat (Assume the fulcrum is at the point where the oar is attached to the boat.)

(e) Sugar tongs

(f) Scissors

(g) Human jaw

(h) Human foot

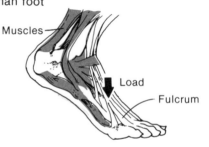

4. To which class of lever does each pulley in Figure 9–4 belong?
5. To which class of lever does the wheel-and-axle in Figure 9–5 belong?

9.4 Torque and the Lever Family

Anyone who has tried to break a short pencil into two pieces knows that the force required is quite large. The force needed to break a long pencil, however, is much smaller. (See Figure 9–7.) We can imagine that the pencil is a lever with the fulcrum at the middle. The total bending effect around the fulcrum is determined by both the force and the length of the pencil.

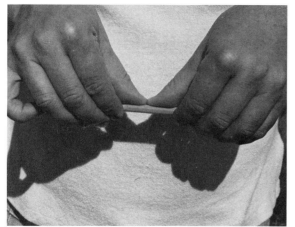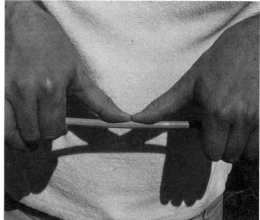

Figure 9–7 A short pencil is harder to break than a long pencil.

A measure of the bending effect around the fulcrum of a lever or similar machine is called **torque**. Torque depends on two factors—the size of the force applied and the distance between the force and the fulcrum. Increasing either quantity increases the torque.

Torque can be calculated using the product of force and distance.

$$T = F \times d$$

In this equation F is the load force or effort force measured in newtons, and d is the perpendicular distance from the force to the fulcrum measured in metres. Thus, torque is measured in newton metres (N·m).

Because torque can be calculated for both the load and the effort, we will use one equation for load torque and another for effort torque. If L stands for load and E for effort, we have:

$$T_L = F_L \times d_L \text{ and } T_E = F_E \times d_E$$

We will call the load distance (d_L) the **load arm** and the effort distance (d_E) the **effort arm**. Both are illustrated in Figure 9–8.

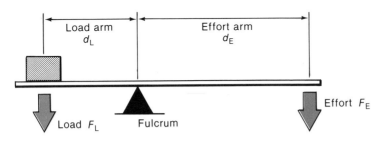

Figure 9–8 Load arm and effort arm for a first-class lever

The equation for torque, $T = F \times d$, should not be confused with the equation for energy transferred, $E = Fd$. The multiplication sign $(\times)$ in $T = F \times d$ will be used to remind you that d is the perpendicular distance from the force to the fulcrum. In $E = Fd$, d is the distance moved by an object influenced by a force. The units for torque and energy are also different. Torque is measured in newton metres and energy in joules.

Sample problem 1: Determine the load torque for the lever in the diagram.

Solution: $T_L = F_L \times d_L$

$\quad\quad\quad = 50 \text{ N} \times 2.0 \text{ m}$

$\quad\quad\quad = 100 \text{ N·m}$

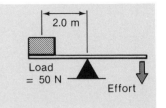

Sample problem 2: Find the effort torque for the pulley in the diagram.

Solution: $T_E = F_E \times d_E$

$\quad\quad\quad = 10 \text{ N} \times 0.05 \text{ m}$

$\quad\quad\quad = 0.5 \text{ N·m}$

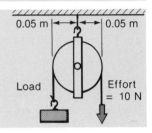

PRACTICE

6. Assume a mechanic is able to exert a 500 N effort on a wrench to loosen a bolt. Calculate the effort torque when the wrench's length is:
    (a) 0.3 m
    (b) 0.5 m

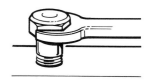

7. Determine the load torque for each simple machine in the diagrams below.

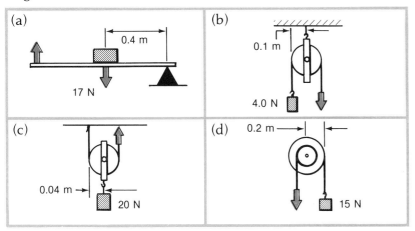

(a)   0.4 m   17 N

(b)   0.1 m   4.0 N

(c)   0.04 m   20 N

(d)   0.2 m   15 N

9.5 Experiment 17: The Law of the Lever

INTRODUCTION

If a lever acts vertically, the force of gravity has to be considered in calculations of torque. To eliminate the effects of gravity, the load and effort forces will be applied horizontally in this experiment.

Two force scales are needed and should be checked for proper adjustment by pulling one against the other. Also, you should learn how to correct for any error that results from holding the scales horizontally.

PURPOSE: To discover the law of the lever using a first-class lever and to check that law using second- and third-class levers.

APPARATUS: metre stick with a hole drilled at the centre; 2 force scales (each to 10 N or more); board, nail; and if #4 is performed, a single fixed pulley, a single movable pulley and a wheel-and-axle

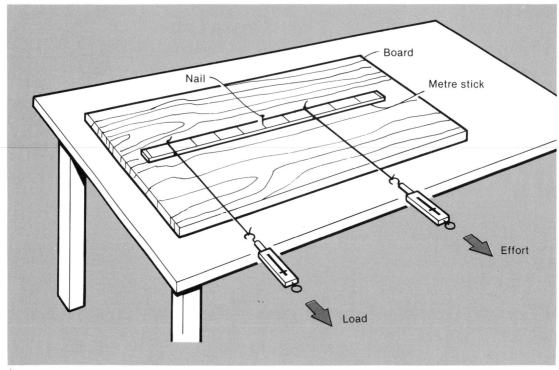

Figure 9–9

PROCEDURE

1. Set up the apparatus as shown in Figure 9–9. The lever should be able to rotate freely around the nail. With the load force at the 10 cm mark and the effort force at the 70 cm mark, exert a force of 4.0 N as a load. Measure the effort force required to keep the lever balanced. Be sure both load and effort are perpendicular to the metre stick. Tabulate your readings in a chart based on Table 9–1.

2. Change the load and effort forces according to the data in the second and third columns of Table 9–1. Complete all the calculations.

   Try to make a general statement (the law of the lever) about the results of the first three columns. Ask your teacher to check your statement before you proceed to # 3.

3. Prove the law of the lever for each situation described in Figure 9–10 and Table 9–1.

4. If the equipment is available, prove the law of the lever for a single fixed pulley, a single movable pulley and a wheel-and-axle arrangement. (Refer to Figure 9–11.) Set up a chart showing your observations and calculations for this procedure.

Table 9-1 Observations and calculations for Experiment 17

Procedure number	1	2	2	3(a)	3(b)	3(c)	3(d)
Class of lever	I	I	I	II	III	I	I
Load force (N)	4.0	8.0	6.0	10.0	4.0	3 4	10.0
Load position (cm mark)	10	30	25	30	10	10 30	10
Load distance (m)	0.4						
Effort force (N)							
Effort position (cm mark)	70	90	90	10	30	90	90
Effort distance (m)	0.2						
Load torque (N·m)							
Effort torque (N·m)							

Figure 9-10

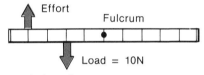

(a) Second-class lever

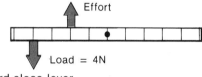

(b) Third-class lever

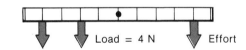

Load = 3N

(c) First-class lever with two loads

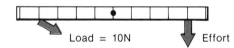

(d) First-class lever with the load not perpendicular to the lever

Figure 9-11

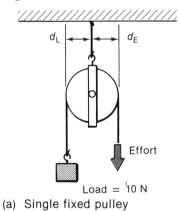

(a) Single fixed pulley

(b) Single movable pulley

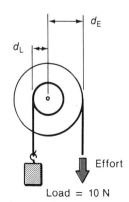

(c) Wheel-and-axle

9.6 Applying the Law of the Lever

The results of Experiment 17 prove that when a lever is balanced, the effort torque equals the load torque. This fact, called the **law of the lever**, can be used to determine the effort force needed to balance or slowly lift a known load. The effort force can be found in the following way:

$$\text{effort torque} = \text{load torque}$$
$$\text{effort} \times \text{effort arm} = \text{load} \times \text{load arm}$$
$$\text{or } F_E \times d_E = F_L \times d_L$$
$$F_E = \frac{F_L \times d_L}{d_E}$$

Sample problem 3: A man wishes to mount his trailer on blocks for the winter. He must lift one corner at a time to have the blocks inserted. If he uses a 3.0 m plank as shown in the diagram, what effort must he apply to lift the corner of the trailer?

Solution: $F_E = \dfrac{F_L \times d_L}{d_E}$

$= \dfrac{3000 \text{ N} \times 0.5 \text{ m}}{2.5 \text{ m}}$

$= 600 \text{ N}$

Effort

LEVER

0.5 m

Load = 3000 N

PRACTICE
8. Calculate the effort force for each situation shown in Figure 9–12.

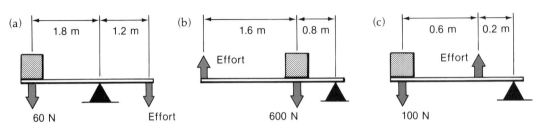

(a) 1.8 m 1.2 m — 60 N — Effort

(b) 1.6 m 0.8 m — Effort — 600 N

(c) 0.6 m 0.2 m — Effort — 100 N

Figure 9–12

9. Rearrange the equation $F_E = \dfrac{F_L \times d_L}{d_E}$ to express:

(a) d_E by itself
(b) F_L by itself
(c) d_L by itself

10. Calculate the unknown quantities:

	F_E (N)	d_E (m)	F_L (N)	d_L (m)
(a)	?	7.5	6.0	2.5
(b)	10	?	50	0.2
(c)	60	3.5	?	0.7
(d)	30	2.0	600	?

9.7 Mechanical Advantage of Machines

In order for a machine to be useful, it must provide some advantage. One of the main advantages of using simple machines is to change the force applied to the load.

Figure 9–13 illustrates a screwdriver acting as a lever to open a paint can. A small effort force moves a large distance to create a large load force that moves a small distance.

Figure 9–13 A small effort force creates a large load force.

The ratio of the load force to the effort force is called the **mechanical advantage** *(MA)* of a machine.

$$MA = \frac{F_L}{F_E}$$

Since mechanical advantage is a ratio of forces, all units divide, so it has no units.

Sample problem 4: Calculate the mechanical advantage of the screwdriver in Figure 9–13 if a 25 N effort force exerts a 125 N load force.

Solution:
$$MA = \frac{F_L}{F_E}$$
$$= \frac{125 \text{ N}}{25 \text{N}}$$
$$= 5$$

Acrobats sometimes use a lever system as illustrated in Figure 9–14. A heavy person jumps on one end of the plank, creating a large effort force over a small distance. A small person (the load) on the other end of the plank moves a larger distance and rises high enough to perform interesting stunts.

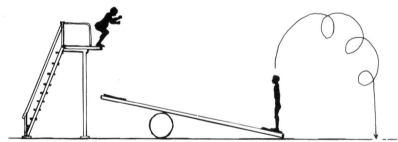

Figure 9–14 A large effort exerts a small load force.

Sample problem 5: Calculate the mechanical advantage of the lever system in Figure 9–14 if a 1000 N effort force exerts a 500 N load force.

Solution:
$$MA = \frac{F_L}{F_E}$$
$$= \frac{500 \text{ N}}{1000 \text{ N}}$$
$$= 0.5$$

Mechanical advantage can also be determined for machines other than levers.

Sample problem 6: A 15 N cart is pulled up an inclined plane with an effort of 5.0 N. Calculate the machine's mechanical advantage.

Solution:
$$MA = \frac{F_L}{F_E}$$
$$= \frac{15 \text{ N}}{5.0 \text{ N}}$$
$$= 3$$

In summary, mechanical advantage can be greater than 1.0, equal to 1.0 or less than 1.0.
(1) If $MA > 1.0$, the load force is increased, and the distance moved is decreased.
(2) If $MA = 1.0$, the load force equals the effort force.
(3) If $MA < 1.0$, the load force is decreased, and the distance moved is increased.

PRACTICE
11. Calculate the mechanical advantage in each case:
   (a) $F_L = 200$ N, $F_E = 80$ N   (b) $F_L = 900$ N, $F_E = 45$ N
   (c) $F_L = 40$ N, $F_E = 160$ N

9.8 Experiment 18: Mechanical Advantage of Simple Machines

INTRODUCTION
The mechanical advantage of any simple machine can be calculated using the relation $MA = F_L/F_E$.

Some measurements in this experiment require that the force scale be inverted. This will cause an error in the measurement. Learn how to correct for this error.

PURPOSE: To discover and compare the mechanical advantages of simple machines.

APPARATUS: metre stick with a hole drilled at the centre; board; nail; 2 force scales (to 10 N or more); 1.0 kg mass (to be used as a 10 N load); pulley systems; wheel-and-axle; inclined plane; cart; block of wood; jackscrew

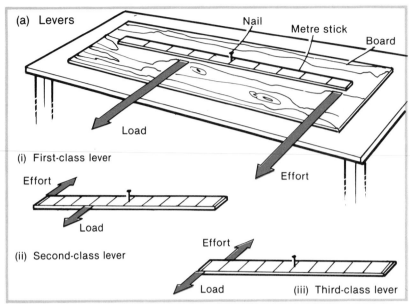

(a) Levers

Nail

Metre stick

Board

Load

(i) First-class lever

Effort

Effort

Load

Effort

(ii) Second-class lever

Load

(iii) Third-class lever

Figure 9–15

PROCEDURE

1. Set up the apparatus as shown in Figure 9–15(a)(i). Exert a 10 N load force at the 30 cm mark and determine the effort force at the 90 cm mark needed to balance the lever. Record the value in a chart (Table 9–2) and calculate the mechanical advantage.
2. Try other positions of the load and effort for a first-class lever. Again calculate the mechanical advantage and record your results.
3. Repeat step 1 for as many situations in Figure 9–15 as possible. In each case, use a load force of 10 N. When you are experimenting with pulleys, try to determine the relationship between the mechanical advantage and the number of strands supporting the load.
4. If time permits, invent and experiment with situations other than those in Figure 9–15.

Table 9–2

Machine	Details	Load force (N)	Effort force (N)	Mechanical advantage
Lever	first-class	10		

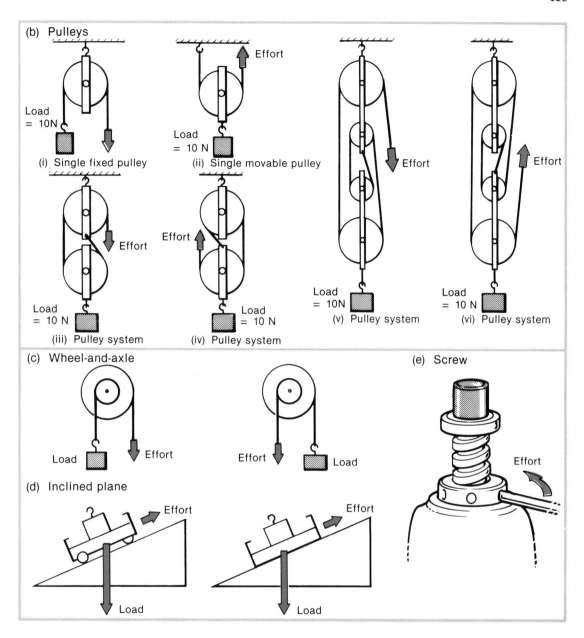

9.9 The Wedge and the Screw

The wedge and screw belong to the inclined plane family. A **wedge** is really a double inclined plane, as shown in Figure 9–16. Knives and axes are examples of the wedge.

Figure 9–16

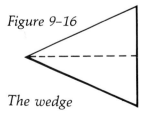

The wedge

A **screw** is actually an inclined plane wrapped around a central shaft. A circular staircase is a simple screw. Wood screws and bottle-cap screws hold things together. A jackscrew lifts a heavy object such as a car. The principle of the screw is also used to make certain measuring devices, such as a micrometer.

9.10 Complex Machines

A **complex machine** is a device made of two or more simple machines. Figure 9–17 illustrates combinations of simple machines used to increase the mechanical advantage of each simple machine.

Figure 9–17 Complex machines

(a) A complex wheel-and-axle arrangement is the basis of operation of the electricity meter shown in the photograph.

(b) How many simple machines make up a pencil sharpener?

9.11 Review Assignment

1. Name the members of the:
   (a) lever family of machines
   (b) inclined-plane family of machines   (9.1)
2. State the main function of each machine shown.   (9.2)

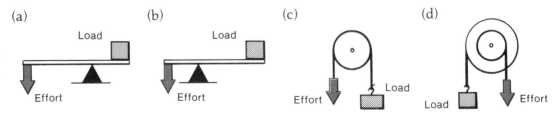

3. State an example in the human body of a:
   (a) third-class lever
   (b) second-class lever   (9.3)
4. Calculate the effort torque on a wheelbarrow if a 300 N effort force is exerted 1.2 m from the fulcrum.   (9.4)
5. What is the load torque on a 0.3 m arm holding a 20 N load? (9.4)
6. State the law of the lever.   (9.5)
7. Use the data given below to determine if the lever is balanced:
   load force = 240 N   load distance = 3.5 m
   effort force = 180 N   effort distance = 4.5 m   (9.5)
8. **Mini-experiment:** Balance a metre stick horizontally on your extended index fingers so that your left index finger is at the 10 cm mark and your right index finger is at the 70 cm mark. Predict the position where your fingers will meet when you gently slide them toward each other. Try it. Use the concept of torque to explain the result.   (9.4, 9.5)
9. A 3.0 m first-class lever has a 600 N load located 0.5 m from the fulcrum.
   (a) Draw a sketch of the lever, showing the effort, load and fulcrum.
   (b) Calculate the effort force needed to lift the load.   (9.6)
10. A wheelbarrow has an 800 N load located 0.6 m from the fulcrum. If a boy needs to exert a 320 N effort to lift the handles of the wheelbarrow, what is the distance from his hands to the fulcrum?   (9.6)
11. In a person's arm, which acts like a third-class lever, the distance from the fulcrum to the muscle (effort) is 4 cm or 0.04 m and the distance from the fulcrum to the hand (load) is 30 cm or 0.3 m. If the muscle can exert an effort of 1500 N, what load can the hand support?   (9.6)

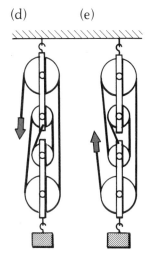

(d)   (e)

12. The distance between the effort and fulcrum of a wheelbarrow is 1.5 m. An effort of 400 N can support a load of 1200 N. What is the distance between the load and the fulcrum? (9.6)
13. A lever has a mechanical advantage of 4.0. What load could be lifted by an effort of 100 N?   (9.7, 9.8)
14. Find the mechanical advantage of each machine shown. (9.7, 9.8)

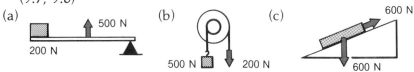

(a)   500 N   (b)   (c)   600 N

200 N   500 N   200 N   600 N

15. In question #14(e) what load could be supported by an effort of 60 N?   (9.7, 9.8)

16. **Class demonstration**

Use two broomsticks and about 5 m of strong cord to set up the arrangement shown in the diagram. The sticks should be parallel and about 40 cm apart. Have two strong persons in the class hold the sticks apart as indicated with the double arrows. Have a third person exert an effort force as shown with the single arrow. Explain the results. (Hint: Relate this demonstration to a system of pulleys.)

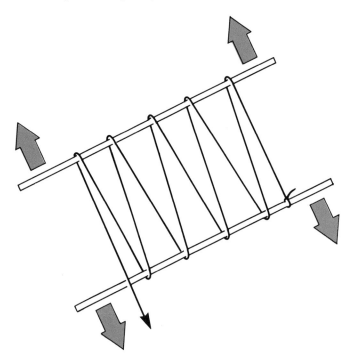

17. Analyse some complex machine in detail. (For example, it is interesting to compare a bicycle and a tricycle. Other examples include a pencil sharpener, typewriter, internal-combustion engine and record player.)

18. Science and technology have provided us with many machines.

(a) List several advantages and disadvantages of using machines in our society.

(b) Assume you must adapt to a new society in which you could use only half as many machines as you now use. Which machines would you discard first?

9.12 Answers to Selected Problems

PRACTICE QUESTIONS

6. (a) 150 N·m
   (b) 250 N·m
7. (a) 6.8 N·m
   (b) 0.4 N·m
   (c) 0.8 N·m
   (d) 3.0 N·m
8. (a) 90 N
   (b) 300 N
   (c) 400 N
9. (a) $d_E = \dfrac{F_L \times d_L}{F_E}$

(b) $F_L = \dfrac{F_E \times d_E}{d_L}$

(c) $d_L = \dfrac{F_E \times d_E}{F_L}$

10. (a) 2.0 N
    (b) 1.0 m
    (c) 300 N
    (d) 0.1 m
11. (a) 2.5
    (b) 20
    (c) 0.25

REVIEW ASSIGNMENT

4. 360 N·m
5. 6.0 N·m
9. (b) 120 N
10. 1.5 m
11. 200 N
12. 0.5 m
13. 400 N

14. (a) 0.4
    (b) 2.5
    (c) 1.0
    (d) 4.0
    (e) 5.0
15. 300 N

IV. Heat Energy

10
Heat Energy and Temperature

GOALS: After completing this chapter you should be able to:
1. Compare ancient theories of heat with modern theory.
2. State the difference between heat and temperature.
3. State the origin of the Celsius temperature scale.
4. Change temperature in degrees Celsius to kelvins and vice versa.
5. State the kinetic molecular theory (KMT) of matter.
6. Use the KMT to explain effects of change of heat energy.
7. List sources of heat energy.
8. Describe and compare the three methods of transferring heat energy —conduction, convection and radiation.
9. Describe methods of preventing heat transfer.
10. Calculate the resistance value of combinations of insulating materials.
11. List examples of applications of heat expansion.
12. Explain effects of the fact that water acts differently than most substances when cooled.
13. Describe medical uses of heat energy.

Knowing the information in this chapter will be especially useful if you plan a career in:
• architecture
• refrigeration
• insulation
• ventilation
• auto air conditioning
• fire protection
• food processing and preserving
• medicine
• chemistry

10.1 The Study of Heat

Heat is a form of energy. It results from the motion of particles.

At least one ancient thinker, a Greek named Democritus (460-370 B.C.), believed that matter was made of tiny particles. He named them atoms (from the Greek word "atomos", which means indivisible). Although he believed atoms in motion caused heat, he could not prove it.

Most other Greeks, including Aristotle (384-322 B.C.), did not agree with Democritus. They thought that heat, or fire, was one of four elements (earth, air, fire and water) making up all matter. To explain water changing into steam, they said that logs (a form of earth) became fire which was added to water. Then the water changed to steam (a form of air).

Although the ancients did not understand much about heat, they made good use of it. For instance, in the third century A.D., a man named Hero, from Alexandria, Egypt, made several devices that used heat. One is called Hero's engine. It is a hollow glass sphere with spouts at the sides, as shown in Figure 10–1. Water in the sphere is heated, producing steam. The steam leaves the spouts in one direction causing the engine to spin in the opposite direction.

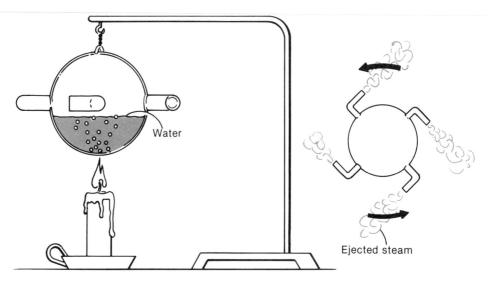

Figure 10–1 Hero's engine

About the end of the 1700s scientists thought that heat was a substance they called **caloric**. Caloric was similar to Aristotle's "fire". It was used to explain many observations. For example, if a hot object was in contact with a cold object, caloric would flow from the hot to the cold object. Also, when caloric was added to an object, it made the object expand.

But the caloric theory ran into trouble when scientists experimented to see if it had any mass. Benjamin Thompson, also known as Count Rumford (1753-1814), performed heat experiments while boring out cannons for the German army. A great amount of heat was produced by friction, especially when the drill was dull. But the total mass of the drill, cannon and metal shavings did not increase or decrease. Since the temperature increased but the mass remained constant, Rumford concluded that caloric had no mass.

Soon after Rumford's discovery, the kinetic molecular theory of matter was developed. It will be discussed in Section 10.4.

Heat is only one of many forms of energy. (Section 8.1 dealt with several other forms.) Studying heat is important because all other forms of energy tend to end up as heat energy. For example, think of what happens to the chemical energy stored in the gasoline used by a car. The gasoline is burned in the car's engine, creating both heat and mechanical energy. The mechanical energy then turns into heat because of friction of the moving parts. Thus, almost all the chemical energy ends up as heat energy.

PRACTICE

1. The action of Hero's engine is based on one of Newton's laws of motion. State which law and use it to explain the spinning of the engine. (Reference: Chapter 7)
2. What was a major weakness of the caloric theory?

10.2 Temperature and Heat

Temperature is a measure of the average energy of motion of individual particles. It increases if the motion of particles increases.

Heat is a measure of the total energy content of the particles of an object. It depends on the object's mass, temperature and type of substance.

For example, compare 1.0 kg of water at 70°C with 100 kg of water at 70°C. The samples have the same temperature, but the 100 kg contains much more heat.

Measuring temperature is important in the study of heat. The **thermometer** is a device used to measure temperature. It works on the principle that a substance expands when heated and contracts when cooled.

The most common type of thermometer is made of a long, thin, glass tube containing mercury or alcohol. As the temperature rises, the liquid expands more than the glass and rises up the tube. Mercury and alcohol are useful in thermometers because of their freezing and boiling temperatures. Mercury freezes at –39°C and boils at 357°C. Alcohol freezes at –117°C and boils at 79°C.

PRACTICE

3. Name three factors that determine the heat content of an object.
4. What type of thermometer would you use in the Arctic in the winter? Why?

10.3 The Celsius and Kelvin Temperature Scales

The temperature scale in everyday use is the Celsius scale. It was invented by Anders Celsius (1701-1744), who lived in Sweden. He based his scale on the most common substance on earth, water. He called the freezing temperature of pure water 0°C and the boiling temperature 100°C. Then he divided the scale into 100 equal parts. Of course, temperatures may go below 0°C or above 100°C.

Besides 0°C and 100°C, there are two other temperatures everyone should recognize. Room temperature is about 20°C and body temperature is 37°C.

Another temperature scale, used mostly by scientists, is the Kelvin scale. It was invented by William Thompson Kelvin, a British scientist (1824-1907). One Kelvin degree equals one Celsius degree; however, the two scales have different zero points. Zero on the Kelvin scale is the coldest temperature possible, known as **absolute zero**. Absolute zero is the same as –273°C. Thus, water freezes at 273 K, which is read 273 kelvins. To change kelvins to degrees Celsius, subtract 273, and to change degrees Celsius to kelvins, add 273.

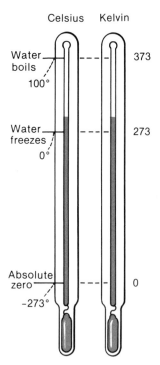

Celsius   Kelvin

Water boils — 100° — 373

Water freezes — 0° — 273

Absolute zero — –273° — 0

Sample problem 1: Change 30°C to kelvins.
Solution: K = °C + 273
= 30 + 273
= 303 K

Sample problem 2: Change 325 K to degrees Celsius.
Solution: °C = K - 273
= 325 - 273
= 52°C

The Celsius and Kelvin scales are compared in Figure 10–2.

Very cold temperatures are useful to scientists because much can be learned about matter when it is at low temperatures. For example, if the gas, helium, is cooled to –269°C (or 4 K), it becomes a liquid, which can be used for special experiments in science laboratories. Very cold temperatures also occur in outer space far away from the sun and other stars. Ordinary thermometers cannot be used to measure such low temperatures. Other types of thermometers, some using electricity, may be used.

Figure 10–2
The Celsius and Kelvin temperature scales

Although there is a lowest possible temperature (–273°C), there is no highest possible temperature. Nuclear explosions, such as those that create energy on the sun, cause temperatures to reach millions of degrees Celsius. Scientists measure those high temperatures by the colour of light given off.

PRACTICE
5. Compare the size of a Kelvin degree with the size of a Celsius degree.
6. Change 50°C to kelvins.
7. State room temperature and body temperature in degrees Celsius and kelvins.
8. Change to degrees Celsius:
   (a) 393 K   (b) 473 K   (c) 173 K

10.4 The Kinetic Molecular Theory of Matter

By 1800 scientists were beginning to accept the theory that matter is made of small particles. They could not see the particles, but they had evidence of their existence.

For example, a Scottish scientist named Robert Brown (1773-1858) discovered a simple way to observe the effects of particles in motion. He added pollen grains to the surface of water and observed the grains using a microscope. The grains followed zigzag patterns without touching each other. Brown concluded that the water particles were constantly in motion and were knocking the grains about. The motion of visible particles being knocked about by invisible particles is now called **Brownian motion**. Figure 10–3 shows an example of Brownian motion in which smoke particles are being knocked about by invisible air particles.

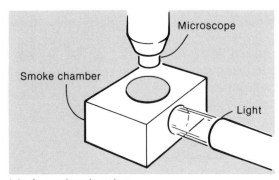

(a) A smoke chamber

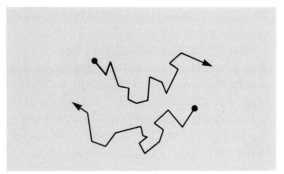

(b) Examples of paths taken by two particles

Figure 10–3 Brownian motion

The modern theory of matter is called the **kinetic molecular theory (KMT) of matter**. "Kinetic" refers to motion. (Recall from Chapter 8 that kinetic energy is energy of motion.) "Molecular" refers to tiny particles called molecules. The KMT states that:

(1) **All matter is made up of small particles called molecules.**
(2) **The molecules are very small compared with the spaces between them.**
(3) **The molecules are constantly in motion, often colliding with each other.**
(4) **The motion of the molecules becomes more violent when the temperature increases.**

In the remainder of this chapter, we will use the KMT of matter to explain many observed results.

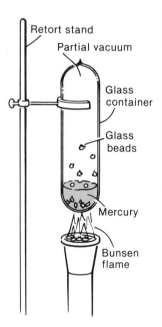

PRACTICE

9. **Class demonstration**

   The diagram shows mercury being heated in an enclosed tube. The heated mercury turns into vapour. The vapour collides with the glass beads, carrying them up the tube. View the action of this demonstration and explain it, using the KMT of matter.

10. When a jar containing a smelly substance, such as ammonia, is opened, the smell quickly spreads throughout the room. Explain how this happens.

10.5 Sources of Heat Energy

The **sun** is the earth's main source of energy. It gives us heat energy directly, as well as light energy, which later changes to heat and other forms of energy.

Chemical reactions, such as the burning of oil, natural gas and coal, produce heat. **Electrical energy** changes into heat for cooking and heating. **Nuclear reactions** give off a great amount of heat that is used to create electricity in nuclear generating stations.

Friction also causes heat. Friction can occur between the surfaces of two objects in contact with each other. It can also occur between the particles of a substance that is being compressed (e.g., a steel bar becomes warm when it is bent back and forth).

Another source of heat is **geothermal energy**, which is stored inside the earth. It is evident when geysers and volcanoes erupt.

Some of our sources of energy are being used up at a fast rate. Scientists are trying to find and improve alternate sources of energy so that our standard of living can be maintained.

10.6 Transferring Heat Energy by Conduction

If a stainless-steel spoon is used to stir soup being heated on a stove, the spoon's handle soon feels warm. Heat energy has been transferred from the soup through the spoon by **conduction**.

The KMT of matter can be used to explain heat conduction in a metal rod. The rod is made of millions of moving molecules. When one end of the rod is heated, molecules at that end gain energy and vibrate more violently. They collide with nearby molecules, causing them to vibrate more. This continues like a chain reaction along the rod. (See Figure 10–4).

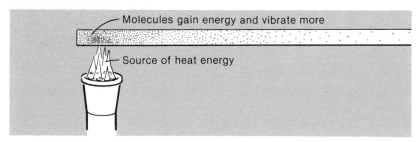

Figure 10-4 Conduction of heat in a metal

Some materials conduct heat better than others. One simple test for conductivity (the ability to conduct heat) is shown in Figure 10–5. Small beads of wax are placed along two equal-sized rods of

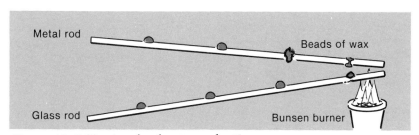

Figure 10-5 Testing for heat conduction

different materials. Then one end of each rod is heated. The beads on the metal rod begin to melt soon, but those on the glass rod do not melt.

The metal rod is called a **conductor**, a material that allows heat energy to transfer through its vibrating molecules. The glass rod is the opposite; it is called an **insulator**.

Table 10–1 compares the conductivity of various materials to the conductivity of air. For instance, water conducts heat 25 times as well as air, and iron 2800 times as well as air.

Table 10–1 Heat Conductivity of Substances Compared to Air

Substance	Conductivity	Substance	Conductivity
Air	1.0	Iron	2800
Cork	1.8	Brass	4600
Water	25	Aluminum	8800
Glass	35	Copper	16 000
Lead	1400	Silver	17 000

A useful application of conductivity that has saved many lives is a safety lamp used in mines. During the Industrial Revolution (about 200 a ago) many miners were killed by explosions in mines when gases were ignited by the flames of their lamps. Sir Humphrey Davy (1778-1829), a British chemist, invented a lamp with a wire mesh surrounding the flame. The wire conducted the flame's heat away so the temperature was not high enough to ignite the gases. Refer to Figure 10–6. (Nowadays, lamps with batteries are used, so the same problem does not exist.)

Figure 10–6 The Davy safety lamp

Wire mesh
(constantly conducts heat
away from flame)

PRACTICE

11. **Class demonstration**
    The conductivities of various metals can be demonstrated using the apparatus shown in the diagram. Place a wax bead at the end of each metal rod. Heat the centre of the apparatus in a Bunsen flame and find how long it takes each bead to melt. List the metals in order of increasing heat conductivity.

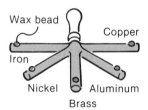

12. Good quality cooking pots are often made with copper bottoms and steel or aluminum sides. Why?

13. Lead and copper are frequently used in radiator cores of automobiles. Why are these good choices?

10.7 Transferring Heat Energy by Convection

Convection is the process in which heated particles of a liquid or gas move from one place to another, carrying energy with them.

The KMT of matter can be used to explain convection. Consider the demonstration apparatus shown in Figure 10–7. The air molecules near the flame are heated. They need more space in which to vibrate, so they spread apart. The warm air becomes less dense and rises through the chimney. As it leaves, it must be replaced with air that comes from the other chimney. The smoke paper makes it possible to see the circulation of air. That circulation is called a **convection current**.

Figure 10–7 Convection apparatus

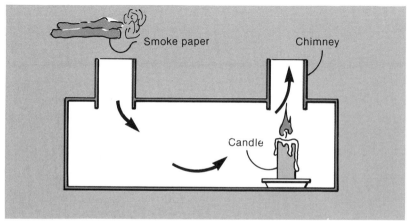

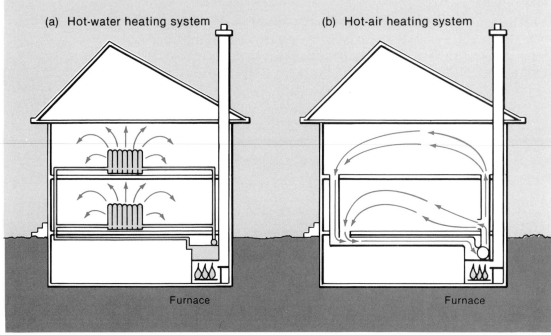

Figure 10–8 Convection currents in homes

Heating systems in homes set up convection currents. Figure 10–8 illustrates two types of heating systems. In diagram (a) water circulates in a convection current carrying heat with it. In diagram (b) air circulates to warm the rooms. In both cases once the material loses its heat to the rooms it becomes more dense and circulates back to the furnace to be reheated.

Large convection currents of air are set up in nature. You may have noticed the effects of sea and land breezes near the shore of a

Figure 10–9 A sea breeze

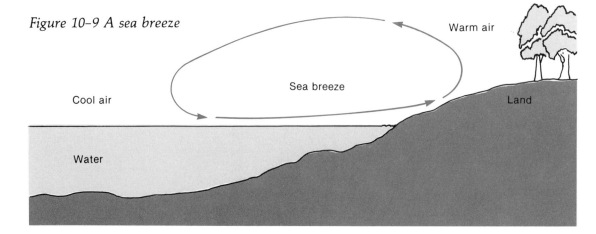

lake. Figure 10–9 shows how a sea breeze is created on a hot, sunny day. The land warms up more quickly than the water, so the air above the land also warms up. That air rises and is replaced by cooler air from above the water. This creates a convection current of air from water to land called a sea breeze.

A land breeze, going from the land toward the water, is noticed at sunset. The land cools down more quickly than the water. Lakes often become quite calm near the shores when a land breeze occurs.

PRACTICE
14. In what kinds of substances (solids, liquids or gases) can convection occur?
15. What happens to the density of a substance when it is heated?
16. Explain the difference between a sea breeze and a land breeze.

10.8 Transferring Radiant Heat Energy

Certain sources of energy, such as light bulbs and the sun, radiate energy outwards. The energy is transferred as a sort of wave motion called **radiation**. Examples of radiant energy include light, microwaves, radar, X rays and radiant heat.

Radiant energy travels at a constant speed (3.0×10^8 m/s) in a vacuum. This is much faster than conduction or convection. Radiant energy travels in straight lines. You may have noticed this fact when standing close to a bonfire. Also, it does not need particles in order to travel. It can travel through the vacuum of outer space. This is much different from conduction or convection, which need particles in order to travel.

What happens when radiant energy strikes an object? The energy is either transmitted, absorbed or reflected.
(1) If the energy is **transmitted**, it simply goes through the object (e.g., light energy goes through glass).
(2) If the energy is **absorbed**, it causes the internal energy of the object to increase (e.g., light energy from the sun absorbed by a dark, dull surface, such as pavement, causes the surface to warm up).
(3) If the energy is **reflected**, it bounces off the object (e.g., light and heat energies bounce off a shiny, smooth surface, such as a mirror).

PRACTICE

17. List three properties of radiant energy.
18. Describe the kind of surface that tends to:
    (a) absorb radiation   (b) reflect radiation
19. Which will melt more quickly on a sunny day in the spring—clean snow or dirty snow? Explain why.

10.9 Preventing Transfer of Heat Energy

Often we want heat energy to be transferred, but sometimes we want to prevent its transfer. In fact, much of North America's valuable energy can be conserved if we learn how to prevent heat transfer from our homes in the winter.

Conduction of heat energy can be prevented by using good insulating materials. Fur is a good insulator, so animals and humans use it to keep warm in the winter. A piece of fur may be at the same temperature as a piece of steel (e.g., −20°C) but when held to your face the fur does not feel cold. That is because it does not conduct the heat away from your face. Of course, you should not consider holding the steel to your face!

Insulating homes and other buildings is important. Proper insulation helps keep heat in during the winter and helps keep heat out during the summer. To prevent conduction, builders use certain materials in the walls and ceilings that have a resistance to heat conduction. The resistance value, or R value, depends on the type of material and its thickness. Table 10–2 lists the R values of several materials.

Table 10–2 Resistance Values of Insulators

Material	Thickness (cm)	Resistance value
Air space	2–10	1.0
Brick	10	0.43
Gypsum board	10	2.6
Plywood	10	5.0
Fibreglass	10	13
Styrofoam	10	20
Polyurethane	10	24

Table 10–2 can be used to find the R values of various thicknesses of a material. For example, 5.0 cm of plywood is R 2.5 and 2.0 cm of plywood is R 1.0. The total R value of a wall or ceiling is the sum of the individual values.

Sample problem 3: Calculate the total resistance value of the wall of a home that has 10 cm of brick, 2.0 cm of plywood, 10 cm of fibreglass and 2.0 cm of gypsum board.

Solution:

10 cm brick	R 0.43
2.0 cm plywood	R 1.0
10 cm fibreglass	R 13
2.0 cm gypsum board	R 0.52
	R 14.95 or R 15

In Table 10–2 notice that the R value of an air space does not seem to depend on the thickness. The important factor is that the air must not be in motion. Such an air space is called "dead". A dead-air space not only prevents conduction, it also prevents convection. Dead-air spaces provide insulation in attics and in double- and triple-glazed windows.

The transfer of radiant heat energy can also be prevented. An example is the **greenhouse effect**. A greenhouse is a glass structure containing plants. Light energy from the sun has short waves that can easily pass through the glass. The plants and other objects absorb the light energy, which then changes into radiant heat energy. The heat energy has long waves that reflect off glass. Thus, the heat energy gets trapped inside the greenhouse, keeping it warm even in winter. (See Figure 10–10.)

Figure 10–10 The greenhouse effect

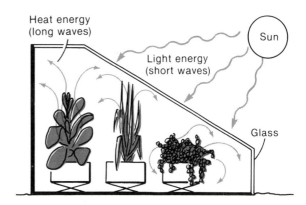

Short waves of light energy pass through the glass. They are absorbed by the plants and changed into radiant heat energy. The heat waves cannot easily pass through glass, so the heat is trapped in the greenhouse.

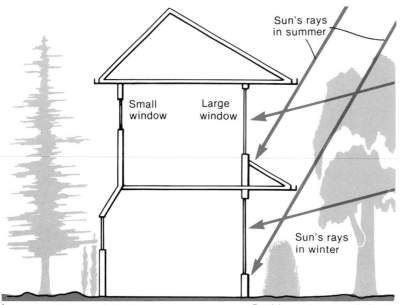

Sun's rays in summer

Small window

Large window

Sun's rays in winter

The diagram illustrates several features of a passive solar-heated home. Other features may include carpets that absorb light energy in the winter, and window shutters that are closed at night to prevent heat loss.

Evergreen trees and shrubs

Deciduous trees and shrubs

Figure 10–11 The basic design of a home having passive solar heating

The greenhouse effect is taken into consideration by architects who design homes having passive solar heating. (Passive means inactive but acted upon. Passive solar heating is much less expensive than active solar heating, in which the sun's energy heats a specially designed apparatus.) Figure 10–11 illustrates the basic design of such homes. They have large windows facing the sunny side, overhanging roofs and deciduous trees, which lose their leaves in the winter. The sun's energy enters the windows in the winter but not in the summer.

PRACTICE

20. State the R value of:
    (a) 20 cm of brick
    (b) 5.0 cm of plywood
    (c) 15 cm of Styrofoam
21. Calculate the R value of a ceiling that has 4.0 cm of gypsum board and 20 cm of fibreglass.
22. Explain how a double-glazed window prevents the transfer of heat.
23. A thermos bottle, like the one shown in the diagram, can keep cold liquids cold and hot liquids hot. Use the information in the diagram to describe how a thermos prevents conduction, convection and radiation.

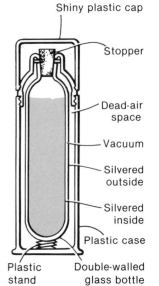

Shiny plastic cap

Stopper

Dead-air space

Vacuum

Silvered outside

Silvered inside

Plastic case

Plastic stand

Double-walled glass bottle

10.10 Expansion Effects of Heat

In almost all situations heating a substance causes it to expand and cooling a substance causes it to contract. (Water is an exception to this between certain temperatures, as you will learn in Section 10.12.)

The KMT of matter can be used to explain expansion and contraction. When a substance is heated, its molecules gain energy and move more quickly. They collide more and push each other apart. Thus, the substance expands. The opposite occurs when a substance cools.

Gases expand readily when heated. The air in a palm glass [Figure 10–12(a)] expands when a hand is placed on one bulb. The expansion forces the coloured water toward the other bulb. In Figure 10–12(b) hot-air balloons rise because the hot air in the balloons expands and becomes less dense than the cool air.

Figure 10–12 Expansion of air

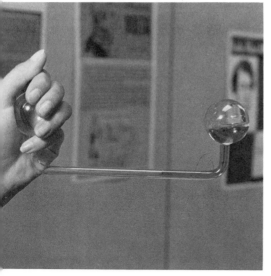

(a) The palm glass
This device can be used to compare the temperatures of various people's hands.

(b) These hot-air balloons are drifting over a golf course near the Bow River in Calgary, Alberta.

Liquids expand when heated. Mercury and alcohol expand uniformly so they are useful in thermometers.

Solids are also subject to expansion and contraction. For instance, spaces between sections of railway tracks are large in the winter and small in the summer. (See Figure 10–13.)

Figure 10–13 Spaces between sections of railway tracks allow for expansion of metals in the summer.

PRACTICE

24. A metal screw cap on a glass jar is too tight to remove. You could use hot water to loosen the cap. Why would this work?

10.11 Experiment 19: Expansion of Solids

INTRODUCTION

Metals expand when heated, but different metals expand different amounts. The difference has useful applications, one of which you will study in this experiment.

PURPOSE: To determine how heat affects the size of metal objects and to observe an application of the expansion of metals.

APPARATUS: ball-and-ring apparatus; compound bar; thermostat and related apparatus; Bunsen burner

PROCEDURE

1. With the ball and ring at room temperature, try to pass the ball through the ring. Now heat the ball vigorously in a Bunsen flame for about two minutes. Again try to pass the ball through the ring.

2. Cool the ball with running water and again try to pass it through the ring.
3. Observe the construction of a compound bar. Heat the bar evenly and describe what happens.
4. Predict what will happen to the hot compound bar when it is held under cold running water. Try it.
5. Set up the thermostat as shown in Figure 10–14. Heat the thermostat's compound bar. Describe the resulting action.

QUESTION

1. Explain how a compound-bar thermostat can control the operation of a furnace in a home.

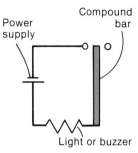

Figure 10–14
The thermostat

10.12 The Special Case of Water

Most substances expand when heated and contract when cooled. Water acts the same way except between certain temperatures. As water cools from 100°C to 4°C it contracts. At 4°C the density of liquid water is greatest because the molecules are closest together. Then as water cools from 4°C to 0°C it acts differently. It expands. This unusual behaviour is very fortunate. It means that in a body of water, such as a lake, the water at the bottom does not go below about 4°C. The colder, less dense water rises to the top, and that is where it freezes. The graph in Figure 10–15 shows how the volume of water changes at temperatures around 4°C.

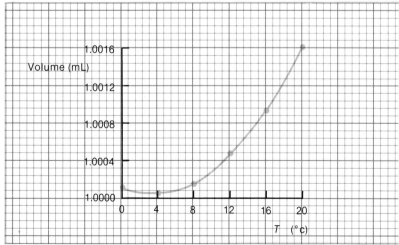

Figure 10–15 The volume of water is a function of temperature.

If water acted like other substances, the coldest water would go to the bottom of the body of water and freeze from the bottom up. In cold weather the entire body of water would then freeze. As you can imagine, life in northern lakes and oceans would not be possible.

PRACTICE

25. If a glass bottle containing pop is placed in a freezer, it might explode. Why?

10.13 Medical Uses of Heat Energy

Medical uses provide interesting applications of both hot and cold temperatures.

Hot baths, often called Roman baths, were used during the time of the Roman Empire to relieve pain. A hot bath relaxes muscles and causes the blood to flow more rapidly. Nowadays, people with painful ailments seek relief at health spas located near hot-water springs around the world.

Hot water provides heat energy to the surface of the human body. To get deeper into the skin, radiation treatment may be used. Radiation from a heat lamp is a form of such treatment. Of course, care must be exercised when applying radiation.

Body temperature is important in medicine. Normal body temperature is 37°C. A slight fever increases that temperature to 38°C. Anyone with a temperature of 40°C should be under a doctor's care.

When a person dies, the body temperature gradually cools to the temperature of its surroundings. By measuring the dead body's temperature, a doctor can determine the time of death within certain limits. Such information can be useful in crime detection.

Heat energy radiating from the diseased area of a body may be used to take a certain kind of picture called a thermograph. A cancerous tumour, for example, is about 1°C warmer than the area near it. So it will show up as a shaded region in a thermograph.

Taking away heat energy is also useful for medical purposes. Cryogenics, the science of making and using cold temperatures, allows the safe storage of rare blood, bone marrow and body tissue. Surgery done at low temperatures prevents both bleeding and pain.

Finally, hearing the wind-chill factor on the weather report helps people decide how to dress in winter. Moving air takes heat away from a body more rapidly than still air. Table 10–3 shows the effect that wind has at three different temperatures. As an example, at a temperature of 0°C, a wind of 10 m/s would cool a body at the same rate as a temperature of –13°C with no wind.

Table 10–3 Wind-chill Factor

Wind speed (m/s)	Actual temperature (°C)			
	20	0	–20	
10	15	–13	–40	} Equivalent temperature (°C) (with no wind)
20	13	–17	–48	

PRACTICE

26. A weather forecaster predicts a low temperature of –20°C and a wind of 10 m/s (almost 40 km/h). What will the wind-chill factor be?

10.14 Review Assignment

1. Distinguish between heat and temperature.　(10.2)
2. Change to kelvins:
   (a) 30°C   (b) 110°C   (c) –200°C　(10.3)
3. Change to degrees Celsius:
   (a) 300 K   (b) 580 K   (c) 173 K　(10.3)
4. Do you think it would be possible for a substance to have no heat whatsoever? Explain your answer.
5. Assume that sometime in the winter you go to the basement of your home in bare feet. You walk across the cement floor onto a rug. The cement and rug are at the same temperature, but the cement feels much colder. Explain why.　(10.6)
6. Describe how a land breeze is set up. Use a diagram and the KMT of matter.　(10.7)
7. What colour of clothing should be worn by a person travelling by camel across a desert? Why?　(10.8)
8. **Class demonstration**
   A radiometer (or Crookes' radiometer) is a device with four vanes connected to a pivot, as shown in the diagram. One side of each vane is painted silver and the other black. Most, but

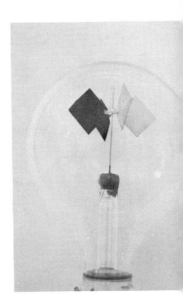

not all, of the air particles have been taken out of the radio-meter.

(a) Predict what will happen when a light is aimed at the vanes.

(b) Check your prediction by experiment.

(c) Explain the results.   (10.8 and your teacher)

9. What is the best method of heat transfer through a:
(a) metal?   (b) vacuum?   (c) liquid?   (10.6, 10.7, 10.8)

10. State which method of heat transfer:
(a) does not require molecules
(b) works because molecules collide with their neighbours
(c) travels at the speed of light
(d) works when molecules circulate in a path   (10.6, 10.7, 10.8)

11. Manufacturers line winter outer wear with the down feathers of geese and ducks. Why?   (10.6, 10.9)

12. Material A has an R value of 6.5 and material B has one of 10.5. Which is the better insulator?   (10.9)

13. Calculate the total R value of a wall with these materials: 10 cm brick, 2.0 cm gypsum board, 5.0 cm Styrofoam (10.9)

14. Explain why metal bridges are made in sections with spaces between the sections.   (10.10)

15. Pyrex glass expands and contracts much less than ordinary glass. Explain why a Pyrex container can be placed from a hot oven into cold water without cracking.   (10.10)

16. A material called invar expands 0.9 mm for each kilometre of length with a temperature increase of 1.0°C. Aluminum expands 25 mm under the same conditions. Which material would make a more accurate measuring tape? Why? (10.10)

17. **Class demonstration**
Add water to a depth of about 2 cm to a can made of thin metal and having a screw cap. (A ditto-fluid can works well.) Heat the water until steam pours rapidly out of the spout. Place the lid tightly on the can and remove the source of heat. Hold the can under cold running water.

(a) What, besides steam, was forced from the can during heating?

(b) Describe the observations after cooling.

(c) Explain what happened inside the can during cooling. (This demonstration also shows that outside the can atmospheric pressure exerts a large force. Refer to Section 3.3 for a discussion of atmospheric pressure.)   (10.10)

18. Two metals, X and Y, are attached in a bimetallic strip and heated. The strip bends as shown in the diagram. Which metal has the lower rate of expansion when heated? (10.11)
19. In what way does the action of water when cooled differ from the action of other substances when cooled? (10.12)

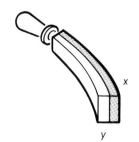

10.15 Answers to Selected Problems

PRACTICE QUESTIONS

6. 323 K
7. 20°C or 293 K; 37°C or 310 K
8. (a) 120°C
   (b) 200°C
   (c) -100°C

20. (a) 0.86
    (b) 2.5
    (c) 30
21. 27
26. -40°C

REVIEW ASSIGNMENT

2. (a) 303 K
   (b) 383 K
   (c) 73 K
3. (a) 27°C
   (b) 307°C
   (c) -100°C
13. 11

11
Measuring Heat Energy

GOALS: After completing this chapter you should be able to:
1. Define specific heat capacity.
2. State and use the SI units of specific heat capacity.
3. Calculate the amount of heat energy gained or lost by an object that experiences a temperature change ($E = mc\Delta T$).
4. Determine the exchange of heat energy when cold and hot water are mixed together.
5. Determine the power rating of a source of heat ($P = \dfrac{E}{t}$).
6. Determine the specific heat capacities of various liquids.
7. List the three states of matter.
8. Name the possible changes of state of matter.
9. Define specific latent heat of fusion (l_F) and specific latent heat of vaporization (l_V), and determine them for water ($l = \dfrac{E}{m}$).
10. State and use the SI units of specific latent heat.
11. Use the kinetic molecular theory of matter to explain the changes of state of matter.
12. Describe applications of changes of state of matter.

Knowing the information in this chapter will be especially useful if you plan a career in:

- refrigeration
- chemistry
- air conditioning

11.1 Specific Heat Capacity

Heat is a measure of the total energy content of the particles of an object. (See Section 10.2.) An object's heat content depends on the:
(1) mass of the object
(2) temperature of the object
(3) substance of which the object is composed

Substance is a factor because different substances have different capacities to hold heat energy. For example, water holds heat energy better than steel. One kilogram of water at 90°C will retain its heat much longer than 1.0 kg of steel at 90°C in the same room.

Water is said to have a higher specific heat capacity than steel. The word specific is used to indicate that we are considering equal masses of substances.

Specific heat capacity, c, is a measure of the amount of energy needed to raise the temperature of 1.0 kg of a substance by 1.0°C. It is measured in joules per kilogram degree Celsius, J/(kg·°C).

The English scientist named James Joule (1818-1889) performed original experiments to find the specific heat capacities of substances. That is why the unit of energy, the joule, is named after him. He discovered that it takes 4200 J of energy to raise the temperature of 1.0 kg of water by 1.0°C. Thus:

$$c_w = 4200 \frac{J}{kg \cdot °C} \text{ where } c_w \text{ is the specific heat capacity of water}$$

This value also means that 1.0 kg of water gives up 4200 J of energy when cooled by 1.0°C.

To calculate the energy gained or lost by an object, we must find the product of the mass m, the specific heat capacity c, and the amount of the temperature change ΔT.

$$E = mc\Delta T$$

Sample problem 1: How much heat energy is needed to raise the temperature of 4.0 kg of water from 3°C to 5°C?
Solution: $E = mc\Delta T$

$$= 4.0 \text{ kg} \times 4200 \frac{J}{kg \cdot °C} \times 2.0°C$$
(units divide)

$$= 33\ 600 \text{ J (or } 3.4 \times 10^4 \text{ J)}$$

Different substances have different specific heat capacities, as shown in Table 11-1.

Table 11-1 Specific Heat Capacities of Common Substances

Substance	$c\ (\frac{J}{kg \cdot °C})$	Substance	$c\ (\frac{J}{kg \cdot °C})$
Water	4200	Glass	840
Alcohol	2500	Iron	450
Ice	2100	Brass	380
Steam	2100	Silver	240
Aluminum	920	Lead	130

Sample problem 2: Calculate the amount of heat energy lost by a 6.0 kg iron rod as it cools from 28°C to 24°C.
Solution: $E = mc\Delta T$

$$= 6.0 \text{ kg} \times 450 \frac{J}{kg \cdot °C} \times 4.0°C$$

$$= 10\ 800 \text{ J or } 1.1 \times 10^4 \text{ J}$$

PRACTICE
1. Calculate the amount of heat energy needed to raise the temperature of:
   (a) 10 kg of water by 5.0°C
   (b) 5.0 kg of alcohol by 6.0°C
   (c) 4.0 kg of lead by 50°C
2. Determine the heat energy lost when:
   (a) 2.0 kg of water cools from 30°C to 22°C
   (b) a 3.0 kg piece of silver cools from 90°C to 20°C
   (c) 2.0 kg of ice is cooled from –3°C to –6°C

11.2 Experiment 20: Measuring Exchange of Heat Energy

INTRODUCTION
When hot and cold water are mixed together, the hot water loses energy to the cold water. In this experiment you will determine whether the heat energy lost by the hot water equals the heat energy gained by the cold water. (You may wish to relate this experiment to the law of conservation of energy in Section 8.6.)

PURPOSE: To determine if heat energy is conserved when hot and cold water are mixed together.

APPARATUS: 2 large Styrofoam cups; thermometer; graduated cyclinder; hot and cold water; stirring rod

PROCEDURE
1. Place 100 mL (0.1 kg) of cold water in one cup and 100 mL of hot water in the other cup. Measure and record the temperature of each sample of water.

2. Add the hot water to the cold water. Stir the mixture and find its final temperature.
3. Calculate the change of temperature and heat energy gained by the cold water.
4. Calculate the change of temperature and heat energy lost by the hot water. Complete the first column of Table 11–2.
5. Compare the heat energy lost with the heat energy gained. Explain any difference.
6. Repeat the above steps using the quantities of hot and cold water indicated in columns 2, 3 and 4 of Table 11–2.

Table 11–2 Observations and Calculations for Experiment 20

Trial number	1	2	3	4	
Volume (mL)	100	150	50	75	
Mass (kg)	0.1				
Starting temp (°C)					
Final temp (°C)					Cold water
ΔT (°C)					
Specific heat [J/(kg·°C)]	4200				
Heat gained (J)					
Volume (mL)	100	50	150	175	
Mass (kg)					
Starting temp (°C)					
Final temp (°C)					Hot water
ΔT(°C)					
Specific heat [J/(kg·°C)]					
Heat lost (J)					

QUESTION
1. Explain how the equipment in this experiment could be changed to obtain better results.

11.3 Power and Heat Energy

Power is the rate of consuming energy. (See Sections 8.7 and 8.8.)
It can be calculated by finding the ratio of energy consumed to the
time.

$$P = \frac{E}{t}$$

In the SI, energy is measured in joules, time in seconds and power
in joules per second or watts.

If a source of heat energy is used to heat a substance, a simple
procedure can be used to find the power of the source. First we
find the energy given by the source to the substance ($E = mc\Delta T$)
in a certain amount of time t. Then we calculate the power
($P = \frac{E}{t}$).

Sample problem 3: A hot plate heats 100 g (0.1 kg) of water
from 20°C to 80°C in 84 s. Calculate the:
(a) heat energy given by the source to the water
(b) power of the source
Solution: (a) $E = mc\Delta T$

$$= 0.1 \text{ kg} \times 4200 \frac{J}{\text{kg} \cdot {}^\circ C} \times 60°C$$

$$= 25\ 200 \text{ J}$$

(b) $P = \frac{E}{t}$

$$= \frac{25\ 200 \text{ J}}{84 \text{ s}}$$

$$= 300 \text{ W}$$

We can also find the energy given by a heat source to a substance if
the power of the source is known. If the equation $P = \frac{E}{t}$ is rearranged to express energy by itself, we have

$$E = Pt$$

Sample problem 4: A 50 W immersion heater warms a sample
of water for 90 s. How much heat energy is given to the water?
Solution: $E = Pt$

$$= 50\frac{J}{s} \times 90 \text{ s} \qquad (50 \text{ W} = 50\frac{J}{s})$$

$$= 4500 \text{ J}$$

PRACTICE

3. A heater raises the temperature of 2.0 kg of water by 2.0°C in 100 s. Find the:
   (a) energy given to the water
   (b) power of the heater
4. A Bunsen burner heats 200 mL (0.2 kg) of water from 10°C to 60°C in 420 s. Determine the:
   (a) energy given to the water
   (b) power of the burner
5. How much heat energy is given by a 200 W burner in 30 s?
6. An electric heater, rated at 80 W, adds heat energy to a sample of water for 2 min (120 s). Find the heat energy given to the water.

11.4 Experiment 21: The Power of a Heater

INTRODUCTION

The instructions for this experiment suggest that a Bunsen burner be used as the heat source. Other sources, such as electric heaters, may be used. The important thing is to learn how to find the power of the source so that the same procedure can be used in Experiments 22, 23 and 24.

PURPOSE: To determine the power of a source of heat energy.

APPARATUS: 250 mL beaker; graduated cylinder; thermometer; stopwatch; Bunsen burner; retort stand; ring clamp; wire support; stirring rod

PROCEDURE

1. Set up the apparatus as shown in Figure 11-1. Add 200 mL (0.2 kg) of cold water to the beaker.
2. Light the Bunsen burner and adjust it to give a constant, powerful flame. Stir the water constantly with the stirring rod. When the temperature reaches 30°C, start the stopwatch.
3. Continue stirring the water. Find the time it takes to heat the water to 70°C.
4. Calculate the heat energy gained by the water ($E = mc\Delta T$).
5. Calculate the power of the burner ($P = \dfrac{E}{t}$).

Figure 11-1

QUESTIONS

1. What factors do you think affect the power rating calculated in this experiment?
2. Explain how the results of this experiment could be improved.

11.5 Calculating Specific Heat Capacity Using Power

Assume you are asked to calculate the specific heat capacity of liquid A. You are given a Bunsen burner and other experimental apparatus. How would you find c_A?

The equation using specific heat capacity is $E = mc\Delta T$. If we rearrange that equation to express c by itself, we have:

$$c = \frac{E}{m\Delta T}$$

Thus, we must find the energy ($E = Pt$) given to the mass m of an object to change its temperature by ΔT.

Sample problem 5: A Bunsen burner, rated at 100 W, is used to heat 0.5 kg of liquid A for 5 min (300 s). The liquid's temperature rises from 25°C to 55°C. Calculate:

(a) the energy given to liquid A
(b) c_A

Solution: (a) $E = Pt$
$= 100 \text{ W} \times 300 \text{ s}$
$= 3.0 \times 10^4 \text{ J}$
or $= 30\ 000 \text{ J}$

(b) $c_A = \dfrac{E}{m\Delta T}$

$= \dfrac{3.0 \times 10^4 \text{ J}}{0.5 \text{ kg} \times 30°C}$

$= 2000 \dfrac{\text{J}}{\text{kg·°C}}$

PRACTICE

7. A 500 W electric heater changes the temperature of 5.0 kg of liquid B by 10°C in 100 s. Find:
   (a) the energy given to liquid B
   (b) the specific heat capacity of liquid B

11.6 Experiment 22: Specific Heat Capacities of Various Liquids

INTRODUCTION

In this experiment you must know the mass of each liquid used. In previous experiments finding the mass of water was easy because the density of water is 1000 kg/m³ or 1000 g/L. Thus, 100 mL of water has a mass of 0.1 kg.

However, different liquids have different densities. So in this experiment a balance is used to find the mass of the liquid.

PURPOSE: To determine the specific heat capacities of various liquids.

APPARATUS: graduated cylinder; triple-beam balance; 150 mL beaker; retort stand; ring clamp; wire support; thermometer; stopwatch; 500 mL beaker; various liquids (cooking oil, ethylene glycol and ethyl alcohol); stirring rod; Bunsen burner

PROCEDURE

1. Set up the apparatus as shown in Figure 11–1, Experiment 21.
2. Determine the power of the burner as you did in Experiment 21. (Use 100 mL or 0.1 kg of water.)
3. Find the mass of the beaker on the balance. Shift the balance's rider to read 100 g more than the mass of the beaker. (For example, if the beaker is 42.5 g, set the balance at 142.5 g.)
4. Add cooking oil to the beaker until the beam is in the balanced position. This means that you have 100 g or 0.1 kg of oil.
5. Place the beaker with the oil on the wire support and begin heating it. Stir the oil constantly with the thermometer.
6. When the temperature reaches 30°C, start the stopwatch. Find the time it takes the temperature to reach 70°C. Remember to keep stirring.
7. Calculate the heat energy gained by the oil ($E = Pt$).
8. Calculate the specific heat capacity of the cooking oil $(c_C = \dfrac{E}{m\Delta T})$.
9. Return the oil to a common container. Clean and dry all the apparatus.
10. Repeat # 3 to # 9 for ethylene glycol. (**Caution: Do not inhale the fumes from hot glycol.**)

11. Repeat #3 to #9 for ethyl alcohol. (**Caution: Alcohol is a fire hazard. Do not let it spill near an open flame.** It should be heated in a water bath in the 500 mL beaker.)

QUESTIONS
1. Compare your answers with those found by other members of the class.
2. The specific heat capacity of ethylene glycol is 2200 J/(kg·°C). Compare your value with this one. Try to account for any difference.

11.7 The States of Matter

Matter may exist in one of three possible states—solid, liquid or gas. A **solid** has a fixed volume and shape. A **liquid** has a fixed volume but no fixed shape. A **gas** has no fixed volume or shape.

A substance may change from one state to another by having heat energy added or taken away. For example, the substance H_2O (water) is solid (ice) at low temperatures. When enough heat energy is added to the solid, it becomes a liquid (water). When more heat energy is added to the liquid, it becomes a gas (steam).

Listed below are the names of the changes of state that may occur:

(1) **Fusion**, or melting, is the change from solid to liquid (e.g., ice melts at 0°C).
(2) **Vaporization**, also called evaporation, is the change from liquid to gas (e.g., water changes to steam when it boils at 100°C).
(3) **Condensation**, or liquification, is the change from gas to liquid (e.g., steam condenses to water at 100°C).
(4) **Solidification**, or freezing, is the change from liquid to solid (e.g., water freezes at 0°C).
(5) **Sublimation** is the change from solid to gas or from gas to solid (e.g., frost to steam or steam to frost).

These changes of state are summarized in Figure 11–2.

Sublimation is not as common as the other changes of state. A substance called dry ice may be used to illustrate sublimation. Dry ice is the solid form of carbon dioxide, a gas at room temperature. The freezing point of dry ice is –79°C, so it will sublimate in a room above that temperature. If a sample of dry ice is exposed to

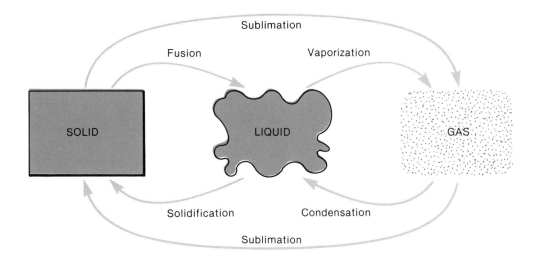

Figure 11-2 Changes of state of matter

the air, it gradually disappears, changing directly from a solid to a gas, without becoming a liquid.

A more interesting demonstration is observed if you place several small chunks of dry ice in a balloon and tie the mouth of the balloon tightly. As the dry ice sublimates, the carbon dioxide gas fills the balloon. (**Caution:** Do not hold the dry ice with bare hands.)

Table 11-3 lists the melting and boiling temperatures of several substances.

Table 11-3 Melting and Boiling Temperatures

Substance	Melting point (°C)	Boiling point (°C)
Aluminum	660	2330
Copper	1080	2580
Hydrogen	−259	−253
Iron	1535	2800
Lead	327	1750
Mercury	−39	357
Silver	691	2190
Water	0	100

PRACTICE

8. Pure water (H_2O) freezes at 0°C and boils at 100°C. At what temperature would each of the following likely occur for H_2O?
   (a) fusion?   (b) vaporization?
   (c) condensation?   (d) solidification?
   (e) sublimation?   (Hint: Think of frost in winter.)
9. Moth balls are made of a chemical that emits a gas at room temperature with a noticeable smell. What change of state occurs as the gas forms?
10. At room temperature (20°C) in what state is:
    (a) aluminum?   (b) hydrogen?   (c) mercury?
    (d) silver?   (Table 11–3)

11.8 Specific Latent Heats of Fusion and Vaporization

A large quantity of heat energy is involved when a substance changes state. Heat energy is required for a change of state from solid to liquid (fusion), or from liquid to gas (vaporization).

The amount of heat energy needed to change 1.0 kg of a substance from a solid to a liquid is called the **specific latent heat of fusion** (l_F). The word specific refers to 1.0 kg of the substance. The word latent means hidden, which refers to the fact that adding heat energy to a substance changing state does not change the temperature.

The equation for the specific latent heat of fusion is:

$$l_F = \frac{E}{m}$$

In the SI heat of fusion is measured in joules per kilogram (J/kg).

Sample problem 6: A total of 4000 J of heat energy will melt a 0.1 kg sample of a metal. What is the metal's specific latent heat of fusion?

Solution: $l_F = \dfrac{E}{m}$

$$= \frac{4000 \text{ J}}{0.1 \text{ kg}}$$

$$= 40\ 000 \text{ J/kg or } 4.0 \times 10^4 \text{ J/kg}$$

The amount of heat energy needed to change 1.0 kg of a substance from a liquid to a gas is called the **specific latent heat of vaporization** (l_v). The temperature remains constant as the substance boils. In equation form:

$$l_v = \frac{E}{m}$$

The units for heat of vaporization are joules per kilogram (J/kg).

Table 11–4 lists the specific latent heats of fusion and vaporization of four substances. Notice that vaporization requires more heat energy per kilogram than fusion.

Table 11–4 Specific Latent Heats of Fusion and Vaporization

Substance	l_F (J/kg)	l_V (J/kg)
Alcohol	1.1×10^5	8.6×10^5
Iron	2.5×10^5	6.3×10^6
Lead	2.3×10^4	8.7×10^5
Silver	1.1×10^5	2.3×10^6

If the specific latent heats are known, they can be used to calculate the heat energy needed to melt or boil away certain amounts of substances. The equation $l = \frac{E}{m}$ can be rearranged to express energy by itself:

$$E = l_F m \text{ and } E = l_V m$$

Sample problem 7: How much heat energy is needed to change 20 kg of solid silver into liquid silver?

Solution: $E = l_F m$

$$= 1.1 \times 10^5 \frac{J}{kg} \times 20 \text{ kg}$$

$$= 2.2 \times 10^6 \text{ J}$$

PRACTICE

11. Calculate the specific latent heat of fusion if 8.4×10^5 J of energy are needed to melt 2.0 kg of a substance.
12. Calculate the specific latent heat of vaporization if 8.4×10^5 J of energy are needed to boil away 0.4 kg of a substance.
13. How much heat energy is required to change 10 kg of:
    (a) liquid silver into gaseous silver at the boiling temperature?
    (b) solid iron into liquid iron at the melting temperature?
    (c) liquid alcohol into gaseous alcohol at the boiling temperature?

11.9 Experiment 23: Specific Latent Heat of Fusion of Ice

INTRODUCTION
As you are performing this experiment, think of possible sources of error. Be sure you keep stirring the ice and water for the entire experiment.

PURPOSE: To determine the specific latent heat of fusion of ice.

APPARATUS: 150 mL beaker; graduated cylinder; mass scale; thermometer; stopwatch; Bunsen burner; retort stand; ring clamp; wire support; 3 regular size ice cubes; stirring rod

PROCEDURE
1. Set up the apparatus as shown in Figure 11–1, Experiment 21. Find the power of the Bunsen burner, using 100 mL (0.1 kg) of water in the 150 mL beaker. Keep the flame constant for the entire experiment. (If you have time to perform Experiment 24, keep the flame constant for it as well.)
2. Cool the beaker, dry it and measure its mass.
3. Add three ice cubes to the beaker and measure the total mass. Calculate the mass of the ice by subtracting the mass of the beaker.
4. With the Bunsen flame on, place the beaker with the ice cubes on the wire support. Start the stopwatch, stir the ice and measure the temperature.
5. Find how long it takes to melt the ice completely. Notice what happens to the temperature as the ice is melting.
6. Calculate the heat energy given to the ice $(E = Pt)$.
7. Calculate the specific latent heat of fusion $(l_F = \dfrac{E}{m})$.

QUESTIONS
1. Describe what happened to the temperature during the melting of the ice.
2. Compare your value with the accepted value of $l_F = 3.3 \times 10^5$ J/kg.
3. Explain any weaknesses in this experiment.

11.10 Experiment 24: Specific Latent Heat of Vaporization of Water

INTRODUCTION

A great amount of heat energy is required to boil away water. (Compare the heats of fusion and vaporization for substances listed in Table 11–4.) For this reason a small amount of water will be used in this experiment.

PURPOSE: To detemine the specific latent heat of vaporization of water.

APPARATUS: 150 mL beaker; graduated cylinder; thermometer; stopwatch; Bunsen burner; retort stand; ring clamp; wire support; stirring rod

PROCEDURE

1. Determine the power of the Bunsen burner in the usual way (Experiment 21).
2. Place 52 mL of water in the beaker and heat it to boiling, stirring constantly.
3. Start the stopwatch as soon as the boiling begins. Notice the temperature of the water as the boiling continues.
4. Measure the time it takes the water to boil almost completely away. Carefully remove the beaker from the heat when there is about 2 mL of water left.
5. Calculate the heat energy given to the water $(E = Pt)$.
6. Calculate the specific latent heat of vaporization of water $(l_v = \dfrac{E}{m}$, where $m = 0.05$ kg$)$.

QUESTIONS

1. Describe what happened to the temperature during the boiling of the water.
2. Compare your value with the accepted value of $l_v = 2.3 \times 10^6$ J/kg.
3. Explain any weaknesses in this experiment.

11.11 Explaining Fusion and Vaporization

Let us summarize what happened to the heat energy in each of the last three experiments involving water.

(1) In Experiment 21 heat energy caused the temperature of liquid water to rise from 30°C to 70°C.
(2) In Experiment 23 heat energy caused the ice to melt, forming liquid water. The heat energy was used for fusion, not for increasing the temperature. (A slight increase in temperature may be noticed toward the end of the experiment.)
(3) In Experiment 24 heat energy caused the water to change to steam. The temperature of the boiling water remained constant.

The KMT of matter can be used to describe why heat energy caused an increase in temperature in Experiment 21. The added heat energy makes the molecules move more violently.

The KMT of matter can also be used to explain fusion and vaporization. Assume we have a quantity of ice at –50°C, and we wish to heat it until the temperature is 150°C. At –50°C the molecules of ice vibrate slowly around a set position. As heat energy is added, the vibrations increase until the temperature reaches 0°C.

At 0°C the heat energy causes the molecules to become free of their set positions. The solid gradually becomes a liquid. The temperature remains constant at 0°C until the ice has melted.

Then more heat energy causes the liquid's temperature to rise from 0°C to 100°C. The molecules move more rapidly as the temperature rises. At 100°C boiling begins. Again the temperature remains constant because the added heat energy goes to free molecules from the liquid water. The molecules that have enough energy gain their freedom and become gas molecules.

When all the water has become steam, the steam molecules can be heated to higher temperatures with added energy. Thus, the steam may be heated to 150°C.

To summarize, when a substance is in one state, adding heat energy increases its temperature. When a substance is changing state (solid to liquid, liquid to gas, or solid to gas), adding heat energy increases the freedom of motion of the molecules but does not increase the temperature.

11.12 Applications of Changes of State of Matter

The freezing point of pure water is 0°C. That freezing point changes when other substances are mixed with water. For example, antifreeze (ethylene glycol) mixed with water makes the freezing point low enough that water in a car's radiator will not freeze in the winter.

Adding salt to water also lowers the freezing point. If the freezing point is lower, the melting point of ice or snow is also lower. That is why salt is spread on major highways that are icy or snowy when the temperature is about –5°C to –10°C.

Changing air pressure affects the boiling temperature of water. Normally, pure water boils at 100°C. However, if the pressure pushing down on the water increases, the temperature must be higher before boiling occurs. Water in a pressure cooker, for example, may have to reach 200°C before it boils. At that high temperature, vegetables would cook much more quickly than at 100°C.

Also, if the air pressure is reduced, as is the case high in the mountains, water boils at a lower temperature. This creates overheating problems for some cars as they drive across mountain ranges.

The fact that water has a high specific latent heat of vaporization (2.3×10^6 J/kg) means that much energy is needed to cause water to evaporate. This explains why a person feels cool when first coming out of a lake or ocean on a hot day. As the moisture evaporates from the body, heat energy is taken away both from the body and from the surrounding air. Sweating involves a similar process to help cool the body.

Finally, water's high specific latent heats of fusion and vaporization help maintain a uniform climate in regions near large bodies of water. Areas by the Atlantic and Pacific coasts experience more moderate climates than areas on the prairies. On the dry prairies there are few large bodies of water to absorb heat energy in the summer or give off heat energy in the winter.

PRACTICE

14. **Student demonstration:** Add room temperature alcohol to one beaker and room temperature water to a second beaker. Place your left index finger into one liquid and your right index

finger into the other. Then hold your fingers in the air.
(a) Which finger feels cooler?
(b) Which liquid evaporates more quickly?
(c) Explain why one finger is cooled more than the other.

11.13 Review Assignment

1. State the SI units used to measure:
   (a) heat energy   (b) specific heat capacity   (c) mass
   (d) change of temperature (e) time   (f) power
   (g) specific latent heat of fusion
   (h) specific latent heat of vaporization
2. Calculate the heat energy lost or gained in each case. (Specific heat capacities are given in Table 11-1, Section 11.1.)
   (a) 5.0 kg of alcohol is warmed from 15°C to 25°C
   (b) 3.0 kg of brass is cooled from 100°C to 60°C
   (c) 2.5 kg of glass is heated from 90°C to 110°C
   (d) 500 g of iron is cooled from 80°C to 0°C   (11.1)
3. In a "perfect" experiment, a sample of hot water is mixed with a sample of cold water. The hot water loses 2.0×10^4 J of heat energy.
   (a) How much energy does the cold water gain?
   (b) What factors would prevent such perfect results in a science laboratory?   (11.2)
4. An electric stove heats 2.0 L (2.0 kg) of water from 15°C to 95°C in 350 s. Calculate the:
   (a) heat energy given to the water
   (b) power of the stove's burner   (11.3)
5. A 1200 W electric frying pan warms a liquid sauce for 4 min (240 s). How much heat energy is given to the sauce? (11.3)
6. A student performs an experiment to find the specific heat capacity of liquid L. The Bunsen burner flame is rated at 200 W for the experiment. The student heats 0.2 kg of the liquid from 25°C to 75°C. The heating takes 120 s.
   (a) Find the energy given by the flame to liquid L.
   (b) Calculate the specific heat capacity of liquid L.
   (c) What liquid might L be?   (11.5 and Table 11-1)
7. State the scientific name for:
   (a) freezing (b) melting   (c) boiling   (11.7)

8. A tray of ice cubes is left for several months in the freezer of a "frost-free" refrigerator. The ice slowly disappears.
   (a) What change of state is gradually taking place?
   (b) Is this a good way to conserve energy?   (11.7)

9. **Class demonstration**
   Crystals form when certain liquids become solids. Phenyl salicylate (also called salol) is a crystal at room temperature. Gently heat a small amount of salol on a glass slide. Watch closely as the liquid cools, forming a crystallized solid. (A microscope will help.) Discuss how the KMT of matter may be used to explain what happens.   (10.4, 11.7, 11.11)

10. Calculate the specific latent heat of fusion in each case:
    (a) 8.0×10^4 J of heat energy is needed to melt 2.0 kg of a certain substance
    (b) 3.3×10^4 J of heat will melt 0.1 kg of ice   (11.8)

11. Calculate the specific latent heat of vaporization in each case:
    (a) 3.0×10^6 J of heat energy boil away 1.5 kg of a liquid
    (b) 2.3×10^5 J of heat energy boil away 100 g (0.1 kg) of water   (11.8)

12. How much heat energy is needed to change:
    (a) 2.0 kg of solid iron into a liquid at the melting temperature?
    (b) 3.0 kg of liquid silver into a gas at the boiling temperature? (Refer to 11.8 and Table 11–4.)

13. Calculate the heat energy that must be taken away from 10 kg of water at 0°C to change it to 10 kg of ice at 0°C. (Use $l_F = 3.3 \times 10^5$ J/kg.)   (11.8)

14. Choose the temperature-time graph that best represents the temperature of H_2O as heat energy is added to it. Explain your choice.   (11.11)

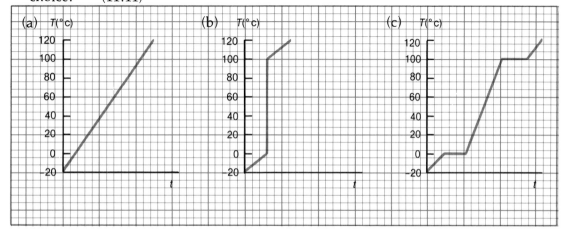

15. Use the KMT of matter to explain why water boils at a lower temperature when the air pressure is reduced.   (10.4, 11.11, 11.12)

11.14 Answers to Selected Problems

PRACTICE QUESTIONS

1. (a) 2.1×10^5 J
   (b) 7.5×10^4 J
   (c) 2.6×10^4 J
2. (a) 6.7×10^4 J
   (b) 5.0×10^4 J
   (c) 1.3×10^4 J
3. (a) 1.68×10^4 J
   (b) 168 W
4. (a) 4.2×10^4 J
   (b) 100 W

5. 6.0×10^3 J
6. 9.6×10^3 J
7. (a) 5.0×10^4 J
   (b) $1000 \dfrac{J}{kg \cdot {}^\circ C}$
11. 4.2×10^5 J/kg
12. 21×10^5 J/kg or 2.1×10^6 J/kg
13. (a) 2.3×10^7 J
    (b) 2.5×10^6 J
    (c) 8.6×10^6 J

REVIEW ASSIGNMENT

2. (a) 1.3×10^5 J
   (b) 4.6×10^4 J
   (c) 4.2×10^4 J
   (c) 1.8×10^4 J
3. (a) 2.0×10^4 J, assuming the energy is conserved
4. (a) 6.7×10^5 J
   (b) 1920 W or 1.9×10^3 W
5. 2.9×10^5 J
6. (a) 2.4×10^4 J
   (b) 2.4×10^3 J
10. (a) 4.0×10^4 J/kg
    (b) 3.3×10^5 J/kg
11. (a) 2.0×10^6 J/kg
    (b) 2.3×10^6 J/kg
12. (a) 5.0×10^5 J
    (b) 6.9×10^6 J
13. 3.3×10^6 J
14. (c)

V. Waves, Sound and Music

12

Vibrations and Waves

GOALS: After completing this chapter you should be able to:
1. Define transverse, longitudinal and torsional vibrations.
2. Define period and frequency and understand how they are related.
3. Given one of period or frequency, calculate the other.
4. Define cycle and amplitude of a vibration.
5. State what factors affect the frequency of a simple pendulum.
6. State the purpose of a wave.
7. State factors that affect the speed of a wave.
8. Describe these parts of a transverse wave: rest line, crest, trough, wavelength and amplitude.
9. Describe these parts of a longitudinal wave: rest position, compression, rarefaction, wavelength and amplitude.
10. Use the universal wave equation $(v = f\lambda)$ to find the speed of a wave, knowing its frequency and wavelength.
11. Describe how constructive interference and destructive interference occur.
12. Draw a diagram to illustrate an interference pattern created by two sources of circular waves in water.
13. State what is meant by mechanical resonance.
14. Describe examples of mechanical resonance.
15. Draw the shape of a standing-wave pattern and from the shape find the wavelength of the waves.

Knowing the information in this chapter will be especially useful if you plan a career in:

• the recording industry
• music (for example, if you plan to become a musician)
• radio
• television
• acoustics
• environmental (sound) pollution
• audio-visual technology

12.1 Vibrations

The topics sound and light have many definitions related to vibrations. Those definitions are more easily understood if you can see what is happening. Thus, we start with vibrations that can be seen. Later we will study the invisible vibrations and waves of sound and light.

A **vibration** is a back-and-forth motion. The main types of vibrations are shown in Figure 12–1.

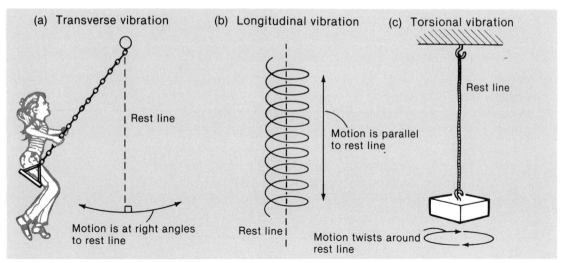

Figure 12–1 Types of vibration

(1) A **transverse vibration** occurs when the object vibrates at right angles to its line of rest (e.g., a child swinging on a swing).

(2) A **longitudinal vibration** occurs when the object vibrates parallel to its line of rest (e.g., a coil spring supporting a car).

(3) A **torsional vibration** occurs when the object twists around its line of rest (e.g., a string supporting an object can be twisted and the object will vibrate around and back).

To study transverse vibrations we will use a pendulum. Consider Figure 12–2. Diagram (a) illustrates the mass and line of rest of a pendulum as well as two quantities, length and amplitude, which may be measured. The **length** of a pendulum is the distance from its top to the middle of the mass. The **amplitude** (A) is the largest distance the mass moves from its rest position. Diagram (b) shows that a **cycle** is a complete vibration of the pendulum.

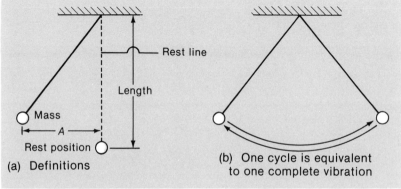

Figure 12–2 The pendulum

Sample problem 1: A child is swinging on a swing. The amplitude of vibration is constant at 1.0 m. How far does the child move in: (a) each cycle? (b) 3 cycles?
Solution:
(a) The child moves 4.0 m in each cycle.
(b) In three cycles the child moves 3 × 4.0 m = 12.0 m.

To study longitudinal vibrations we will consider a mass hung on the end of a spring, as shown in Figure 12–3. Diagram (a) shows the spring and mass at rest. In diagram (b) the mass has been lifted up. The amplitude *(A)* is the greatest distance from the rest position. Then the mass is released and drops to its lowest position in diagram (c). The cycle is complete when the mass returns to its top position, as in diagram (d).

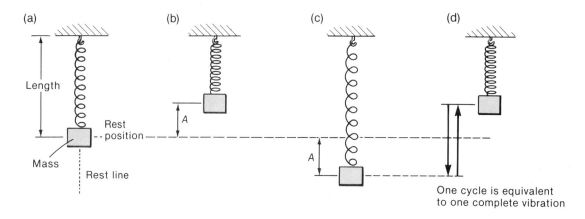

Figure 12–3 The spring

Torsional vibrations can be demonstrated by using a wave machine, such as the one shown in Figure 12–4. In that machine, a central bar twists, causing the rods attached to it to vibrate in transverse waves.

PRACTICE
1. What type of vibration is illustrated by each of the following?

(a)   (b)

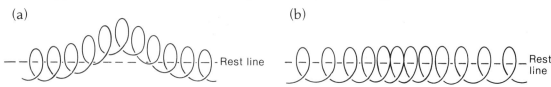

Figure 12–4 A wave machine

2. For the pendulum shown in the diagram, state the:
   (a) type of vibration
   (b) amplitude
3. The diagrams show a mass at rest on a spring and then vibrating. State the:
   (a) type of vibration   (b) amplitude

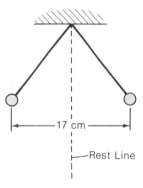

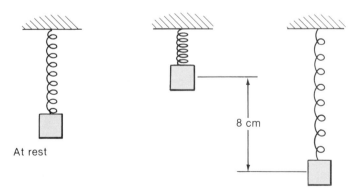

4. The amplitude of a certain pendulum is 10 cm. How far does the mass move in five cycles?
5. The amplitude of vibration of a mass on a spring is 3.0 cm. How far does the mass move in two cycles?

12.2 Frequency and Period

A vibrating object has a frequency and period. The **frequency** *(f)* is the number of cycles that occur in a specific amount of time.

$$f = \frac{\text{number of cycles}}{\text{total time}}$$

In the SI frequency is measured in cycles per second or hertz (Hz). This unit is named after the German physicist, Heinrich Hertz (1857-1894).

The **period** *(T)* is the amount of the time required for one cycle of vibration.

$$T = \frac{\text{total time}}{\text{number of cycles}}$$

The SI unit of period is seconds per cycle, or simply seconds.

Sample problem 2: A ball bounces 16 times in 8.0 s. Calculate the ball's: (a) frequency (b) period

Solution: (a) $f = \dfrac{\text{number of cycles}}{\text{total time}}$

$$= \frac{16 \text{ cycles}}{8.0 \text{ s}}$$

$$= 2.0 \text{ cycles/s or } 2.0 \text{ Hz}$$

(b) $T = \dfrac{\text{total time}}{\text{number of cycles}}$

$$= \frac{8.0 \text{ s}}{16 \text{ cycles}}$$

$$= 0.5 \text{ s/cycle or } 0.5 \text{ s}$$

Since frequency is measured in cycles per second and period in seconds per cycle, they are reciprocals of each other.

$$T = \frac{1}{f} \text{ and } f = \frac{1}{T}$$

Sample problem 3: Find the period if $f = 4.0$ Hz.

Solution: $T = \dfrac{1}{f}$

$$= \frac{1}{4.0 \text{ Hz}}$$
$$= 0.25 \text{ s}$$

Sample problem 4: Find the frequency if $T = 0.2$ s.

Solution: $f = \dfrac{1}{T}$

$ = \dfrac{1}{0.2 \text{ s}}$

$ = 5.0 \text{ Hz}$

PRACTICE

6. In a movie projector, 1800 frames are displayed every 60 s. Calculate the frequency and period for this situation.
7. Calculate the period of vibration if the frequency is:
   (a) 2.0 Hz   (b) 20.0 Hz   (c) 0.4 Hz
8. Find the frequency of an object that vibrates with a period of:
   (a) 5.0 s   (b) 0.75 s   (c) 0.01 s

12.3 Experiment 25: The Pendulum

INTRODUCTION

A pendulum swings with a regular period so it is useful as a device to measure time. In fact, Galileo Galilei made the first pendulum clock in 1581. First he used his pulse to discover the regular period of vibration of a lamp hanging in a church in Pisa, Italy. Then he performed laboratory experiments similar to the one you will perform now. He wanted to determine what factors affect the period and frequency of a swinging pendulum.

PURPOSE: To learn what happens to the frequency of a pendulum when you change its amplitude, mass and length.

APPARATUS: retort stand; clamp; string; stopwatch; metre stick; metal masses (50 g, 100 g and 200 g)

PROCEDURE

1. Set up a pendulum of length 100 cm using a 50 g mass. Remember that the length is measured to the middle of the mass.
2. Give the pendulum an amplitude of 10 cm. Measure the time for 20 complete cycles. Calculate the frequency. Enter the values in a chart similar to that shown in Table 12–1.

Table 12–1 Observation Chart for Experiment 25

Procedure #	Length (cm)	Mass (g)	A (cm)	Time for 20 cycles (s)	f (Hz)
2	100	50	10		

3. Repeat #2 using amplitudes of 20 cm and 30 cm. Determine what happens to the frequency of vibration when you change the amplitude.
4. Determine what happens to the frequency of vibration when you change the mass. Keep the length at 100 cm and the amplitude at 10 cm. Use masses of 100 g and 200 g. Be sure the length of the pendulum is always measured to the middle of the mass.
5. Determine what happens to the frequency of vibration when you change the length of the pendulum. Keep a constant amplitude of 10 cm and a constant mass of 50 g. Use lengths of 80, 60, 40 and 20 cm.

QUESTION
1. A pendulum clock, like the one shown in Figure 12–5, is running slower than normal. What should be done to increase its frequency?

Figure 12–5
A pendulum clock

12.4 Transfer of Energy

Energy can be transferred from one place to another by several methods. Electrical energy is transferred by the repelling force of tiny particles in a wire. Mechanical energy, such as the kinetic energy of a hammer, is transferred by exerting a force on an object. Heat energy may be transferred through particles by conduction and convection, as explained in Chapter 10. Another means of transferring energy is by **waves**.

There are many types of waves. Water waves are common. On oceans and lakes they are created by wind forces. The energy from water waves can wear away rocks along shores, creating interesting formations as well as small pebbles and grains of sand. See Figure 12–6.

Waves on water are easily seen, as are waves on ropes and coiled springs (Figure 12–7). For this reason, we will study waves on ropes, springs and water first. Then we will study waves, especially sound waves, that are invisible.

Figure 12-6 *The village in the photograph is located on the west coast of Portugal.*

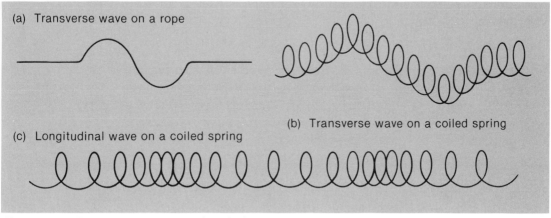

(a) Transverse wave on a rope

(b) Transverse wave on a coiled spring

(c) Longitudinal wave on a coiled spring

Figure 12–7 *Waves on ropes and coiled springs*

Another form of energy transfer by waves is called **radiation**. It is different from the other types of energy transfer because it does not need particles. Heat energy, for example, may be transferred by radiation. (See Chapter 10.) Light energy is transferred by radiation. Radio waves and X rays are other examples of radiation. We will study more about such waves after we have learned about waves we can see and hear.

12.5 Experiment 26: Pulses Travelling along a Coiled Spring

INTRODUCTION

A pulse (—⌒—or —⌣—) is simply half a wave (—⌒⌣—). The knowledge gained by studying pulses on a spring can be applied to all types of waves.

This experiment is not difficult to perform, but the conclusions are very important to the study of waves.

The distances indicated in the instructions are meant for short springs. If long ones are used, double all distances.

PURPOSE: To study the action of pulses moving along a coiled spring.

APPARATUS: coiled spring (such as a Slinky toy); piece of masking tape; piece of paper; stopwatch; metre stick
Caution: Do not overstretch the springs and do not let go of a stretched spring.

PROCEDURE

1. Attach the masking tape to a coil near the middle of the spring. Stretch the spring along a smooth surface (the floor) to a length of 2.0 m. With one end of the spring held rigidly, use a rapid sideways jerk to create a transverse pulse at the other end. See Figure 12–8(a). Describe the motion of the particles of the spring. (Hint: Watch the tape attached to the spring.)
2. With the same setup as in #1, use a rapid forward push to create a longitudinal pulse along the spring. Refer to Figure 12–8(b). Again describe the motion of the particles of the spring.
3. Stand a folded piece of paper on the floor close to the middle of the spring, as shown in Figure 12–8(c). Use energy transferred by a transverse pulse to knock the paper over. Describe where the energy came from and how it got to the paper.
4. Stretch the spring to a length of 1.0 m. Measure the time for a

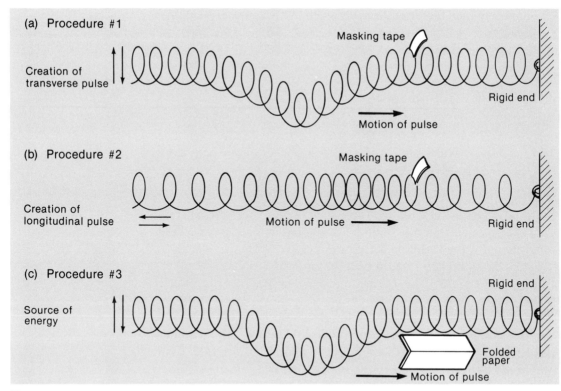

(a) Procedure #1

Creation of transverse pulse

Masking tape

Rigid end

Motion of pulse

(b) Procedure #2

Creation of longitudinal pulse

Masking tape

Motion of pulse

Rigid end

(c) Procedure #3

Source of energy

Rigid end

Folded paper

Motion of pulse

Figure 12-8

transverse pulse to travel from one end of the spring to the other **and back again**. Repeat the measurement several times for accuracy while trying to keep the amplitude constant. Calculate the speed of the transverse pulse along the spring ($v = \frac{d}{t}$).

5. Determine whether the speed of the pulse in #4 depends on the size or shape of the pulse.
6. Repeat #4, using a stretch of 2.0 m, then 3.0 m.
7. If different types of springs are available, find the speed of a transverse pulse along them.

QUESTIONS

1. Based on the observations in this experiment, discuss whether the following statements are true or false:
   (a) The purpose of a pulse or wave is to transfer energy from one place to another.
   (b) Energy may move from one end of a spring to the other.
   (c) The particles of a spring move from one end of a spring to the other with the energy.

2. Choose which factors listed below affect the speed of a pulse in a material.
   (a) type of material
   (b) condition of the material (e.g., stretching a spring changes its condition)
   (c) amplitude of the pulse

12.6 Periodic Waves

Periodic waves are created by a source of energy vibrating at some constant frequency. Both periodic transverse waves and periodic longitudinal waves may be set up on a coiled spring.

Figure 12–9 indicates the important parts of a transverse wave. Figure 12–10 shows the corresponding parts of a longitudinal wave. A transverse wave consists of a **crest** above the rest line and a **trough** below the rest line. A longitudinal wave consists of a **compression** where the particles are close together and a **rarefaction** where the particles are spread apart or rarefied. The **amplitude** (A) is the largest distance from the rest line or rest position. The **wavelength** (λ) is the length of one wave. (The symbol λ is taken from the Greek alphabet. It is spelled lambda.) As shown in the diagrams, there are several ways of measuring wavelength.

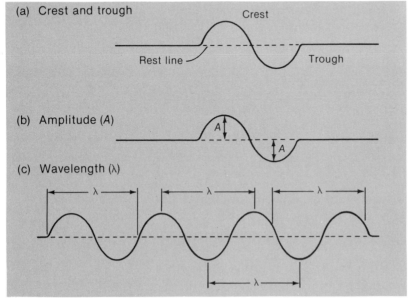

Figure 12–9 Transverse waves

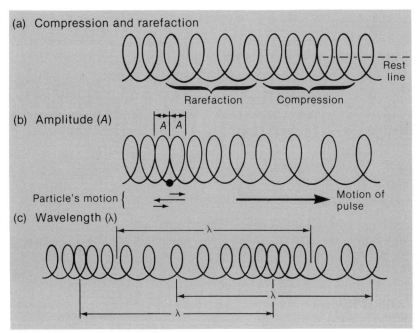

Figure 12–10 Longitudinal waves

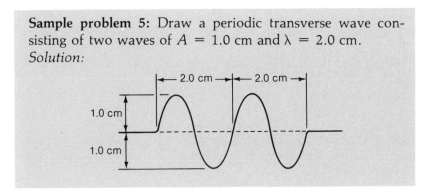

Sample problem 5: Draw a periodic transverse wave consisting of two waves of $A = 1.0$ cm and $\lambda = 2.0$ cm.
Solution:

Periodic waves are caused by vibrations. A vibrating source of waves has a frequency and a period, as you learned in Section 12.2. If the frequency of the vibration increases, the wavelength decreases. This can be demonstrated using a rope or a coiled spring.

PRACTICE

9. Measure the amplitude and wavelength of the periodic transverse wave in Figure 12–9(c).

10. Measure the wavelength of the periodic longitudinal wave in Figure 12–10(c).

11. Draw a periodic transverse wave consisting of two waves such that $A = 0.5$ cm and $\lambda = 4.0$ cm.
12. What happens to the wavelength of a periodic wave if the frequency decreases?

12.7 The Universal Wave Equation

You learned in Experiment 26 that the speed of a pulse or wave depends on the material and the condition of the material. For a pulse on a spring, the speed can be found using $v = \dfrac{d}{t}$. For a periodic wave, however, another equation should be derived.

Consider diagram (a) in Figure 12–11. It shows a set of train cars travelling to the right. If each car is 20 m long and takes 4.0 s to pass point X, then the train's speed is

$$v = \frac{d}{t}$$
$$= \frac{20 \text{ m}}{4.0 \text{ s}}$$
$$= 5.0 \text{ m/s}$$

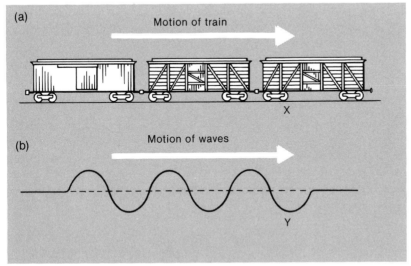

Figure 12–11 Comparing train cars and waves

Now compare the train cars to the periodic wave in diagram (b). If each 20 m wave was created in a period of 4.0 s, then the speed of the periodic wave past point Y is the ratio of the wavelength to the period.

$$v = \frac{d}{t}$$

$$= \frac{\lambda}{T}$$

$$= \frac{20 \text{ m}}{4.0 \text{ s}}$$

$$= 5.0 \text{ m/s}$$

This equation for speed ($v = \frac{\lambda}{T}$) may also be written $v = (\frac{1}{T})\lambda$. Since $f = (\frac{1}{T})$, we can now write an equation for speed in terms of frequency and wavelength.

$$v = f\lambda$$

This last equation is called the **universal wave equation**. It is used to find the speed of all periodic waves whose frequency and wavelength are known. Units involving the universal wave equation are shown in the example that follows.

Sample problem 6: Find the speed of a wave in water if $f = 4.0$ Hz and $\lambda = 2.5$ m.

Solution: $v = f\lambda$

$\qquad = 4.0$ Hz $\times$ 2.5 m

$\qquad = 4.0 \frac{\text{cycles}}{\text{s}} \times 2.5 \frac{\text{m}}{\text{cycle}}$ (see note below)

$\qquad = 10.0$ m/s

Note: If frequency is stated in cycles per second and wavelength in metres per cycle, you can see how the cycles divide, leaving metres per second, the correct unit of speed.

Sample problem 7: A periodic source of waves creates a wave of $\lambda = 3.2$ cm every 0.5 s. Calculate the:
(a) frequency of the waves
(b) speed of the waves

Solution:

(a) $f = \frac{1}{T}$

$\qquad = \frac{1}{0.5 \text{ s}}$

$\qquad = 2.0$ Hz

(b) $v = f\lambda$

$\qquad = 2.0$ Hz $\times$ 3.2 cm

$\qquad = 6.4 \frac{\text{cm}}{\text{s}}$

216

PRACTICE
13. Calculate the speed in each situation:
    (a) $f = 12$ Hz, $\lambda = 2.5$ m
    (b) $f = 60$ Hz, $\lambda = 0.05$ m
    (c) $f = 200$ Hz, $\lambda = 0.4$ m
14. Calculate the speed if:
    (a) $T = 3.0$ s, $\lambda = 12$ m
    (b) $T = 0.4$ s, $\lambda = 8.0$ m
    (c) $T = 0.02$ s, $\lambda = 1.6$ m
15. Rearrange the equation $v = f\lambda$ to solve for:
    (a) frequency
    (b) wavelength
16. Calculate the unknown quantities in the chart below:

	v (m/s)	f (Hz)	λ (m)
(a)	?	25	0.3
(b)	300	?	0.5
(c)	20	400	?

17. A 3.5 m sound wave is moving with a speed of 350 m/s. Calculate the:
    (a) frequency of the sound
    (b) period of vibration of the source of the sound

18. A wave machine creates water waves in a tank with a frequency of 7.5 Hz. If the waves are travelling at a speed of 45 cm/s, calculate the wavelength of the waves.

12.8 Interference of Pulses and Waves

You have studied the action of a single pulse travelling along a spring. What happens if a pulse moving in one direction meets a pulse moving in the opposite direction? The pulses interfere with each other for an instant. The interference of transverse pulses can be observed using a spring or wave machine. Interference also occurs for longitudinal pulses but is not as easily observed.

Two types of interference can occur. **Destructive interference** occurs when a crest meets a trough. If the crest and trough are equal in size, they destroy each other for an instant, then continue

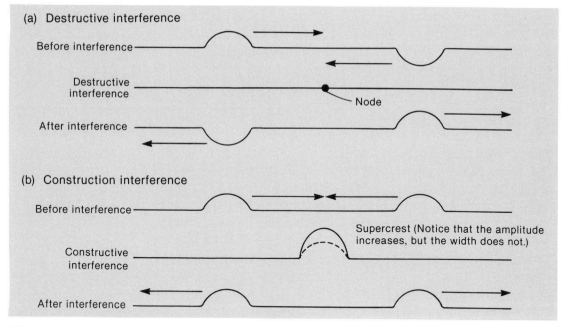

Figure 12–12 Interference of transverse pulses

in their original directions. The point where total destruction occurs is called a **nodal point** or **node**. Figure 12–12(a) illustrates destructive interference for transverse pulses. A node is produced with longitudinal pulses where a compression meets a rarefaction of equal size.

   Constructive interference occurs when pulses build each other up to a larger size. This can occur if a crest meets a crest causing a **supercrest**, or a trough meets a trough causing a **supertrough**. Figure 12–12(b) shows a supercrest resulting when two transverse crests meet. Constructive interference can also occur for longitudinal pulses.

PRACTICE
19. State whether the interference is constructive or destructive:
   (a) a crest meets a trough
   (b) a supertrough is formed
   (c) a compression meets a compression
   (d) a node is produced
20. Draw a series of diagrams, similar to those in Figure 12–12(b), showing the creation of a supertrough when two transverse pulses meet. Draw each trough 2.0 cm wide with an amplitude of 1.0 cm.

12.9 Viewing Waves on Water

Studying pulses on coiled springs is a good way to start the topic of waves. However, a spring has only one dimension, length. A surface of water has two dimensions, length and width, so studying water waves is the next step. This will help in the study of sound waves travelling in three-dimensional air. (Air spaces have length, width and depth.)

A **ripple tank**, shown in Figure 12–13(a), is a device used in science laboratories to study waves in water. It is a raised, shallow tank with a glass bottom. For most experiments the tank is level and contains water to a depth of about 10 mm. Periodic straight waves are produced by a motor connected to a straight bar. Periodic circular waves are produced when the motor is connected to a point source.

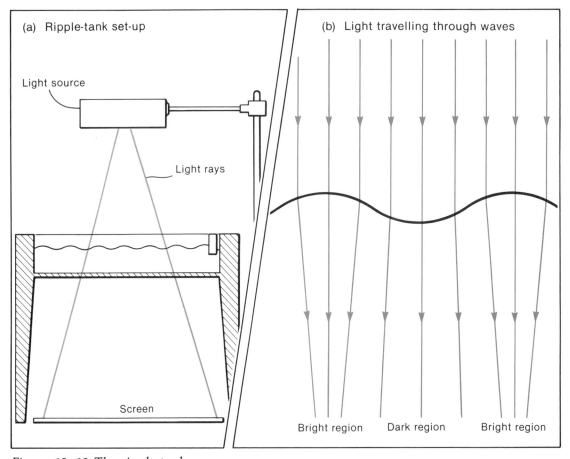

Figure 12–13 The ripple tank

A light source is used to allow the transverse water waves to be easily seen. The light source, held by a stand above the water, sends light through the water to a screen below the tank. The crest of each wave acts like a magnifying glass, focussing the light to a bright region. The trough of each wave spreads the light out, making a dark region. The bright and dark regions appear on the screen, as shown in Figure 12–13(b).

12.10 Experiment 27: Interference of Periodic Waves in a Ripple Tank

INTRODUCTION
If two periodic sources produce circular waves in the same material, such as water in a ripple tank, an interference pattern occurs. When a crest from one source meets a trough of equal size from the other source, a nodal point results. Many supercrests and supertroughs are also formed. In such an interference pattern the nodal points form distinct lines called **nodal lines**.

In the study of sound in Chapter 13, an interference pattern can be observed in the classroom using sound waves. That pattern resembles the one viewed in this experiment, although the types of waves are different.

PURPOSE: To observe and learn how to draw an interference pattern that occurs on water using two point sources of equal frequency.

APPARATUS: ripple tank and related apparatus; two-source ripple-tank motor; retorts and clamps

PROCEDURE
1. Place water in the ripple tank to a depth of about 10 mm. Be sure the tank is level.
2. Set up the motor and place the two point sources so they are 4.0 cm apart and slightly into the water. Operate the motor at a medium frequency with the sources bouncing up and down together. Describe in detail the resulting interference pattern. A sketch in your notebook will help.
3. Predict, then determine the effect on the pattern when you increase the distance between the sources but keep the frequency constant.

4. Predict, then determine the effect on the pattern when you increase the frequency of the motor but keep the distance between the sources constant.
5. Empty the tank and dry it completely.

QUESTIONS
1. To learn how the interference pattern in this experiment is created draw the diagram described below. Figure 12–14 shows how to start the diagram.
   (a) The diagram is most easily drawn on a piece of metric graph paper turned sideways. Near the middle of the sheet, mark two points 6.0 cm apart to represent the sources of waves.
   (b) Draw a circle with a radius of 2.0 cm from each source. Then draw circles having radii of 4.0 cm, 6.0 cm, and so on, until the page is filled. The circles represent crests of waves with λ = 2.0 cm.

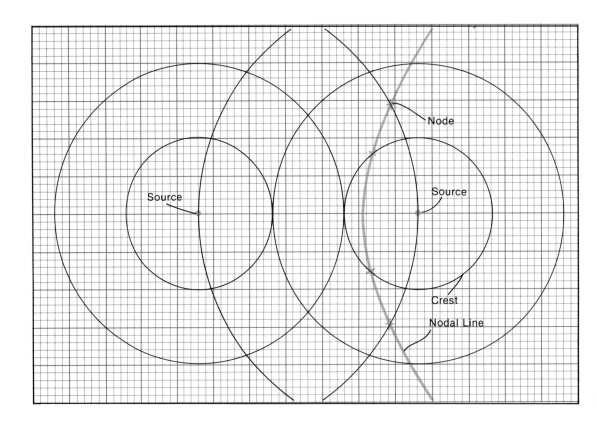

(c) Since the troughs are midway between the crests, you should be able to locate many nodal points. Place an "X" at each nodal point. Once you discover the pattern of nodal points, draw in the nodal lines. (Only one nodal line is shown in Figure 12–14.)

(d) Label the diagram completely.

2. Relate the pattern in question #1 to the pattern observed in this experiment.

12.11 Mechanical Resonance

Any object that vibrates will do so with its largest amplitude if it is vibrating at its own natural frequency. This natural or **resonant frequency** is the frequency at which a vibration occurs most easily. When the vibration is mechanical the natural vibration is called **mechanical resonance**. If you understand the examples of mechanical resonance that follow, you should be able to understand resonance of sound waves.

A pendulum of a certain length has its own resonant frequency (Section 12.3). A playground swing, which acts like a long pendulum, also has its own resonant frequency. If you are pushing someone on a swing, the amplitude of vibration can be built up by pushing at the correct instant in each cycle. In other words, your pushing frequency equals the resonant frequency of the swing.

If a car is stuck in snow, it can be rocked back and forth at the resonant frequency of the system. This motion builds up the amplitude, helping the car get out of the snow.

Another example of resonant frequency was discovered by military leaders in previous centuries. If the soldiers in an army marched across a small bridge in unison, the amplitude of vibration of the bridge built up. If the frequency of the soldiers' steps was the resonant frequency of the bridge, the vibration could possibly break the bridge. To prevent this, the soldiers were told to "break step" as they crossed bridges.

A spectacular example of mechanical resonance was the disaster that caused the collapse of a bridge in the state of Washington, U.S.A., in 1940. A bridge called the Tacoma Narrows Bridge was suspended by huge cables across a valley. On a windy day soon after its opening, the bridge began vibrating at its resonant frequency. At first the bridge vibrated as a transverse wave. Then

(a) The centre span of the bridge is shown vibrating torsionally before collapse.

(b) The vibrations caused the bridge to collapse.

Figure 12–15 The Tacoma Narrows Bridge

one of the suspension cables came loose and the entire 850 m centre span of the bridge vibrated torsionally. The vibrations were so great that the bridge collapsed. Refer to Figure 12–15.

12.12 Standing Waves—A Special Case of Both Interference and Resonance

If periodic transverse waves travel in opposite directions on a spring, rope or wave machine, a standing-wave pattern can be set up. The pattern has supercrests, supertroughs and nodes. The nodes stay standing in the same position, so the formation is called a **standing-wave pattern**. (Standing waves produce the sound in wind instruments, as you will learn in Chapter 14.)

A standing-wave pattern is an example of both interference and resonance. To understand why, set up such a pattern yourself. Tie one end of a long rope to a rigid support. Send periodic waves toward the rigid end. Those waves will reflect back and **interfere** with the ones you are sending. This interference causes nodes and loops. A **loop** is a position of largest amplitude. It is actually a combination of a supercrest and supertrough.

Try to create patterns having the shapes shown in Figure 12–16. As you are creating the patterns, notice that there is only one frequency that produces each shape. That frequency is the **resonant frequency** of the system. Try producing patterns with three, four or more nodes between the ends.

*Figure 12–16
Standing waves*

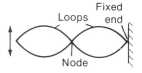

(a) Low frequency; long wavelength; zero nodes between the ends

(b) Medium frequency; shorter wavelength; one node between the ends

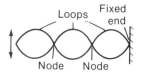

(c) Higher frequency; still shorter wavelength; two nodes between the ends

(c) This is a view of the rebuilt bridge. What structural changes have been made to the new bridge?

Assume that in Figure 12–16 the distance from the rigid end to the source of vibrations is 3.0 m. The standing wave in diagram (b) is one wavelength long, so its wavelength is 3.0 m.

Sample problem 8: What are the wavelengths of the standing waves in diagrams (a) and (c) of Figure 12–16?

Solution: (a) The standing wave is half a wavelength:
$$\tfrac{1}{2}\lambda = 3.0 \text{ m}$$
$$\therefore \lambda = 6.0 \text{ m}$$
  (c) The standing wave is 1½ wavelengths:
$$\tfrac{3}{2}\lambda = 3.0 \text{ m}$$
$$\therefore \lambda = 2.0 \text{ m}$$

From these examples, we can conclude that the distance from one node to the next in a standing-wave pattern is ½λ. Check this fact in Figure 12–16. We can also conclude that for a given length of rope only certain wavelengths and frequencies create standing waves. These ideas will be discussed further in the study of sound waves.

PRACTICE
21. What is the wavelength of the standing waves shown in the diagram?
22. In a certain standing-wave pattern, the distance from one node to the next is 2.0 m. What is the wavelength of the standing waves?

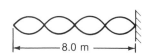
8.0 m

12.13 Review Assignment

1. State whether the vibration indicated is transverse or longitudinal:
   (a) a tree sways in the wind
   (b) a sewing-machine needle moves up and down
   (c) a child bounces on a pogo stick   (12.1)
2. A pendulum mass moves 16 cm in one cycle. What is the amplitude of vibration?   (12.1)
3. Calculate the period of each of these motions:
   (a) a child, while skipping, jumps off the ground 80 times in 120 s
   (b) a pulse beats 25 times in 15 s
   (c) a man shovels at a rate of 15 shovelfuls per minute. (12.2)
4. Calculate the frequency of the following:
   (a) an automatic gun fires 10 bullets at a target every 0.5 s
   (b) a tuning fork vibrates 21 times in 0.1 s
   (c) a recording timer creates 3600 dots in one minute. (12.2)
5. How are frequency and period related to each other? (12.2)
6. Calculate the period if the frequency is:
   (a) 5.0 Hz
   (b) 500 Hz   (12.2)
7. Calculate the frequency if the period is:
   (a) 10 s
   (b) 0.25 s   (12.2)
8. A pendulum is shortened. State what happens to its:
   (a) frequency
   (b) period   (12.3)
9. What is the main function of a wave?   (12.4)
10. Name the type of wave that results when the particles of an object vibrate:
    (a) parallel to the rest line
    (b) at right angles to the rest line   (12.1, 12.4)
11. Name two factors that affect the speed of a wave in a material. (12.5)
12. Measure the amplitude and wavelength of the wave shown in the diagram.   (12.6)

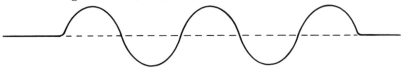

13. What is the universal wave equation?   (12.7)
14. A source of sound sends 500 Hz sound waves through water. The wavelength of the waves is 3.0 m. What is the speed of the sound in the water?   (12.7)
15. The wavelength of a certain sound is 0.3 m. If the speed through a certain material is 240 m/s, calculate the frequency of the sound waves.   (12.7)
16. What is the wavelength of a sound that travels at a speed of 340 m/s if its frequency is 510 Hz?   (12.7)
17. Under what conditions can a node occur during the interference of:
    (a) transverse pulses?
    (b) longitudinal pulses?   (12.8)
18. What happens to the frequency of a periodic wave if the wavelength decreases?   (12.6, 12.10)
19. Describe in your own words the meaning of mechanical resonance.   (12.11)
20. Describe two examples of mechanical resonance other than those given in Section 12.11.
21. **Class demonstration:** Set up the apparatus shown in the diagram. Place the retort stands about 50 cm apart. Connect the stands with a tight string. Suspend two pendulums of equal length (about 30 cm) and equal mass (50 g). Set one pendulum swinging at right angles to the horizontal string. Watch carefully for several minutes. Explain in detail what you observe.   (12.11)

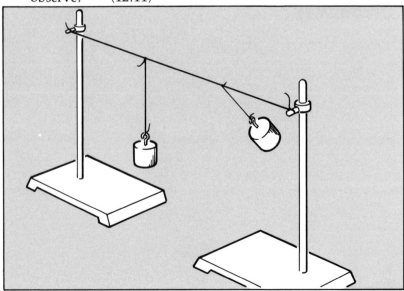

226

22. A 6.0 m rope is used to create standing waves. Draw a diagram of the shape of the standing wave when it has a wavelength of:
(a) 12.0 m   (b) 6.0 m   (c) 3.0 m   (12.12)

12.14 Answers to Selected Problems

PRACTICE QUESTIONS

2. (b) 8.5 cm
3. (b) 4.0 cm
4. 200 cm
5. 24 cm
6. $f = 30$ Hz, $T = 0.03$ s
7. (a) 0.5 s
   (b) 0.05 s
   (c) 2.5 s
8. (a) 0.2 Hz
   (b) 1.3 Hz
   (c) 100 Hz
13. (a) 30 m/s
    (b) 3.0 m/s
    (c) 80 m/s

14. (a) 4.0 m/s
    (b) 20 m/s
    (c) 80 m/s
15. (a) $f = \dfrac{v}{\lambda}$
    (b) $\lambda = \dfrac{v}{f}$
16. (a) 7.5 m/s
    (b) 600 Hz
    (c) 0.05 m
17. (a) 100 Hz
    (b) 0.01 s
18. 6.0 cm
21. 4.0 m
22. 4.0 m

REVIEW ASSIGNMENT

2. 4 cm
3. (a) 1.5 s
   (b) 0.6 s
   (c) 4 s
4. (a) 20 Hz
   (b) 210 Hz
   (c) 60 Hz

6. (a) 0.2 s
   (b) 0.002 s
7. (a) 0.1 Hz
   (b) 4 Hz
14. 1500 m/s
15. 800 Hz
16. 0.67 m

13
Sound Energy and Hearing

GOALS: After completing this chapter you should be able to:
1. State what produces sound energy.
2. Describe how sound energy is transmitted from one place to another.
3. Determine the speed of sound in air experimentally.
4. Given the temperature of air, find the speed of sound in air

$$[v = 332 \ \tfrac{m}{s} + (0.6 \ T) \ \tfrac{m}{s}].$$

5. List factors that affect the speed of sound in various materials.
6. Describe what is heard when listening to the interference of sound waves.
7. Calculate the beat frequency when sounds of two different frequencies are heard together.
8. Describe how beats can be used to tune musical instruments.
9. Describe examples of resonance in sound.
10. Explain how a human ear transmits sound to the brain.
11. Define audible range and recognize its average value for human hearing.
12. Define infrasonic and ultrasonic frequencies.
13. Describe applications of ultrasonic frequencies.

Knowing the information in this chapter will be especially useful if you plan a career in:
• the recording industry
• music (for example, if you plan to become a musician)
• radio
• television
• acoustics
• environmental (sound) pollution
• audio-visual technology

13.1 The Production of Sound Energy

Sounds we hear are described in many ways. Leaves rustle, lions roar, babies cry, birds chirp, corks pop—the list is long. The energy that creates those sounds comes from vibrating objects.

Some vibrations that make sound can be seen. If you pluck a guitar string or strike a low-frequency tuning fork, vibrations are observed. If you watch the low-frequency woofer of a loudspeaker system, you can see it vibrating.

Some vibrations that create sound cannot be seen. When you speak, for example, parts of your throat vibrate. When you make a sound by blowing into an empty pop bottle, the air molecules in the bottle vibrate.

All sound energy is produced by vibrations, both visible and invisible. Definitions and ideas related to vibrations were studied in Chapter 12.

PRACTICE

1. State what vibrates to produce the sound from a:
   (a) banjo (b) drum   (c) coach's whistle
2. **Class demonstration:** Strike a 256 Hz tuning fork and touch the prongs to the surface of some water. Describe what happens and why.

13.2 The Transmission of Sound Energy

In an electric bell, sound is created by a vibrating metal arm striking a metal sounder. Such a bell is shown in a jar in Figure 13-1. The jar is connected to a vacuum pump. As the pump sucks the air out of the jar, sound from the bell becomes more difficult to hear. If all the air is taken out, no sound can be heard, even though the metal arm may be seen to vibrate. Thus, sound energy needs a material to travel through. It cannot travel in a vacuum.

Air is the most common material that transmits sound energy to our ears. Sound energy travels through air by means of longitudinal waves. (Remember in the coiled-spring experiment, Section 12.5, that the energy from your hand was transmitted by wave action along the spring.) Longitudinal waves have compressions where the molecules are close together and rarefactions where the molecules are spread apart. (See Section 12.6.)

Although air molecules are invisible, the effect of their motion can be seen. Figure 13-2 shows a way of demonstrating compressions and rarefactions of air. A piece of cardboard, in line with a set of lit candles, is waved back and forth. Compressions force the flames away from the cardboard and rarefactions force the flames toward the cardboard.

In a physics laboratory a common source of sound energy is a tuning fork, illustrated in Figure 13-3. A tuning fork has two prongs. When one prong is struck from the side with a rubber hammer, both prongs move together, as shown in diagram (a). This causes a compression of air molecules between the prongs, as shown in diagram (b). At the same instant there are rarefactions outside the prongs.

Figure 13-1 Bell in a vacuum

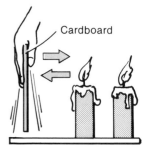

Figure 13-2
Illustrating compressions and rarefactions of air molecules

At the next instant, as shown in diagram (c), the prongs are spreading apart. This causes a rarefaction between the prongs and compressions outside them. The air molecules vibrate back and forth at the same frequency as the tuning fork. They transfer the sound energy from the source to the listener by colliding with each other as shown in diagram (d).

Diagram (e) shows how longitudinal sound waves can be represented by transverse waves. Transverse waves are often used in diagrams because they are easier to draw than longitudinal waves.

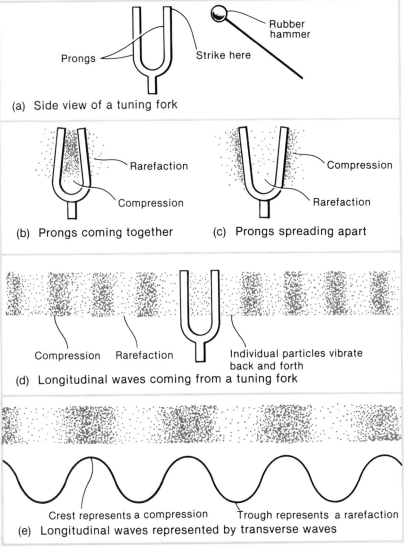

(a) Side view of a tuning fork

(b) Prongs coming together   (c) Prongs spreading apart

(d) Longitudinal waves coming from a tuning fork

(e) Longitudinal waves represented by transverse waves

Figure 13–3 The tuning fork and longitudinal waves

3. Describe what happens to air molecules during:
   (a) compression   (b) rarefaction

13.3 Experiment 28: Measuring the Speed of Sound in Air

INTRODUCTION

During a thunder storm, lightning is noticed first. Several seconds later, thunder is heard. A similar phenomenon occurs at track meets when the timer of a 100 m sprint stands at the finish line and watches for a puff of smoke from the starter's pistol. Shortly after the smoke is seen, the sound of the gun is heard. The light travels very fast (3.0×10^8 m/s), but the sound travels slowly enough to have its speed measured by students.

PURPOSE: To determine the speed of sound in air.

APPARATUS: tape measure; 2 blocks of wood; stopwatches; thermometer

PROCEDURE

1. Choose an outside location to perform the experiment. The distance between the source of the sound and the listeners should be at least 100 m. Measure that distance.
2. Measure and record the air temperature. This value will be used in Section 13.4.
3. Have one student create a loud sound by banging one block on

Figure 13-4

100 m or more

the other, as shown in Figure 13–4. Find the time it takes the sound to reach the listeners after the boards are seen to touch.

4. Repeat # 3 several times to find an average value.

5. Calculate the speed of sound in air using the equation $v = \dfrac{d}{t}$.

QUESTIONS

1. Describe any weaknesses in this experiment.

2. A pistol is used to start a 500 m race along a straight track. A puff of smoke is seen and 1.5 s later the sound is heard at the finish line. What is the speed of the sound?

3. Assume that during a storm, thunder was heard 8.0 s after lightning was seen. If the speed of sound in air was 350 m/s, what was the distance from the lightning to the observer? (Hint: Rearrange the equation $v = \dfrac{d}{t}$ to solve for distance.)

13.4 Temperature and the Speed of Sound in Air

The speed of sound in air is 332 m/s at 0°C. If the air temperature increases, the speed increases because the air molecules move more rapidly. (The speed of sound in air depends on other factors that will not be mentioned here.)

For every degree Celsius rise in temperature, the speed of sound in air increases by 0.6 m/s. Thus, at 1°C, the speed of sound is 332 m/s + 0.6 m/s = 332.6 m/s. In general, the speed of sound in air can be found using the equation:

$$v = 332 \, \frac{m}{s} + (0.6 \, T) \, \frac{m}{s} \quad \text{where } T \text{ is the air temperature in } °C$$

Sample problem 1: Calculate the speed of sound in air when the temperature is 16°C.

Solution: $v = 332 \, \dfrac{m}{s} + (0.6 \, T) \dfrac{m}{s}$

$\qquad = 332 \, \dfrac{m}{s} + (0.6 \times 16) \dfrac{m}{s}$

$\qquad = 332 \, \dfrac{m}{s} + 9.6 \, \dfrac{m}{s}$

$\qquad = 341.6$ m/s or 342 m/s

If the temperature drops below 0°C, the speed of sound in air is less than 332 m/s.

PRACTICE

4. Calculate the speed of sound in air when the temperature is:
   (a) 3°C
   (b) 20°C
   (c) –2°C
5. In Experiment 28 you recorded the outside air temperature. Use that temperature to calculate the speed of sound in the air. Compare this value with the one found in the experiment.

13.5 The Speed of Sound in Various Materials

The speed of sound in a material depends on the spacing of the molecules and how easily the molecules move. For example, sound travels faster in water than in air because the water molecules are closer together and transmit sound energy more easily.

Table 13–1 lists the speed of sound in various materials. Scientists use these values to study the structure of the earth, search for oil and minerals, and locate objects beneath the surface of the sea. (This will be discussed further in Section 13.10.)

Table 13–1 The Speed of Sound in Common Materials

State	Material	v ($\frac{m}{s}$) at 0°C
Gas	Carbon dioxide	258
	Oxygen	317
	Air	332
Liquid	Alcohol	1240
	Sea water	1470 (depends on salt content)
	Fresh water	1500
Solid	Pine wood	3300
	Maple wood	4100
	Steel	5000

PRACTICE

6. **Mini-experiment:** Hold a ticking stopwatch about 1 m from your ear. Listen. Then touch a metre stick to one ear and have

your partner hold the ticking stopwatch to the other end. Describe what happens and why.

7. A 21-gun salute is about to be given by a navy ship anchored 3000 m from shore. A swimmer near shore sees a puff of smoke from the gun, quickly pops her head under water, and listens. In 2.0 s she hears the sound of the gun and then lifts her head above the water. In another 6.6 s she hears the sound that has come through the air from the same shot. Find the speed of the sound in the:
(a) water (b) air

13.6 Experiment 29: Interference of Sound Waves

INTRODUCTION

Destructive interference occurs for longitudinal waves when a compression meets a rarefaction. **Constructive interference** occurs when a compression meets a compression or when a rarefaction meets a rarefaction. (See Section 12.8.)

A two-source interference pattern, which included **nodal lines**, was created in water in a ripple tank (Section 12.10). A similar pattern will be created for sound. However, the nodal lines may not be distinct in a room where the walls reflect sound easily.

In section 13.2 the vibrations of a tuning fork were described. The fork was viewed from the side. Figure 13–5 shows longitudinal waves spreading out from a tuning fork when viewed from the top. This diagram will help explain the creation of the interference pattern near a tuning fork.

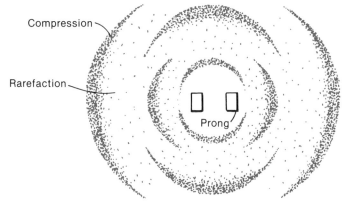

Figure 13–5 Top view of a tuning fork

PURPOSE: To observe and explain interference patterns of sound in air.

APPARATUS: tuning fork; demonstration setup for a two-source interference pattern (frequency generator and two loudspeakers)

PROCEDURE
1. Strike the tuning fork on a rubber stopper or your elbow. Hold the fork vertically near your ear and **slowly** rotate it. Listen carefully for loud and soft sounds and have your partner help you locate their exact positions.
2. Draw a diagram of the top view of the tuning fork showing the positions of the loud and soft sounds.
3. Place two loudspeakers 1.0 m apart, facing the same direction. Connect them to the same frequency generator. Set the frequency of the sound to 800 Hz. Walk slowly back and forth in front of the speakers, listening for the loud and soft sounds. Have several students line up along nodal lines to observe the interference pattern.

QUESTIONS
1. Use a diagram similar to Figure 13–5 to illustrate how interference occurs near a tuning fork.
2. Explain how the interference pattern in water in a ripple tank relates to the sound interference pattern in Procedure # 3.
3. Explain how this experiment shows that sound energy travels by means of waves.

13.7 The Production of Beats

In Section 13.6, interference of sound waves was created by using only one frequency at a time. When sounds of different frequencies are heard together, interference can also occur. This is especially true if the two frequencies are almost the same. For example, if a 256 Hz tuning fork is heard with a 250 Hz note from a frequency generator, a loud and soft interference pattern occurs. The pattern of loud and soft sounds is called the production of **beats**. Once you hear a demonstration of beats, you will find it easy to recognize them.

Beat frequency is the number of beats heard per second. It is

found by subtracting the lower frequency from the higher frequency. In the example above, the beat frequency is
256 Hz – 250 Hz = 6 Hz.

> **Sample problem 2:** Two 384 Hz tuning forks are sounded together and no beats are heard. Then a metal clip is attached to a prong of one fork and again the forks are sounded together. This time a beat frequency of 4 Hz is heard. What is the new frequency of the fork with the clip?
> *Solution:* The frequency of the fork must be either 4 Hz higher or 4 Hz lower than 384 Hz. Adding extra mass to a tuning fork makes it more difficult for the prongs to vibrate, so the frequency must be 4 Hz lower. Thus, it is 380 Hz.

An application of beats is the tuning of a musical instrument such as a piano. A note on the piano is sounded with the corresponding tuning fork. The tension in the piano strings can be adjusted until no beats are heard.

PRACTICE
8. State the beat frequency when the following pairs of frequencies are heard together:
   (a) 202 Hz, 200 Hz   (b) 341 Hz, 347 Hz   (c) 1003 Hz, 998 Hz
9. A 512 Hz tuning fork is sounded with a second tuning fork and a beat frequency of 7 Hz is heard. State two possible frequencies of the second tuning fork.

13.8 Resonance in Sound

No doubt you have heard the high-pitched squeal produced when a person runs a moist finger around the lip of a long-stemmed glass. The frequency of such a sound is the natural frequency for the glass. That natural frequency is called **resonant frequency**, as you learned in Section 12.11. The resonant frequency of a long-stemmed glass can be changed by adding water to the glass.

Mechanical resonance (Section 12.11) creates a large, easy amplitude at a natural frequency. Short pendulums have high frequencies and long pendulums have low frequencies.

Sound resonance also creates a large amplitude of vibration. For

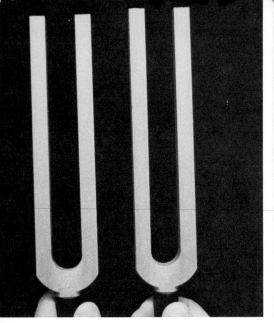

(a) Tuning forks   (b) Mounted tuning forks

Figure 13–6 Demonstrating resonance in sound

example, tuning forks vibrate at their own resonant frequencies. Short tuning forks have high frequencies and long ones have low frequencies.

Two identical tuning forks may be used to show how an object can be made to vibrate at its resonant frequency. Obtain two 256 Hz tuning forks. Strike one and hold it close to the other, as shown in Figure 13–6(a). After about 15 s, stop the first fork from vibrating and listen to the second one. It is vibrating because it has picked up the vibration from the first fork, which has the same resonant frequency. If the forks are of different resonant frequencies, the transfer of energy does not occur.

The demonstration is more effective if mounted tuning forks of equal frequency are used. See Figure 13–6(b).

Transfer of energy at resonant frequencies can also be shown if a piano is available. Suppress the right (sustaining) pedal to free all the strings in the piano. Sing a certain note loudly into the piano. Listen for the sound of the strings that vibrate in resonance with your voice.

Resonance of sound in musical instruments will be studied in more detail in Chapter 14.

PRACTICE
10. (a) What happens to the resonant frequency of an object when its length increases?
   (b) On what facts do you base your answer to part (a)?

13.9 Hearing and the Human Ear

Sound waves that reach our ears cause sensations that result in nerve signals being sent to the brain. The brain then interprets what we have heard.

Our ears are very important and sensitive organs. They react to a wide range of frequencies. With the proper equipment your teacher can check the frequency response for everyone in the class. A frequency generator is connected to a loudspeaker. The frequencies of sound may range from quite low to more than 25 000 Hz (25 kHz), depending on the speaker.

All the frequencies you can hear make up your **audible range**. Most students have an audible range of about 20 Hz to 20 000 Hz. The ear is most sensitive to frequencies between 1000 and 3000 Hz.

Figure 13–7 The human ear

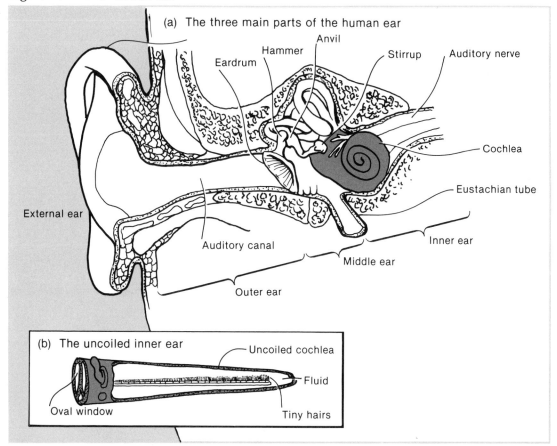

(a) The three main parts of the human ear

Eardrum · Hammer · Anvil · Stirrup · Auditory nerve · Cochlea · Eustachian tube · External ear · Auditory canal · Inner ear · Middle ear · Outer ear

(b) The uncoiled inner ear

Uncoiled cochlea · Fluid · Oval window · Tiny hairs

Often older people are unable to hear high frequencies. Young people who expose their ears to loud sounds for long periods of time also may lower their audible range.

The human ear is complex, but a short description of it will help you understand how we hear. The human ear consists of three parts—the outer ear, middle ear and inner ear. Refer to Figure 13–7(a).

The **outer ear** consists of the **external ear** that we can see, and the **auditory canal**. The external ear funnels the sound waves into the auditory canal. The auditory canal directs the longitudinal sound waves to the **eardrum**. The eardrum, which is about the thickness of a hair, connects the outer ear to the middle ear.

The **middle ear** consists of three tiny bones called the hammer, anvil and stirrup. The vibrating eardrum transfers its energy to these tiny bones. The bones act like a system of levers to transfer energy to the **oval window**. The oval window joins the middle ear to the inner ear.

The **eustachian tube** joins the middle ear to the throat. It allows air pressure to become equal on both sides of the eardrum. Your ears "pop" when the air pressure equalizes whenever you are ascending or descending in elevators or aircraft.

The **inner ear** consists of a coiled tube called the **cochlea** and the **auditory nerve**. The cochlea is filled with fluid and has thousands of tiny hairs of varying lengths. See Figure 13–7(b). Vibrations from the oval window force special hairs to vibrate. Different lengths of hair have different resonant frequencies. The vibrating hairs change energy of vibrations into electrical energy. The electrical energy is transferred through the auditory nerve to the brain. The brain then records what you have heard.

It is estimated that about one person in twenty in North America is either deaf or hard of hearing. Deafness may be caused if signals cannot travel through the auditory nerve to the brain. There is no cure for such deafness.

Deafness may also be caused by damage to the eardrum or the middle ear. This problem may be solved by an operation or by the use of a hearing aid. A hearing aid transmits energy through the skull to the inner ear. The inner ear then acts in the normal manner.

PRACTICE

11. The tiny hairs in the cochlea have different lengths. Which lengths (short or long) vibrate at high frequencies? Explain why you think so.

13.10 Infrasonics, Ultrasonics and Echo Finding

The average human audible range is from 20 Hz to 20 000 Hz. Frequencies lower than 20 Hz are called **infrasonic**. (Infra means lower than.) If we could hear frequencies lower than 20 Hz, we would often be bothered by sounds in and around us. For example, some sounds inside the body have a frequency of about 10 Hz. You can imagine how annoying it would be to hear spurts of blood every time your heart took a beat.

Frequencies higher than 20 000 Hz are called **ultrasonic**. (Ultra means higher than.) A dog whistle creates sound that is ultrasonic. Dogs' ears are sensitive to frequencies higher than we can hear. Dogs are not the only animals that hear ultrasonic sounds. Several others are listed in Table 13–2.

Table 13–2 Audible Ranges

Animal	Audible range (Hz)	Animal	Audible range (Hz)
Human	20 to 20 000	Bat	1000 to 120 000
Dog	15 to 50 000	Porpoise	150 to 150 000
Cat	60 to 65 000	Robin	250 to 210 000

Some animals, such as the bat, navigate and hunt using ultrasonics. The bat sends out high-frequency sounds that reflect off objects. The reflected sounds return to the bat and allow it to tell what is in the way. This explains why the bat can navigate in darkness as well as in light.

Equipment is available that uses reflection of ultrasonic sounds in water. A process called echo finding is used to determine the depth of water below a ship or to locate a school of fish. Ultrasonic sounds are sent from the ship, as shown in Figure 13–8. They reflect off the object and return to the ship. An instrument measures the time for the signal to return. Then the equation $v = \dfrac{d}{t}$ can be used to find either the speed of the sound or the distance it travelled.

Figure 13–8 Echo finding using ultrasonics

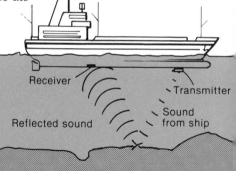

Sample problem 3: A ship is anchored where the depth of water is 120 m. A sound signal is sent to the bottom of the lake and returns in 0.16 s. What is the speed of the sound in water?
Solution: The distance travelled by the sound is 240 m.

$$v = \frac{d}{t}$$

$$= \frac{240 \text{ m}}{0.16 \text{ s}}$$

$$= 1500 \text{ m/s}$$

Ultrasonic sounds have several other applications. They can help find flaws in metal products. They are used to clean electronic parts of watches and surgical instruments. In medicine they are used to detect brain damage and certain cancers. They are also used to study the growth of unborn babies. Ultrasonic sounds are much less dangerous than high-energy X rays for these medical applications.

PRACTICE

12. Ultrasonic sound is sent to the ocean floor 360 m below a submarine. The sound reflects back to the submarine in 0.5 s. Find the speed of the sound in water.
13. The speed of sound in a fresh-water lake is 1500 m/s. Ultrasonic sound is sent from the surface of the water to the bottom of the lake. It returns in 0.2 s. How deep is the lake? (Hint: $d = vt$)

13.11 Review Assignment

1. State what produces sound energy.   (13.1)
2. What vibrates to produce the sound from:
   (a) an acoustic guitar?
   (b) an electric doorbell?
   (c) a stereo system?   (13.1)
3. Describe how sound energy is transferred from one place to another.   (13.2)
4. Why does sound energy not travel in a vacuum?   (13.2)
5. There is no air on the moon. How can astronauts on the moon hear each other speak?
6. At a certain instant a tuning fork is creating compressions outside the prongs. What is being created between the prongs?   (13.2)

7. At a baseball game a physics student with a stopwatch sits behind the centre-field fence marked 136 m. He starts the watch when he sees the bat connect with the ball. He stops the watch when he hears the resulting sound. The time is 0.4 s. How fast is the sound energy travelling?   (13.3)

8. At a speed of 345 m/s in air, how far can sound travel in 6.0 s?   (13.3)

9. State the speed of sound in air when the temperature is:
    (a) 25°C   (b) 8°C   (c) −10°C   (13.4)

10. Explain why sound energy travels faster in warm air than in cold air. (Hint: Consider the motion of the individual air molecules. If you studied the kinetic molecular theory of matter in Section 10.4, some ideas there may also help.)   (13.4)

11. State in which type of substance (solid, liquid or gas) sound seems to travel:
    (a) fastest   (b) slowest   (13.5 and Table 13–1)

12. Sound in a certain substance travels 3750 m in 2.5 s.
    (a) What is the speed of the sound in the substance?
    (b) According to Table 13–1, what is the substance?   (13.5)

13. How far can sound travel in steel in 1.5 s?   (13.5)

14. Describe the sound heard when beats are produced.   (13.7)

15. Find the beat frequency when the following pairs of frequencies are sounded together:
    (a) 256 Hz, 261 Hz   (b) 512 Hz, 508 Hz   (13.7)

16. What are the possible beat frequencies when these tuning forks are available: 256 Hz, 259 Hz, 251 Hz.   (13.7)

17. Explain how the production of beats could be used to tune a guitar string.   (13.7)

18. Describe two examples of resonance in sound.   (13.8)

19. State the meaning of:
    (a) human audible range   (b) ultrasonic
    (c) infrasonic   (13.9, 13.10)

20. State your own audible range.   (13.9)

21. Describe how a bat uses sound energy to find its way around.   (13.10)

22. Reflecting sound is used to find the speed of sound in water. A signal is sent from a ship to the floor of the ocean 420 m below the ship. The reflected signal is received 0.6 s later. What is the speed of the sound in the water?   (13.10)

23. Ultrasonic sound is used to locate a school of fish. The speed of sound in the ocean is 1450 m/s and the reflection of sound reaches the ship 0.1 s after it is sent. How far is the school of fish from the ship?   (13.10)

13.12 Answers to Selected Problems

PRACTICE QUESTIONS
4. (a) 333.8 m/s
   (b) 344 m/s
   (c) 330.8 m/s
7. (a) 1500 m/s
   (b) 349 m/s
8. (a) 2 Hz
   (b) 6 Hz
   (c) 5 Hz
9. 505 Hz, 519 Hz
12. 1440 m/s
13. 150 m

REVIEW ASSIGNMENT
7. 340 m/s
8. 2070 m
9. (a) 347 m/s
   (b) 336.8 m/s
   (c) 326 m/s
12. (a) 1500 m/s
13. 7500 m
15. (a) 5 Hz
    (b) 4 Hz
16. 3 Hz, 5 Hz, 8 Hz
22. 1400 m/s
23. 72.5 m

14

Music and Musical Instruments

GOALS: After completing this chapter you should be able to:
1. Describe the scientific differences between noise and music.
2. State the three characteristics of musical sounds and describe what each characteristic depends on.
3. Describe the two main musical scales.
4. Given the frequency of one musical note, calculate the frequency of a note one octave from it.
5. State the units used to measure the loudness of sound.
6. Describe the dangers of loud sounds.
7. Describe how changing the shape of a sound wave changes the quality of the musical sound.
8. State how tension, length and diameter affect the frequency of a vibrating string.
9. State the two main parts of stringed instruments.
10. List various types of stringed instruments.
11. Describe how resonance of sound is created in vibrating columns of air.
12. Calculate the wavelength of a sound that resonates in an air column knowing the distance from one loud sound to another.
13. Calculate the speed of sound in air given the frequency and wavelength ($v = f\lambda$).
14. State four ways of making air in wind instruments vibrate.
15. List various types of wind instruments.
16. List various types of percussion instruments.
17. Describe the functions of the main parts of the human voice.
18. Describe the difference between electrical and electronic instruments.
19. Describe how certain factors affect the acoustics of rooms and auditoriums.

Knowing the information in this chapter will be especially useful if you plan a career in:
• the recording industry
• music (for example, if you plan to become a musician)
• radio
• television
• acoustics
• environmental (sound) pollution
• audio-visual technology
• medicine

14.1 Noise and Music

Noise is sound that is unpleasant or annoying. **Music** is sound that is pleasant and harmonious.

The difference between noise and music depends somewhat on

the judgment of the individual. However, there is also a scientific difference.

In order to study noise and music, we use our ears as well as an instrument called an oscilloscope, which is illustrated in Figure 14-1. Longitudinal sound waves in air can be changed to transverse waves displayed on the oscilloscope screen. If you listen to a sound as you watch its wave, you will learn how the pleasantness of a sound changes when the shape of the wave changes.

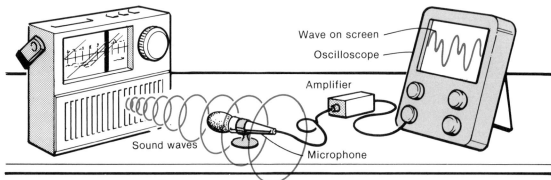

Figure 14-1 Using an oscilloscope to "see" sound

Figure 14-2 illustrates the shapes of waves that appear on an oscilloscope screen when various sounds are heard. Diagram (b) shows that the shape of noise waves is not smooth or regular.

Figure 14-2 Shapes of sounds on an oscilloscope screen

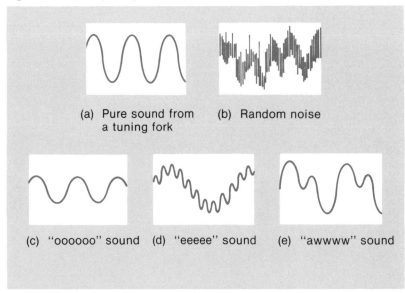

The next three sections discuss in detail the three main characteristics of musical sounds—pitch, loudness and quality. Oscilloscope displays of the waves of sound from musical instruments will help you distinguish noise from music scientifically.

14.2 Pitch and Musical Scales

If you are near a pond on a summer evening, you might hear crickets chirping and bullfrogs croaking. The sounds are easy to distinguish. Sound from crickets has a high pitch, and sound from bullfrogs has a low pitch.

A pitch wheel, shown in Figure 14-3, is a device that shows how pitch depends on the frequency of vibration. It consists of a set of three equal-sized wheels. Each wheel has a different number of teeth. As the wheels spin, a piece of paper is held up to each wheel in turn. The paper vibrates with the lowest frequency when it touches the wheel with the least teeth. This creates the sound of lowest pitch. The other wheels create sounds of higher pitch. Thus, pitch increases as frequency increases.

As the frequency of a sound changes, the sound wave also changes. This can be illustrated, using a frequency generator connected to an oscilloscope. Resulting waves are shown in Figure 14-4. It is clear that as the frequency increases the wavelength decreases.

Figure 14-4
Oscilloscope displays of frequency and wavelength

Figure 14-3 The pitch wheel

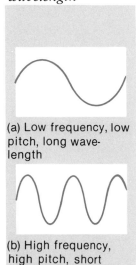

(a) Low frequency, low pitch, long wave-length

(b) High frequency, high pitch, short wave-length

There are two main musical scales that we will study. One scale, which we will call the **scientific scale**, is used in science laboratories. Tuning-fork frequencies are often named using this scale. For example, 256 Hz is the note "middle C" and 512 Hz is "high C". Table 14–1 lists eight common frequencies on the scientific scale.

Table 14–1 Some Frequencies on the Scientific Scale

Note	C	D	E	F	G	A	B	C
Frequency (Hz)	256	288	320	341	384	427	480	512

In Table 14–1, notice that 512 Hz is exactly twice the frequency of 256 Hz. The two frequencies sound pleasant when heard together. They are exactly one **octave** apart. If one note is an octave above another, its frequency is twice as high as the first note. For instance, a frequency of 1024 Hz is one octave above 512 Hz.

Appendix E at the end of the book describes more about the scientific scale. Music enthusiasts will benefit from studying that appendix.

The other musical scale, which we will call the **musicians' scale**, is based on a frequency of 440 Hz. That frequency is the note A above middle C on the piano. An octave below that A has a frequency of 220 Hz and an octave above has a frequency of 880 Hz. Figure 14–5 shows part of the piano scale, including the notes and their frequencies. This is the scale used to tune most musical instruments. Appendix E describes the mathematical details of how the frequencies for this scale are calculated.

The standard frequency of A = 440 Hz is used in many countries throughout the world. To allow the public to check for that frequency, government agencies broadcast it by telephone and radio. For example, if you dial the telephone number 1-303-499-7111 (at your own expense!) you will be given the standard time in Greenwich, England, and a standard frequency broadcast at intervals of exactly 1.0 s.

PRACTICE

1. On what does the pitch of a musical sound depend?
2. State the frequency of a note one octave above a note of:
   (a) 200 Hz   (b) 320 Hz   (c) 580 Hz
3. State the frequency of a note one octave below a note of:
   (a) 200 Hz   (b) 320 Hz   (c) 580 Hz
4. State the frequency of a note two octaves above a note of:
   (a) 300 Hz   (b) 1000 Hz   (c) 256 Hz

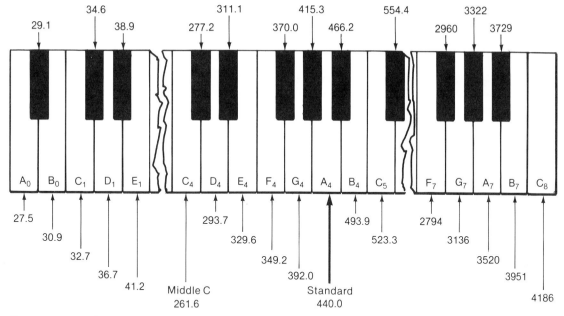

Figure 14–5 The musicians' scale illustrated on a piano keyboard

14.3 Loudness of Sounds

Of course there is a difference in loudness between a soft whisper and the roar of nearby thunder. The loudness we hear depends mainly on the amplitude of vibration of the source of sound and the distance from the source to our ears. Figure 14–6 shows how the amplitude changes when the loudness changes, at a constant frequency.

Loudness is measured in the unit bel. The bel (B) is named after Alexander Graham Bell, the man who invented the telephone. (Bell was born in Scotland in 1847. He worked in Canada and the United States. He died in Nova Scotia in 1922.)

A sound of zero bels (0 B) is called the **threshold of hearing**, which is the softest sound a healthy ear can hear. A sound of 1 B is 10 times as loud as a sound of 0 B. A sound of 2 B is 100 times as loud as a sound of 0 B, and so on. The **threshold of pain** occurs at 12 B. Above that loudness, sound becomes painful and probably harmful.

In Table 14–2, the first column lists the loudness of sounds from 0 B to 12 B. The second column lists the same sounds in decibels

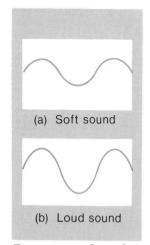

(a) Soft sound

(b) Loud sound

Figure 14–6 Sounds of different loudness on an oscilloscope

(dB) which are units often used on instruments that measure loudness. Since "deci" means one-tenth, 1 B = 10 dB. The third column states the number of times the sound is louder than the threshold of hearing. The last column lists examples of sounds of different loudness.

Table 14-2 Loudness of Sounds

Loudness (B)	Loudness (dB)	Value of loudness compared to threshold of hearing	Example
0	0	1	Threshold of hearing
1.0	10	10	Empty church on quiet street
2.0	20	100	Average whisper at 1 m
3.0	30	1000	Library reading room
4.0	40	10 000	Inside car with engine on
5.0	50	100 000	Quiet restaurant
6.0	60	1000 000	Conversation at 1 m
7.0	70	10 000 000	Machinery in factory
8.0	80	100 000 000	Noisy street corner
9.0	90	1000 000 000	Loud hi-fi in average room
10.0	100	10 000 000 000	Close to stage at rock concert
11.0	110	100 000 000 000	Jet taking off at 60 m
12.0	120	1000 000 000 000	Threshold of pain

From Table 14-2 you can tell that a sound of 120 dB is a million million times as loud as the softest sound (0 dB) we can hear. This means that the human ear is sensitive to a great range of loudnesses.

Listening to loud sounds for long periods of time can cause hearing loss. We use the term **noise pollution** to describe the effect of excess noise in our modern society. Governments try to prevent noise pollution by setting standards of noise levels on streets and in places of work. Also, ear protection must be provided for people who work where the loudness is greater than about 80 dB. See Figure 14-7.

Figure 14–7 Ground personnel working near airplanes must have ear protection.

PRACTICE

5. On what does the loudness of a sound depend?
6. State the number of times the first sound is louder than the second:
   (a) 4 B, 3B   (b) 8 B, 6 B   (c) 100 dB, 70 dB
   (d) 120 dB, 80 dB
7. A 50 dB sound is increased in loudness 1000 times. What is the new loudness?

14.4 Quality of Musical Sounds

Assume you hear the note middle C equally loudly from a piano, a violin and a trumpet. The three sounds have equal pitch and equal loudness. However, they sound different. The difference is in their **quality**.

A beginning music student will create poor-quality sound on a musical instrument. An experienced player using the same instrument will create high-quality sound. A small portable radio produces sound of poor quality compared to sound from an expensive high-fidelity sound system. Thus, the quality of musical sounds can differ greatly.

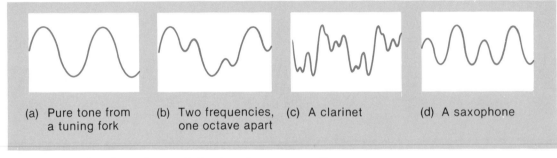

(a) Pure tone from a tuning fork

(b) Two frequencies, one octave apart

(c) A clarinet

(d) A saxophone

Figure 14-8 Quality of sound displayed on an oscilloscope

The quality of a sound means how pleasant the sound is. Scientifically, the quality depends on the shape of the sound waves. This can be demonstrated on an oscilloscope using sound waves from various musical instruments. Figure 14-8 shows sound waves of different quality. The pure tone in diagram (a) is a boring sound. The other diagrams show waves of higher quality.

PRACTICE

8. On what does the quality of a musical sound depend?

14.5 Experiment 30: The Pitch of Vibrating Strings

INTRODUCTION

Stringed instruments were invented thousands of years ago. The sounds made by the strings depended on various factors that you will control in this experiment.

Stringed instruments, such as the guitar, violin, piano and banjo, are made with various lengths and types of strings. Not only do the lengths and types differ, but the tension or tightness of the strings also differs.

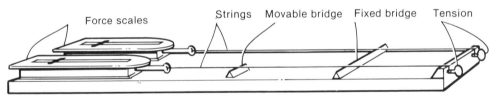

Figure 14-9 A sonometer

PURPOSE: To find how the pitch of a vibrating string depends on its length, tension and diameter.

APPARATUS: sonometer with 2 or more strings (see Figure 14–9); metre stick

PROCEDURE
1. Adjust the tension in the thin wire to 30 N. Pluck the string as shown in Figure 14–10(a). Listen to the pitch produced.
2. Place the movable bridge midway between the ends of the string. Pluck the string as shown in Figure 14–10(b). Compare the pitch of this shorter string to the pitch heard in # 1.

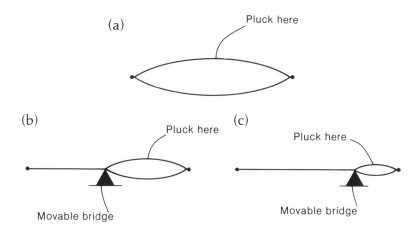

Figure 14–10

3. Place the bridge at a position one-quarter of the way along the wire as shown in Figure 14–10(c). Again pluck the string and compare the pitch to that in # 1 and # 2.
4. Using the full length of the string and a tension of 30 N, pluck the string and listen. Now change the tension to 35 N, listen to the sound, and compare the two pitches. Use other tensions and compare the sounds.
5. Set all the strings on the sonometer to equal lengths and tensions. Compare the diameters of the strings. Pluck the strings and compare their pitches.

QUESTION
1. A note sounded on a guitar is flat; in other words, its frequency is lower than it should be. What must be done to increase the frequency?

14.6 Stringed Instruments

Stringed instruments consist of two main parts—the **vibrator** and the **resonator**. The vibrator is the string. The resonator is the case, box or sounding board that the string is attached to. A string by itself does not give a loud or pleasant sound. It must be attached to a resonator to improve its loudness and quality. Even a tuning fork has a louder and better sound if its handle is touched to a desk or a wall.

Stringed instruments can be played by plucking, striking or bowing. The quality of sound is different in each case. The quality also depends on what part of the string is plucked, struck or bowed. For example, a string plucked in the middle has a pure sound, as shown in Figure 14-11(a). Plucking at other positions changes the shape of the wave and thus the quality. This is illustrated in diagrams (b) and (c).

Figure 14-11 Changing the quality of sound of a vibrating string

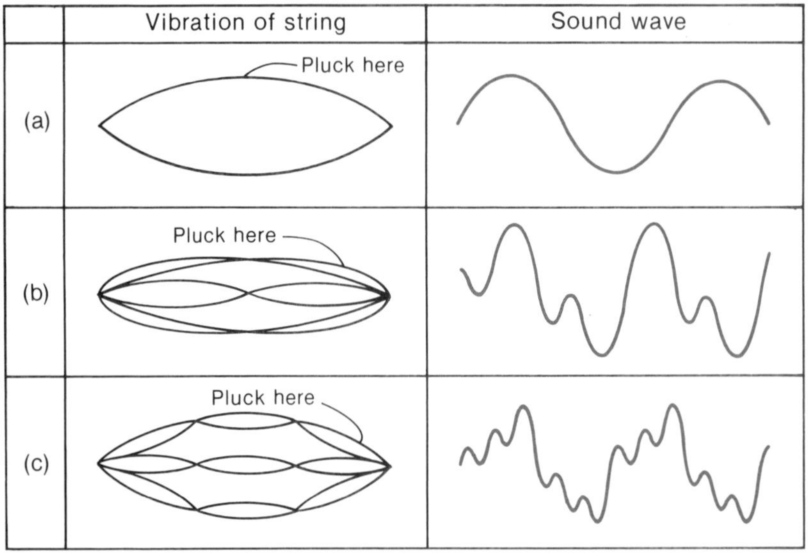

	Vibration of string	Sound wave
(a)	Pluck here	
(b)	Pluck here	
(c)	Pluck here	

Stringed instruments that are **plucked** include the harp, banjo, guitar, mandolin (Figure 14-12) and ukulele. The harp is a complex instrument. It has 46 strings. The other stringed instruments are all similar to each other. They have 4 to 8 strings as well as frets to guide the placing of fingers. The lower notes have thick wires with low tension. The higher notes have thin wires with more tension.

The main stringed instrument that is **struck** is the piano. A piano key is connected by a system of levers to a hammer that strikes the string or strings of a certain note. (See Figure 14–13.) A modern piano has 88 notes with a frequency range from 27.5 Hz to 4186 Hz. The short, high-tension wires produce high-pitch notes. The long, thick wires produce low-pitch notes. The sounds from the strings are increased in loudness and quality by the wooden sounding board of the piano.

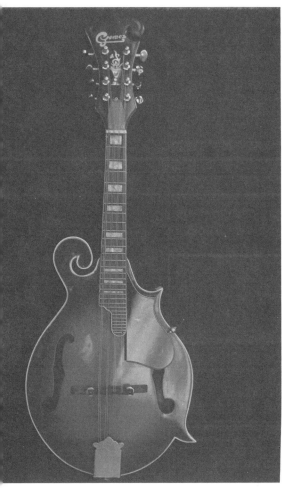

Figure 14–12 The mandolin

Figure 14–13 Inside an upright piano

Stringed instruments that are usually **bowed** belong to the violin family. This family consists of the violin (Figure 14–14), viola, cello and bass. The bows are made with dozens of fine fibres that are rubbed with rosin to increase friction between them and the strings. Each instrument has four strings and wooden sounding boards at the front and back of the case. The members of the violin family have no frets. Thus, the pitch can be changed gradually, not necessarily in steps as in the guitar family. (The only other instrument capable of changing pitch gradually is the trombone.) Again, the smallest instrument has the highest-pitch sounds and the largest has the lowest-pitch sounds.

Stringed instruments do not give out a great amount of energy. That is why a large number of violins are needed in an orchestra compared with the number of drums or trumpets.

14.7 Vibrating Columns of Air

A sound can be made by blowing into an empty pop bottle. The air inside the bottle vibrates, creating sound. When water is added to the bottle, the pitch of the sound changes. The shorter the column of vibrating air, the higher the frequency. The frequency of the vibrating air is its natural or **resonant frequency**.

Resonance occurs in a column of air when a standing wave fits neatly into the column. Standing waves result from interference. (See Section 12–12.) A device that illustrates resonance of sound in a column of air is shown in Figure 14–15. The flask can be raised or lowered to adjust the level of water in the tube. This changes the length of the air column. If a tuning fork, say 512 Hz, is sounded and held near the open end of the air column, resonant sounds will be noticed as the length of the air column changes.

To understand how resonant sounds are caused by standing waves, consider Figure 14–16. It shows waves of sound in a column of air that is open at one end. (The waves shown are transverse only because they are easier to draw.) The air molecules in the tube cannot vibrate easily at the closed end, where the water is located. Thus a node exists there. At the open end, the vibration of air molecules is easy, so a loop occurs there. Resonance occurs only when a wave of sound fits into the tube so that a node occurs at the closed end and a loop at the open end.

In Figure 14–16 the second column, (b), is half a wavelength longer than the first column. Also, the third column is half a wavelength longer than the second column. This means that the distance from one maximum sound to another in an air column is one-half the wavelength ($\frac{1}{2}\lambda$) of the sound. The distance from one

Figure 14–14
The violin has no frets. Compare this to the mandolin shown in Figure 14–12.

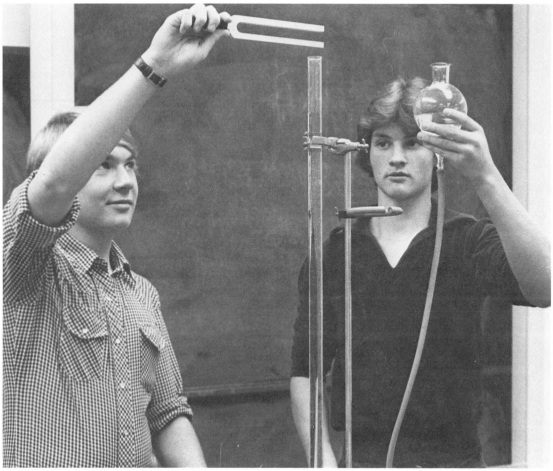

Figure 14–15 Resonance apparatus

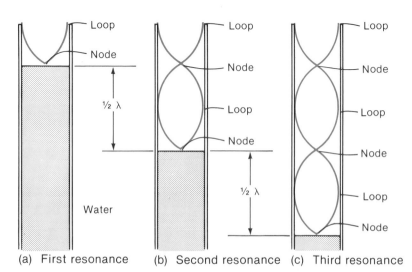

Loop

Node

½ λ

Loop

Node

½ λ

Loop

Node

Loop

Node

Loop

Node

Water

Node

Loop

Node

Figure 14–16 Standing-wave patterns in columns of air closed at one end

(a) First resonance   (b) Second resonance   (c) Third resonance

node to another is also ½λ. (This was also true for standing waves on ropes, studied in Section 12.12.)

> **Sample problem 1:** A vibrating tuning fork is held near the mouth of a column filled with water. The water level is lowered until the first loud sound is heard. Then the water level is lowered another 18 cm and a second loud sound is heard. What is the wavelength of the sound from the tuning fork?
>
> *Solution:* ½λ = 18 cm
> ∴λ = 36 cm or 0.36 m

A similar calculation can be made for columns of air that are open at both ends. Figure 14–17 shows why the distance from one maximum to the next is again ½λ. In these columns loops occur at both ends because air molecules vibrate easily there.

Figure 14–17 Standing-wave patterns in columns of air open at both ends

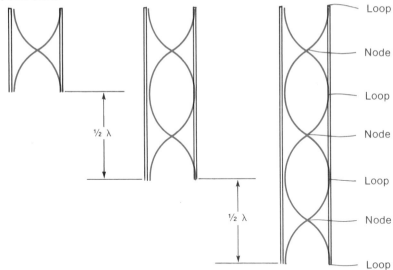

(a) First resonance   (b) Second resonance   (c) Third resonance

These ideas will be applied to an experiment involving columns of air. Then they will be used to explain wind instruments, which create sound because of vibrating air in columns.

PRACTICE

9. What vibrates to create sound in a column of air?
10. In an air column closed at one end the distance from one loud sound to the next is 22 cm. What is the wavelength of the sound producing resonance?

11. In an air column open at both ends the distance between loud sounds is 60 cm. What is the wavelength of the sound?
12. **Class demonstration**: Listen to the sound of water being added to a graduated cylinder. What happens to the pitch of the sound as the water is added? Explain why.

14.8 Experiment 31: Measuring the Wavelength of Sound in a Column of Vibrating Air

INTRODUCTION

The distance between resonant or loud sounds in air columns can be used to find the wavelength of a sound. (See Section 14–7.) If the frequency of the sound is known, the universal wave equation $(v = f\lambda)$ can be used to find the speed of sound in air. If the air temperature (T) is known, the equation $v = 332$ m/s + $(0.6\ T)$ m/s can be used to check the speed found using $v = f\lambda$. If the experiment is a success, the speeds should be the same.

As you perform this experiment, notice that low-frequency sounds need long columns of air and high-frequency sounds need short columns. This relates closely to what you will learn about wind instruments in Section 14.9.

PURPOSE: To use resonance of sound in air columns to find the wavelengths of sounds.

APPARATUS: 3 tuning forks of known frequency (between 384 Hz and 2000 Hz); resonance apparatus (Figure 14–15); rubber stopper; thermometer; metre stick; column of air open at both ends
Caution: Do not allow the vibrating metal tuning fork to touch the glass tube because it will shatter easily.

PROCEDURE

1. Adjust the water level in the long glass tube by raising the supply flask by hand until the water is near the top of the tube.
2. One person should strike the first tuning fork and hold it close to the mouth of the tube. Remember the **Caution** above. Another person should slowly lower the level of the water. When the first resonant sound occurs, a third student should mark the level of the water.

Raise and lower the water around that level until you are sure of it. Then measure and record the length of the air column.

3. Repeat #2 for the second resonant sound and a third if there is one.
4. Calculate the wavelength of the resonant sound produced.
5. Use the wavelength and frequency to find the speed of the sound in air.
6. Measure the air temperature. Use it to find the speed of sound in air. Compare this answer to the one in #5.
7. Repeat the entire procedure, using tuning forks of different frequencies.
8. Repeat the procedure, using a column of air open at both ends, if one is available.

14.9 Wind Instruments

All wind instruments have columns of vibrating air molecules. The frequency of vibration of the air molecules depends on the length of the column and whether the column is open or closed at the ends. As is the case with all vibrating objects, the large instruments create low-frequency sounds and the small instruments create high-frequency sounds.

In some wind instruments the length of each air column is fixed. This is true of a pipe organ. However, in most wind instruments, such as the trombone, the length of the air column can be changed.

To cause the air molecules to vibrate, something else must vibrate first. There are four general ways of causing air molecules to vibrate in wind instruments.

(1) **Air reed instruments:** Air is blown across or through an opening. The moving air sets up a turbulence inside the column of the instrument. Examples are the pipe organ [Figure 14–18(a)], flute [Figure 14–18(b)], piccolo, recorder and fife. The flute and piccolo have keys that are pressed to change the length of the air column. The recorder and fife have side holes that must be covered with fingers to change the length of the air column.

(2) **Single-membrane reed instruments:** Moving air sets a single reed vibrating. This in turn sets the air in the instrument vibrating. Examples include the saxophone (Figure 14–19), clarinet and bagpipe. Again, the length of the air column is changed by holding down keys or covering side holes.

(a) This beautiful pipe organ is located in a famous church in Czestochowa, Poland.

(b) The flute

Figure 14–18 Air reed instruments

(3) **Double-mechanical reed instruments:** Moving air causes a set of two reeds to vibrate against each other. These vibrations cause air in the instrument to vibrate. Examples include the oboe (Figure 14–20), English horn and bassoon. Keys are pressed to change the length of the air column.

(4) **Lip reed instruments** (also called **brass instruments**): The player's lips function as a double reed. They vibrate, causing air in the instrument to vibrate. The air does not escape

Figure 14–19
The saxophone

Figure 14–20
The oboe

Figure 14–21
The trumpet

through side holes as in the other wind instruments. The air must travel all the way through the brass instrument. Examples are the bugle, trombone, trumpet (Figure 14–21), French horn and tuba. The length of the air column is changed by pressing valves or keys that add extra tubing to the instrument.

PRACTICE

13. From the pairs of instruments listed, choose the one that would have a higher pitch. (It will be helpful to discuss in class the sizes of the instruments.)
    (a) piccolo, flute
    (b) oboe, English horn
    (c) bassoon, English horn
    (d) tuba, trumpet

14.10 Percussion Instruments

Percussion means the striking of one object against another. Percussion instruments are usually struck by a firm object such as a hammer, bar or stick. These were the first musical instruments invented because they are fairly easy to make. (Even doctors use percussion when they tap a patient's ribs and listen for sounds that indicate either clear or congested lungs.)

Percussion instruments can be put in the following categories:

(1) **Indefinite pitch:** These instruments are meant for special effects or to keep the beat of the music. Examples include the triangle (Figure 14–22), bass drum and castanets.

(2) **Definite pitch:** These instruments have bars or bells of different sizes. When struck, the bars or bells produce their own resonant frequencies. Examples are the tuning fork, orchestra bells, marimba (Figure 14-23), xylophone and carillon.

Some instruments, such as the accordion and harmonica, are difficult to classify as one type of instrument. The accordion and harmonica use moving air to set reeds into vibration. However, they do not have resonating columns of air, so they are not usually called wind instruments. They are better classified as percussion instruments in which air knocks against reeds, causing them to vibrate. See Figure 14–24.

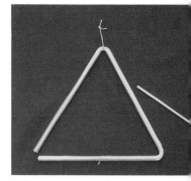

Figure 14–22 The triangle

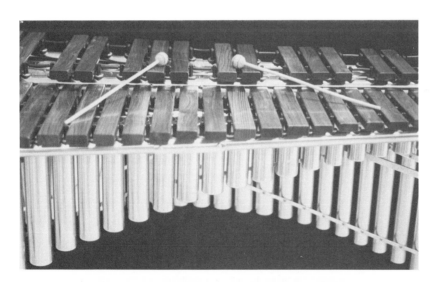

Figure 14–23 The marimba

Figure 14–24 The harmonica

14.11 The Human Voice

The human voice is a fascinating instrument. The main parts of the body that help create sound are shown in Figure 14–25(a). Diagram (b) shows how the flow of air from the lungs causes sound.

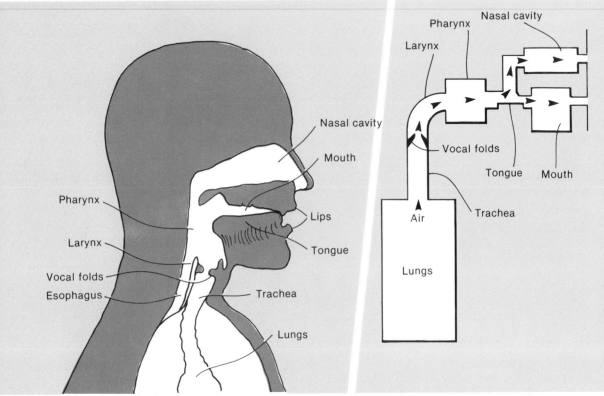

Figure 14–25 The human voice

The voice consists of three main parts:
(1) the **source** of air (lungs)
(2) the **vibrators** (vocal folds)
(3) the **resonators** (lower throat or pharynx, mouth and nasal cavity)

To create most sounds, air from the lungs causes the vocal folds to vibrate. The vocal folds are two bands of skin that act like a double reed. The loudness is controlled by the amount of air. The frequency is controlled by muscular tension as well as the size of the vibrating parts. As usual, the larger instruments have lower resonant frequencies, so, in general, male voices are lower in frequency than female voices. Refer to Table 14–3.

Table 14–3 Approximate Frequency Ranges of Singers

Type of singer	Frequency range (Hz)	Type of singer	Frequency range (Hz)
Bass	82 – 294	Alto	196 – 698
Baritone	110 – 392	Soprano	262 – 1047
Tenor	147 – 523		

The quality of sound from the voice is controlled by the parts in the resonating cavities such as the lips, tongue, mouth and nasal cavity. You can see an interesting demonstration of this by holding a microphone to your throat and then to your mouth while making the same sound. If the signals are viewed on an oscilloscope screen, the effect of the resonating cavities can be seen.

Of course, the quality of sound may also be improved by proper training. Good singers can control such effects as vibrato and tremolo. **Vibrato** is a slight changing of frequency and **tremolo** is a slight changing of amplitude.

14.12 Electrical Instruments

Electrical instruments are made of three main parts—a **source** of sound, a **microphone** and a **loudspeaker**. At hockey or football games the announcer's voice directs sound energy into a microphone. The microphone changes sound energy into electrical energy. The electrical energy then causes vibrations in a loudspeaker.

Many of the musical instruments discussed in the previous sections can be made into electrical instruments by adding a microphone and a loudspeaker. This is often done with stringed instruments that normally give out low amounts of energy. A microphone can be attached directly to the body of the instrument. In some cases the design of the instrument is changed. An electric guitar, for example, may have a body that is solid rather than hollow. See Figure 14–26.

Loudspeakers are important in determining the quality of sound from an electrical instrument. A single loudspeaker does not have

Figure 14–26 An electric guitar

the same frequency range as our ears. A set of two or three loud-speakers must be used to obtain both quality and frequency range. Table 14–4 lists three common sizes of loudspeakers used in electrical sound systems.

Table 14–4 Loudspeakers

Name	Approximate size (cm)	Frequency range (Hz)
Woofer (low-range)	25 – 40	25 – 1000
Squawker (mid-range)	10 – 20	1000 – 10 000
Tweeter (high-range)	4 – 8	3000 – 20 000

14.13 Electronic Instruments

No doubt you have heard of musical or electronic synthesizers. They are instruments that produce vibrations using electronics rather than normal sources such as stringed, wind and percussion instruments. The electronic parts that create the vibrations are called transistors and vacuum tubes.

An electronic instrument consists of four main parts:
(1) The **oscillator** creates the vibrations.
(2) The **filter circuit** selects the frequencies that are sent to the mixing circuit.

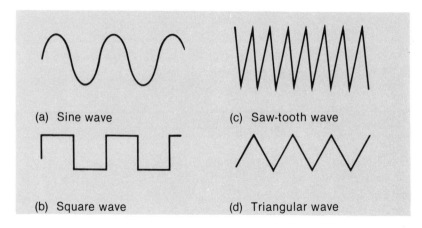

(a) Sine wave

(c) Saw-tooth wave

(b) Square wave

(d) Triangular wave

Figure 14-27 The waves shown can be used to create various other waves.

(3) The **mixing circuit** adds various frequencies together to produce the final sound.
(4) The **amplifier and speaker system** makes the sound loud enough to be heard.

Synthesizers and electronic organs are common electronic instruments. They can control the shape of their sound waves. Thus, sound can be produced to resemble the sound of almost any musical instrument. The basic shapes of waves that are used to make other waves are shown in Figure 14-27.

Synthesizers and electronic organs can also control the **attack** and **decay** properties of the sound. When a sound is first heard, its attack may be sudden, delayed or overshot. Refer to Figure 14-28(a). When a sound comes to an end, its decay may be slow, fast or irregular. See Figure 14-28(b).

(a) Growth patterns

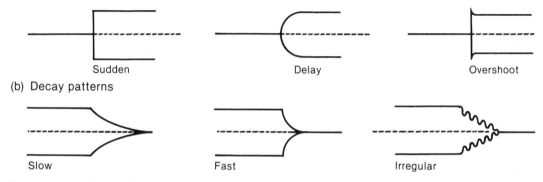

Sudden

Delay

Overshoot

(b) Decay patterns

Slow

Fast

Irregular

Figure 14-28 Control patterns

14.14 Acoustics

Some people claim that their singing voices are better in the shower than anywhere else. If that is true, it is because of all the reflections of sound in a small room.

The qualities of a room that determine how well sound is heard in the room are called **acoustics**.

The acoustics of a room depend on the shape of the room, what is in the room and what is on the walls, ceiling and floor. Sounds in a large, empty room are hollow and poor in quality. When rugs and furniture are added to the room, the acoustics improve.

In auditoriums and theatres, special designs must be used to help improve acoustics, especially for listening to music. The walls and ceiling must be designed to provide a good balance of reflection and absorption of sound. Figure 14–29 shows an auditorium with many features to improve acoustics.

Figure 14–29 This 2300-seat concert hall is in the National Arts Centre in Ottawa, Ontario. The design takes into consideration both acoustics and beauty.

(a) A recording or sound-testing studio is designed to absorb sound.

(b) The Hollywood Bowl in Hollywood, California, is designed to direct sound from the stage to the audience. This outdoor theatre seats over 18 000 people.

(c) This outdoor theatre is located in Tallin, Estonia. The shell is designed to direct sound from a 30 000-member choir toward the audience. In the background is the Baltic Sea.

Figure 14–30 Special acoustics effects

(d) This 25 000-seat outdoor theatre was built in Ephesus, Turkey, by the Romans about 2000 a ago. The theatre has excellent acoustic properties.

Sometimes construction is needed for special purposes. This is shown in Figure 14–30. Diagram (a) shows the design of a recording or sound-testing studio. Diagram (b) shows a band shell in a park directing sound to an outdoor audience. Diagram (c) shows an open-air theatre designed especially for choirs. Diagram (d) shows a 25 000 seat open-air theatre built by the Romans in Turkey almost 2000 a ago. A person who stands at the centre of the stage and speaks with ordinary loudness can be heard everywhere in the theatre. Perhaps the Romans could not explain acoustics scientifically the way we can today but they certainly could design theatres with excellent sound characteristics.

This is only the beginning of the fascinating topic of sound—an important type of energy that surrounds us at all times.

14.15 Review Assignment

1. Which of the following descriptions best suits each oscillo-scope display shown: pure sound; noise; high-quality sound? (14.1)

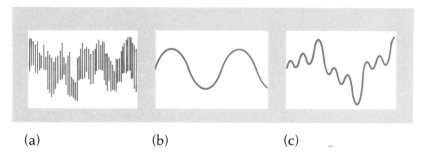

   (a)   (b)   (c)

2. (a) Name the three characteristics of musical sounds.
   (b) State what each characteristic depends on.   (14.1 to 14.4)
3. List the following frequencies in order of decreasing pitch: 512 Hz, 768 Hz, 384 Hz, 256 Hz, 420 Hz   (14.2)
4. Which sound has the higher pitch?   (14.2)

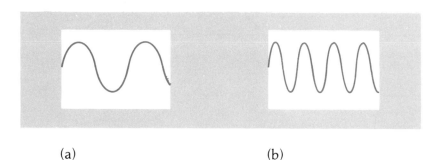

       (a)   (b)

5. What is the frequency of middle C on the:
   (a) scientific scale?
   (b) musicians' scale?   (14.2)
6. Use Table 14-1 and Figure 14-5 to calculate the frequency of a note one octave above:
   (a) F on the scientific scale   (c) C_1 on the musicians' scale
   (b) B on the scientific scale   (d) G_4 on the musicians' scale   (14.2)

7. A certain note has a frequency of 1000 Hz. Calculate the frequency of a note:
   (a) one octave above it
   (b) one octave below it
   (c) two octaves above it
   (d) two octaves below it   (14.2)
8. One person has a threshold of hearing of 10 dB and another has a threshold of hearing of 30 dB. Which person has better hearing?   (14.3)
9. How many times louder is a 7.0 B sound than a 4.0 B sound?   (14.3)
10. Describe practical ways to reduce noise pollution at a busy intersection in a city.   (14.3)
11. Compare the qualities of the sounds shown in the diagrams. (14.4)

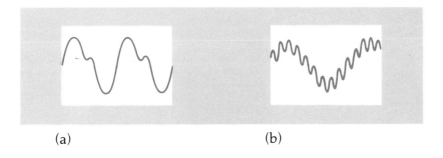

(a)   (b)

12. State what happens to the frequency of a vibrating string when:
    (a) the tension increases but the length remains constant
    (b) the length increases but the tension remains constant
        (14.5)
13. What are the two main parts of stringed instruments? (14.6)
14. Give an example of a stringed instrument that:
    (a) has frets
    (b) has no frets
    (c) is struck   (14.6)
15. Describe two ways of changing the quality of sound of a stringed instrument.   (14.6)
16. The distance from one loud sound to another in a column of air is 0.27 m. What is the wavelength of the sound?   (14.7)
17. The frequency of the sound resonating in the air column in question 16 above is 650 Hz. Calculate the speed of the sound in air.   (14.8)

18. (a) Name four methods of getting air to vibrate in wind instruments.
    (b) State an example of an instrument that uses each method in (a) above.   (14.9)
19. What is an example of a percussion instrument that has:
    (a) an indefinite pitch?
    (b) a definite pitch?   (14.10)
20. In section 14.6 a piano was described as a stringed instrument. Do you think it could also be described as a percussion instrument? Why or why not?
21. Discuss in class whether the human voice might be classified as a stringed, wind or percussion instrument.   (14.11)
22. Electrical and electronic instruments have different sources of vibrations. Explain the difference.   (14.12 and 14.13)
23. Give an example of an:
    (a) electrical instrument
    (b) electronic instrument   (14.12, 14.13)
24. Discuss in class the acoustics of both your physics classroom and the school's auditorium. In each case consider:
    (a) what has been done to provide good acoustics
    (b) what could be done to improve the acoustics   (14.14)
25. To review the most important concepts in the topics of sound, answer these questions:
    (a) What is needed to cause sound?
    (b) By what means does sound energy get transferred from one place to another?
    (c) If the frequency of a sound wave increases, what happens to its wavelength?
    (d) What is the universal wave equation?
    (e) Given the temperature of air, how do you find the speed of sound in air?
    (f) How is interference of sound waves produced?
    (g) What is the human audible range?
    (h) What are the characteristics of musical sounds and what do they depend on?
    (i) How does the size of a musical instrument relate to its frequency range?

14.16 Answers to Selected Problems

PRACTICE QUESTIONS

2. (a) 400 Hz
   (b) 640 Hz
   (c) 1160 Hz
3. (a) 100 Hz
   (b) 160 Hz
   (c) 290 Hz
4. (a) 1200 Hz
   (b) 4000 Hz
   (c) 1024 Hz

6. (a) 10
   (b) 100
   (c) 1000
   (d) 10 000
7. 80 dB
10. 44 cm
11. 120 cm

REVIEW ASSIGNMENT

7. (a) 2000 Hz
   (b) 500 Hz
   (c) 4000 Hz
   (d) 250 Hz

9. 1000 times
16. 0.54 m
17. 351 m/s

VI. Electricity and Electromagnetism

15

Static Electricity

Knowing the information in this chapter will be especially useful if you plan a career in:

• pollution control (for example, use of static electricity)
• electricity
• telecommunications
• laboratory technology

15.1 The Force of Electricity

Everyone has felt little electric shocks from touching certain objects after walking across a rug. Everyone has also seen a lightning bolt during a thunderstorm. Both the electric shock and lightning result when a build-up of electric charge is transferred from one place to another. The rapid transfer of charge is called a **discharge**. Refer to Figure 15–1.

The build-up of electric charges on an object is called **static electricity**. The word static means at rest. Thus, static electricity is not the type of electricity that flows through a wire. Rather it builds up

Figure 15-1 Static-electricity discharge in the classroom
The device on the left creates a build-up of electric charge. The
charge then jumps to the sphere on the right.

on an object and may then discharge (i.e., jump to another object).

The electric charge that builds up and causes shocks and lightning also creates a force that is easily observed. If you rub a balloon with a cloth, the balloon will stick to a wall or ceiling. If you stroke a brush through clean, dry hair, the hair is attracted to the brush. If you remove clothes from a hot dryer, the clothes stick together. In these examples the force of attraction is called the **force of electricity**.

The force of electricity was first studied and described in Greece about 600 B.C. by a man named Thales. He started with a substance called amber, a hard, yellowish resin from dead trees. He rubbed the amber with fur and discovered that the amber attracted feathers and pieces of straw. Because the Greek word for amber is **elektron**, modern words related to electricity stem from that discovery.

In the seventeenth century scientists finally advanced the study of electricity by experimenting. One of the most famous scientific experiments was performed by an American inventor, Benjamin Franklin (1706-1790). He wanted to discover if lightning was similar to the static electricity created by rubbing a piece of amber. During a thunderstorm he connected a kite to one end of a long rope and a large metal key to the other end. He held onto the rope using a silk cloth and flew the kite in the storm. (See Figure 15–2.) He discovered that lightning had the same effect as static electricity in the laboratory, except it was strong enough to knock him over. If he had not used the silk cloth as insulation against electricity, he likely would have been killed!

Figure 15–2 Benjamin Franklin's famous experiment

It was Benjamin Franklin who introduced the idea of positive and negative electric charges. He called the charge on amber rubbed with fur **negative**, and we still use that term today.

Since the time of Franklin, much has been discovered about electricity. Today so much of what we do depends on electricity that we tend to take this amazing force for granted.

PRACTICE
1. List four examples of static electricity given in this section.
2. List two examples of static electricity other than those given in this section.

15.2 Atoms and Types of Charges

In order to explain the action of electric charges we must have an idea of what matter is made of.

All matter is composed of very small particles called **atoms.** Each atom has three important types of particles. **Protons** and **neutrons** exist close together in the central part, or nucleus, of the atom. Around the nucleus travel tiny particles called **electrons.** Refer to Figure 15–3.

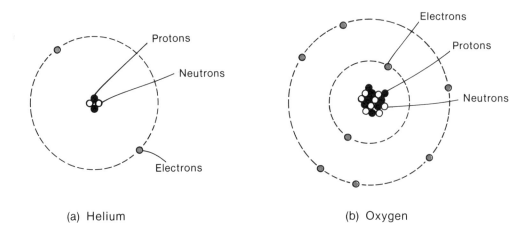

(a) Helium   (b) Oxygen

Figure 15–3 Atoms of helium and oxygen

All the atoms of one element, such as oxygen, are the same. However, different elements have atoms with different numbers of protons, neutrons and electrons. For instance, an oxygen atom is different from a helium atom (Figure 15–3).

Electrons have the same type of charge as a piece of amber rubbed with fur. Therefore, electrons have a **negative** charge. Protons have the opposite charge, **positive.** The size of charge on one electron equals the size of charge on one proton—only the sign is different. A neutron has a **neutral** charge, in other words, no charge at all.

Protons and neutrons are strongly held together and massive, so they do not contribute to the flow of charge in solids. Electrons are light and move easily, so they are reponsible for the build-up or flow of electric charge.

These basic ideas will be used to explain many events observed in the topic of electricity.

PRACTICE

3. State the type of charge on each of the following:
   (a) a proton
   (b) a neutron
   (c) an electron
   (d) a nucleus
   (e) an atom having an equal number of protons and electrons
   (f) an atom having more electrons than protons
4. A piece of amber rubbed with fur becomes negatively charged. What particles must have been deposited on the surface of the amber?

15.3 Experiment 32: The Laws of Static Electricity

INTRODUCTION

The **laws of static electricity** may be called the **law of repulsion** and the **law of attraction.** Repulsion means repelling or pushing apart. Attraction is the opposite—it means pulling together. Both the pushing and pulling result from the force of electricity.

In this experiment two types of plastic are used. One type, **acetate**, is clear. The other type, **polythene**, is white. When polythene is rubbed with wool or cat's fur, the polythene takes on a negative charge. This fact is used to determine all other types of charges. (In some laboratories ebonite or vinylite is used instead of polythene, and glass is used instead of acetate. Glass should be rubbed with silk.)

An **electroscope** is a device, such as a plastic rod, that can be used to store a known charge. That charge is used to find the type and size of a charge that exists on some other object.

PURPOSE: To prove the laws of repulsion and attraction in static electricity.

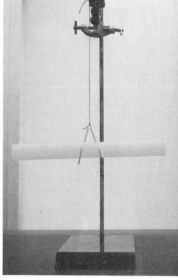

Figure 15-4

APPARATUS: retort stand with clamp and support; 2 polythene and 2 acetate strips; wool cloth, comb; plastic pen

PROCEDURE

1. Suspend the polythene strip as shown in Figure 15-4. Hold the suspended strip in the middle and rub both ends with wool. Rub

one end of the other polythene strip with wool and hold it close to one end of the first strip. Try this a few times and describe your observations.

2. Again rub the suspended polythene with wool. Now hold the wool close to the polythene. Describe what happens.

3. Rub an acetate strip with wool and hold the strip close to the suspended, charged polythene. Describe the effect. What kind of charge is on the acetate? Again try holding the wool close to the polythene.

4. Suspend the acetate from the retort stand. Repeat #1 and #2, using two acetate strips.

5. Rub a polythene strip with wool and hold the strip close to the suspended, charged acetate. Describe what happens.

6. Suspend the polythene strip once again from the retort stand. Charge the polythene strip so it can be used as an electroscope to test other charges. Use the electroscope to determine the size (small, medium or large) and type (positive or negative) of charge on:
(a) a comb rubbed through your hair
(b) a plastic pen rubbed on a piece of clothing

QUESTIONS

1. State the laws of repulsion and attraction.
2. When polythene is rubbed with wool, what type of particle (proton, neutron or electron) must be deposited on the polythene? How do you know?
3. A certain electroscope is charged positively. A small charged object is brought close to the electroscope. The electroscope is repelled. What kind of charge is on the object?

15.4 Explaining the Build-up of Static-Electricity Charges

In Experiment 32 you created a charge on a strip of polythene (or similar substance) by rubbing it with wool. The charge on the polythene was negative. We know that electrons have negative charges, so we are now able to explain how the polythene attained a negative charge.

Before rubbing occurs, the polythene has a neutral charge (equal number of protons and electrons). The wool also has a neutral charge. See Figure 15–5(a). The polythene atoms have a strong at-

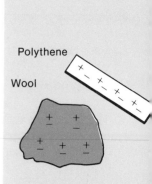

Polythene

Wool

(a) Neutral charges before rubbing

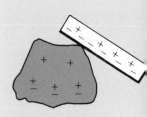

(b) Rubbing causes electrons to go from the wool to the polythene

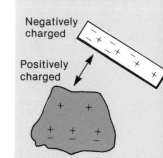

Negatively charged

Positively charged

(c) Opposite charges attract

Figure 15–5 Charging wool and polythene

traction for electrons while the wool atoms do not. When the wool and polythene are rubbed together, the rubbing (or friction) causes heat energy. The heat energy given to the electrons in the wool atoms helps some electrons escape from the wool. Those electrons are attracted to the polythene. This creates a negative charge on the polythene and a positive charge on the wool. See Figure 15–5(b).

When the wool and polythene are held close together, they attract each other because opposite charges attract. Refer to Figure 15–5(c).

The atoms of different substances have different abilities to attract electrons. For example, rubber atoms attract electrons more than wool or fur atoms. Table 15–1 lists several substances in the order in which their atoms attract electrons. The list is called the **static-electricity series**.

Table 15–1 The Static-Electricity Series

+	
Acetate	atoms have a poor attraction for electrons
Glass	
Wool	
Cat's fur	
Silk	
Aluminum	
Cotton	
Polythene	
Rubber	
Gold	atoms have a large attraction for electrons
–	

Sample problem 1: What types of charges result on cotton and rubber when they are rubbed together?
Solution: According to Table 15–1, rubber atoms have a larger attraction for electrons than cotton atoms. Thus, the rubber will become negatively charged and the cotton positively charged.

PRACTICE

5. When acetate and wool are rubbed together, the acetate becomes positively charged.
   (a) What type of charge is on the wool?
   (b) Do the electrons escape from the acetate to the wool, or from the wool to the acetate?
6. State the types of charges that result when each pair of substances is rubbed together:
   (a) rubber, silk   (c) glass, cotton
   (b) silk, glass   (d) cat's fur, gold

15.5 Conductors and Insulators

All matter is composed of atoms that contain electrons. But not all atoms have electrons that can move easily from one atom to the next. A material whose electrons do not travel easily is called an **electrical insulator.** The opposite kind of material, an **electrical conductor,** allows the transfer of electrons easily.

A simple test can be made to determine whether a substance is a conductor or an insulator. Figure 15–6 shows an object to be tested with one end touching an electroscope that has a negative charge. Another negative charge is brought close to the opposite end of the test object. If the material is a conductor, its electrons will be repelled toward the electroscope and the electroscope will swing away. If the material is an insulator, nothing will happen.

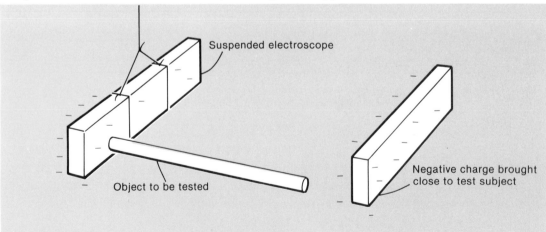

Figure 15–6 Testing for conductors and insulators

Table 15–2 lists several conductors and insulators. Notice that metals are good electrical conductors.

Table 15–2 Electrical Conductors and Insulators
(in alphabetical order)

Conductors	Insulators
Aluminum	Air (dry)
Copper	Amber
Gold	Glass
Iron	Paper
Nickel	Rubber
Silver	Silk
	Wool

To deposit a charge on a metal conductor, the conductor must be insulated from its surroundings. The metal sphere in Figure 15–7(a) is resting on an insulating stand. Such a sphere can be charged negatively or positively. In diagram (b) it has a negative charge. Once it is charged, you can discharge it by touching it with your hand. The charge travels from the sphere through your body to the earth. The process of conducting a charge to the earth is called **grounding** (symbol ⏚). See diagram (c). Grounding a small metal sphere may be compared to pouring a cup of water into the ocean. It makes a lot of difference to the cup but no difference to the ocean.

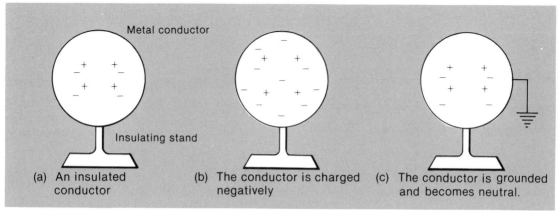

(a) An insulated conductor

(b) The conductor is charged negatively

(c) The conductor is grounded and becomes neutral.

Figure 15–7 Charging and discharging an insulated conductor

An important application of grounding is used during aircraft refuelling, when there is a danger that a static electricity spark may cause the fuel to explode. To prevent this danger, the aircraft and the fuel-hose nozzle are grounded before refuelling begins. See Figure 15–8.

PRACTICE

7. Were the materials you used in Experiment 32 insulators or conductors?
8. What kind of charge results on an insulated conductor that is grounded?

Figure 15–8 An airplane must be grounded during refuelling to prevent a static-electricity discharge.

15.6 Distribution of Charges on Insulators and Conductors

If one end of a polythene strip is rubbed with wool, only that end of the polythene becomes charged. Polythene is an insulator, so the electrons do not spread throughout it. Rather they stay piled up at one location, as shown in Figure 15–9.

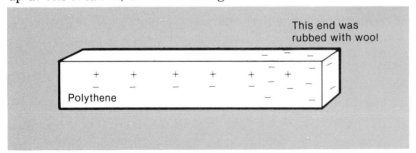

Figure 15–9 Charge distribution on an insulator

Charges act differently on conductors. If a metal sphere is charged negatively, the electrons spread evenly over the outside surface of the sphere. If the conductor has a pointed shape, the electrons tend to repel each other toward the point. Negative charges on various shapes of conductors are shown in Figure 15–10.

In Figure 15–10(c) the electrons could easily be repelled off the pointed end. In general, a pointed conductor is very useful in allowing a charge to pass to or from an object. An important application of this principle is the **lightning rod**.

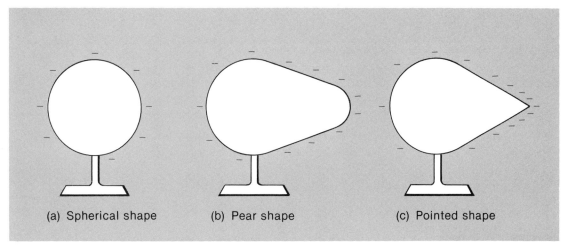

Figure 15–10 Charge distribution on metal conductors

A lightning rod helps prevent lightning from striking a house or barn. The rod is a pointed metal conductor, placed higher than any other part of the building. (See Figure 15–11.) Whether the charge travels from the ground to the clouds or from the clouds to the ground, it can drain quickly through the conductor. This prevents the lightning from starting a fire.

Figure 15–11 The lightning rod

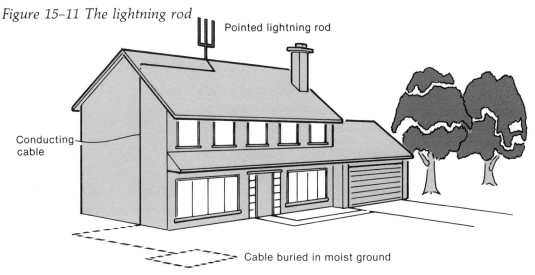

PRACTICE

9. From which object listed below would electrons jump most easily? Why?
   (a) a pointed insulator   (c) a pointed conductor
   (b) a spherical insulator   (d) a spherical conductor

15.7 Charging Conductors by Contact

There are two ways of depositing a charge on an insulated conductor—by contact and by induction. To charge by contact, all you have to do is touch a charged object, such as a polythene strip, to the conductor. This is shown in Figure 15–12.

In considering Figure 15–12, remember that only the electrons move in conductors. Notice that the charge on the conductor is the same as the original charge on the strip.

Charging a conductor by induction is described in Experiment 33.

PRACTICE

10. A conductor is charged by contact using a positive strip. What type of charge results on the conductor?

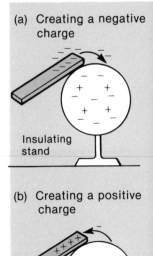

(a) Creating a negative charge

Insulating stand

(b) Creating a positive charge

Figure 15–12
Charging by contact

15.8 Experiment 33: Induction and Charging Conductors by Induction

INTRODUCTION

The pushing and pulling forces of electricity can separate charges on a conductor. This process is called **induction**. It is illustrated in Figure 15–13. In diagram (a), a negatively charged strip is brought close to a neutral conductor. Electrons on the conductor are forced (induced) to the far end of the conductor. In diagram (b) a positively charged strip attracts the electrons in a neutral conductor.

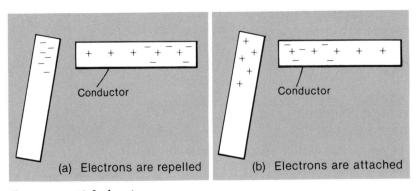

(a) Electrons are repelled

(b) Electrons are attached

Figure 15–13 Induction

Induction can be used to deposit a charge onto a conductor. The ways of doing this will be described in the experiment.

A special device called an **electrophorus** is used in this experiment. It consists of a smooth metal plate, which rests on a rough insulating surface. (See Figure 15–14.) The electrophorus can be used to store a large static-electricity charge.

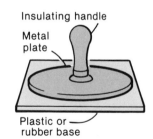

Insulating handle

Metal plate

Plastic or rubber base

Figure 15–14 The electrophorus

PURPOSE: To learn how to induce charges on objects and how to charge a metal conductor by induction.

APPARATUS: 2 polythene strips; 1 acetate strip; retort stand and clamp; wool cloth; stream of water; 2 metal spheres mounted on insulating stands; electrophorus set

PROCEDURE

1. Charge a polythene strip negatively by rubbing it with wool. Hold it close to a smooth stream of water from a tap. Describe what happens.
2. Repeat #1, using an acetate strip charged positively by rubbing it with wool. Again describe what happens.
3. Suspend a polythene strip from the retort stand. Charge the strip to act as an electroscope. What type of charge is on the electroscope?
4. Place two metal spheres together and neutralize them by touching them with your hand. Bring a charged polythene strip close to one end of the pair as shown in Figure 15–15(a). Remove the sphere that is farthest from the polythene strip. Predict the type of charge on each sphere. Check your prediction using the charged electroscope. This is one way of charging conductors by induction.
5. Charge a polythene strip and bring it close to a neutralized mounted sphere. Ground the sphere as shown in Figure 15–15(b). Predict the type of charge remaining on the sphere. Check your prediction using the charged electroscope. This is another way of charging conductors by induction.

(a) By separation

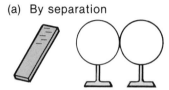

Charged polythene strip

(b) By grounding

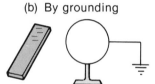

Charged polythene strip

Figure 15–15 Charging by induction

6. Rub the polythene base of the electrophorus with wool. Place the metal plate on the charged base. Touch the metal briefly. Predict the type of charge on the metal plate. Check your prediction using the charged electroscope. A spark may result if you ground the electrophorus plate.

QUESTIONS

1. Choose which of the following statements about charges are true:

(a) positive attracts positive
(b) positive attracts neutral
(c) positive attracts negative
(d) negative attracts negative
(e) negative attracts neutral

2. Draw a series of diagrams to explain how charging by induction occurred in Procedure # 4.

15.9 Using Diagrams to Explain Charging by Induction

In Experiment 33, Procedure # 5, you charged a metal conductor by induction starting with a negatively charged polythene strip. The charge that resulted on the metal was positive. We will use diagrams to explain how the conductor became positively charged.

Consider Figure 15-16. In diagram (a) the metal sphere is neutral, having an equal number of positive and negative charges. In diagram (b) the negative polythene is brought close to the sphere. Electrons on the surface of the conductor are repelled far away. The conductor is then grounded in diagram (c). Some electrons leave the conductor and are repelled to the ground. Finally, in diagram (d), the result is a positive charge on the conductor. We say the metal has been charged positively by induction.

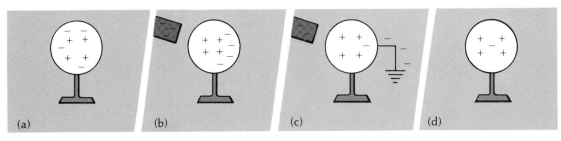

(a)   (b)   (c)   (d)

Figure 15-16 Charging a metal sphere by induction

Notice in Figure 15-16 that the charge on the conductor is opposite to the charge on the polythene strip. If the original charge had been positive, the final charge on the conductor would be negative. Thus, when charging by induction, the final charge on the conductor is opposite to the type of the original charge.

Several other observed results may be explained using a series of diagrams similar to Figure 15-16.

PRACTICE

11. A conductor is charged by induction using a positively charged strip. What type of charge results on the conductor?
12. Draw a series of diagrams to show how to charge a single metal sphere by induction starting with a positively charged strip. (Hint: Only the electrons move. When the sphere is grounded, electrons travel **from** the ground **to** the conductor.)
13. Draw a series of diagrams illustrating how the electrophorus was charged in Experiment 33, Procedure # 6.

15.10 Static-Electricity Generators

A static-electricity generator is a device that creates a large build-up of charge. The charge can be used for demonstrations and practical purposes.

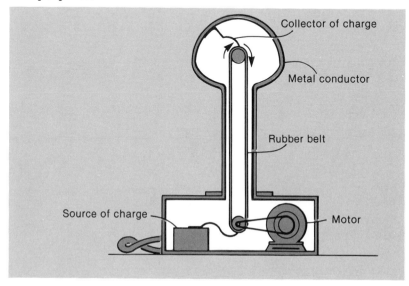

Figure 15-17 The Van de Graaff generator

A common type of static-electricity generator is called the Van de Graaff generator, named after its designer. Refer to Figure 15–17. The source of charge at the bottom of the machine gives energy to millions of electrons to free them from atoms. The electrons are carried by a rubber belt (an insulator) to the top of the machine. The electrons are distributed rapidly and evenly over the outside surface of the metal conductor.

The Van de Graaff generator or similar device can be used to demonstrate many phenomena in static electricity. Some of these are described in the next section.

PRACTICE

14. What type of charge ends up on the outside of the Van de Graaff generator described in this section?

15.11 Demonstrations of Static Electricity

Viewing the demonstrations of static electricity described here will help you understand applications of static electricity discussed in Section 15.12.

Caution: Only the teacher should operate the generator. Proper grounding procedure should be used.

(1) **The principle of the lightning rod:** A needle or pin can be used by the teacher to discharge the generator. Large sparks are prevented because the electrons leak continually to the ground.

(2) **Electric wind:** A pointed piece of metal is attached to the metal sphere of the generator. Electrons stream rapidly from the pointed metal. The effect can be noticed if a lit candle is placed in the way of the electric wind.

(3) **Flying thread and/or flying fur:** Pieces of thread are taped to the generator or a piece of fur is placed on top of it. When the generator is turned on, the results are interesting.

(4) **Flying hair:** A person with clean dry hair stands on an insulating stand and holds onto the generator. The electrons are discharged through the hair. See Figure 15–18(a).

(5) **Arcing:** A discharge sphere is placed 3 or 4 cm from the main generator. When the room lights are out, sparks can be seen where the electrons jump from one sphere to the other. Light,

(a) Flying hair   (b) Charging a Leyden jar

Figure 15–18 Demonstrations of static electricity

heat and sound are created by the arc. Arc welding is based on this principle.

(6) **Pinwheel:** When electrons are repelled off the ends of the pinwheel, Newton's third law of motion (the "action-reaction" law) comes into effect.

(7) **Smoke condenser:** Smoke is placed into the device and the generator is turned on. The smoke particles become charged by induction. Some industries use this principle in their large smokestacks.

(8) **Volta's hailstorm:** Small styrofoam spheres are covered with metallic paint, a good conductor. The spheres undergo rapid charging and discharging.

(9) **Leyden jar:** (See the caution note below.) Two conducting sheets are insulated from each other [Figure 15–18(b)]. A large static-electricity charge can be placed onto the jar. The jar has the capacity to hold a charge. Such a device is called a **capacitor**. Figure 15–19 illustrates how the Leyden jar becomes charged [diagram (a)] and discharged [diagram (b)].

Caution: A spark from the Leyden jar can be extremely dangerous. Be sure to follow the safety precaution recommended by the manufacturer.

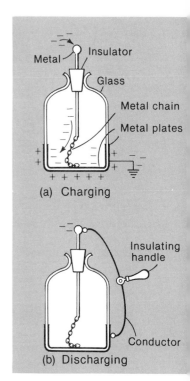

Figure 15–19 The Leyden jar

PRACTICE

15. Explain why the flying fur in demonstration (3) was repelled from the generator.

16. Explain why the pinwheel in demonstration (6) is a good example of Newton's third law of motion, discussed in Section 7.10.

15.12 Applications of Static Electricity

Static electricity is often harmful and sometimes useful. This section tells how to reduce the unwanted effects of static electricity. Then it presents useful applications that have not yet been mentioned.

Static electricity can be a nuisance. For example, it may cause clothes to cling to the body, and it may cause scratchy sounds during the playing of a record. Sprays have been developed to reduce this type of static. Adding moisture to the air also helps.

Explosions and fires may be caused by a static-electricity discharge. In hospital operating rooms explosive gases (anesthetics) are sometimes needed. To prevent spark discharges, the floor and the footwear worn by the doctors and nurses must be good conductors of electricity.

Static electricity can also be a help. One simple use is in the wrapping of food. Static-electricity forces help plastic wrap cling to food. This keeps the food fresh longer.

Many large industries use static electricity to help prevent air pollution. A device called a precipitator (Figure 15–20) works as follows. A metal electrode is negatively charged. Polluted gases pass by the electrode. Particles in the gases become charged negatively by induction. They are then attracted to the positively charged collectors. If the particles are liquid, they simply drain off. If the particles are solid, the collectors must be shaken once in a while to force the particles to fall down.

Figure 15–20 The static-electricity precipitator

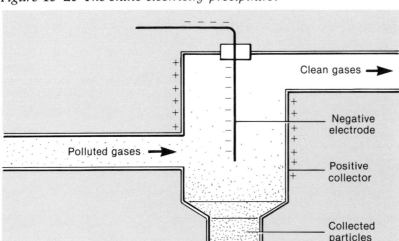

An application that has gained interest in recent years is the negative-ion generator. An **ion** is a charged particle. An ion generator sends negative electrons into the air, creating charged particles of air. Experiments have shown that these charged particles help make people feel energetic. Several such devices are now used in cars, homes and offices.

There are many other applications of static electricity. The details will not be described here, but a list of some of those applications will indicate their importance.

(1) cleaning air in homes using static electricity
(2) separating minerals from ores
(3) separating unwanted particles from grain seeds
(4) separating shells from nuts
(5) applying certain spray paints
(6) coating short fibres on rugs and the insides of musical instrument cases
(7) making expensive sandpaper
(8) spraying insecticides on plants
(9) photocopying
(10) studying living cells
(11) making microphones with high-frequency response
(12) arc welding

It is obvious that static electricity is important. However, its applications are not as common as the applications of current electricity, discussed in the next chapter.

PRACTICE

17. Explain why the conducting floor and footwear of doctors and nurses in an operating room help prevent sparks.

15.13 Review Assignment

1. Define and give two examples of the force of electricity. (15.1)
2. What is meant by static-electricity discharge? (15.1)
3. Draw a diagram of an atom having three protons, three neutrons and three electrons. (Show two electrons close to the nucleus and the third electron farther away.) (15.2)
4. What kind of charge does the atom in #3 above have? (15.2)
5. (a) Which particle is the most important in explaining static-electricity events?
   (b) What kind of charge does that particle have? (15.2)

6. Describe a method to prove the law of:
   (a) repulsion
   (b) attraction   (15.3)
7. Objects A and B are rubbed together. Object A loses electrons to object B.
   (a) What kind of charge is on A?
   (b) What kind of charge is on B?
   (c) What happens if A is brought close to a negatively charged electroscope?   (15.2, 15.3, 15.4)
8. When hard rubber and wool are rubbed together, the rubber becomes negatively charged. Explain how this happens, starting with neutral pieces of rubber and wool.   (15.4)
9. What kinds of charges result on each substance when the following pairs of substances are rubbed together? (Refer to Table 15–1, Section 15.4.)
   (a) gold, rubber
   (b) aluminum, rubber
   (c) aluminum, silk
10. What is the difference between electrical conductors and electrical insulators?   (15.5)
11. What is meant by the term "grounding" in static electricity? (15.5)
12. The diagrams below show insulated metal conductors that are charged negatively. Show how electrons are distributed on the conductors.   (15.6)

(a)

(b)

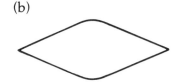

13. The diagram below shows a neutral metal conductor. A negatively charged strip is brought close to the right end of the conductor. Draw a diagram in your notebook, showing the distribution of charges on the conductor.   (15.8)

14. State the final charge on a conductor that is:
    (a) charged by contact using a negative strip
    (b) charged by contact using a positive strip
    (c) charged by induction using a negative strip
    (d) charged by induction using a positive strip   (15.7, 15.8)

15. The diagrams below show a step-by-step procedure for charging two metal conductors by induction. Redraw the diagrams in your notebook, showing the distribution of charges in each case.   (15.8, 15.9)

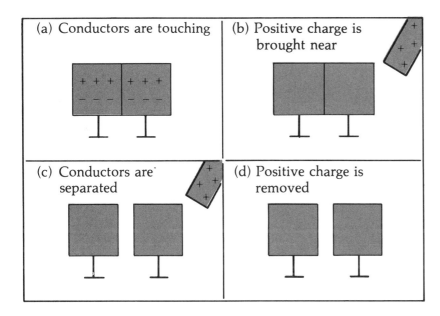

16. Static electricity shocks are noticed more in the dry winter months than in the moist summer months. What can be done to the air to reduce shocks in the winter?   (15.12)

15.14 Answers to Selected Problems

PRACTICE QUESTIONS

3. (a) positive
   (b) neutral
   (c) negative
   (d) positive
   (e) neutral
   (f) negative
4. electrons

5. (a) negative
6. (a) rubber becomes negative, silk positive
8. neutral
10. positive
11. negative

REVIEW ASSIGNMENT

4. neutral
5. (a) the electron
7. (a) positive
   (b) negative

9. (a) gold becomes negative, rubber positive
14. (a) negative   (c) positive
    (b) positive   (d) negative

16

Current Electricity

Knowing the information in this chapter will be especially useful if you plan a career in:

• electricity
• telecommunications
• laboratory technician
• power distribution (linemen, etc.)
• electronics
• computers
• astronomy

16.1 Electrons in Motion

The study of current electricity began about two hundred years ago when an Italian doctor experimented with the legs of dead frogs. The doctor, named Luigi Galvani (1737-1798), noticed that the frog's legs jerked when they were touched with two different metals at the same time. He thought that something inside the legs had created electricity, which had caused the jerking movement.

Another Italian, a scientist named Count Alessandro Volta (1745-1827), heard about Galvani's experiments and began his own research. Volta proved it was the metals, not the frog's legs, that created the electricity. He was able to create electricity by simply placing two different metals into a fruit and connecting the metals with a wire. Such a setup, now called a **voltaic cell**, is shown in Figure 16-1. The instrument that indicates the flow of electrons is called a **galvanometer**. It was named after Luigi Galvani.

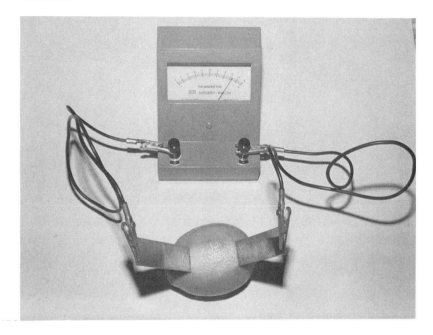

Figure 16-1 A simple voltaic cell

The type of electricity created by a voltaic cell is called current electricity. **Current electricity** is a continuous movement of electrons in a path. This is different from static electricity (studied in Chapter 15) in which the electrons pile up on an object. (In this book we are concerned with the motion of negative electrons. Other scientists, especially those studying chemistry, are also concerned with the motion of positive charges.)

The electrons that move in current electricity get energy from some source. Then they repel each other through a path, such as a wire, and give their energy to some object. Of course, there are thousands of devices that use electrical energy. Our lives would be very different without this important type of energy.

PRACTICE
1. State the type of charge on the particles that travel in current electricity.

16.2 Sources of Electrical Energy

People have learned how to change several kinds of energy into electrical energy. The source of energy depends greatly on what the electricity is used for. The electricity may be used to start a car, listen to a portable radio, operate computers on a satellite, heat a home and so on.

Following is a description of five types of energy that can be changed into electrical energy. Scientists are constantly searching for new ways to perform this important task.

(a) Voltaic cell

(1) **Chemical Energy** A chemical reaction can give energy to electrons to free them from atoms. When a wire is connected to a source, the electrons have a path through which to travel. We will mention three kinds of cells that change chemical energy into electrical energy.

   (a) **The voltaic cell:** This source, invented by Alessandro Volta and discussed in the previous section, has two different metal strips placed in a liquid chemical. One strip becomes charged negatively and the other positively. Electrons repel each other from the negative strip through a wire to the positive strip [Figure 16-2(a)]. The electrons give energy to whatever is in their path until the liquid chemical or metal strips are used up.

(b) Primary cell (in the form of a "dry" cell)

   (b) **The primary cell or dry cell:** This source is similar to a voltaic cell. It consists of two different metals and a chemical paste. Like the voltaic cell, it cannot be recharged. However, it is more convenient because it is small and "dry". [Refer to Figure 16-2(b).] Primary cells are used to operate portable devices such as watches, calculators, radios and electronic toys.

   (c) **The secondary cell or storage cell:** Again, this source has two different metals and a chemical substance. The important advantage of the secondary cell is that it can be recharged. This is done by using an outside source of electricity. Secondary cells are used in cars, trucks and camera-flash sets. [See Figure 16-2(c).]

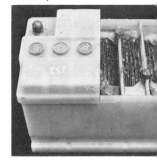

(c) Secondary cell

*Figure 16-2
Chemical sources of electrical energy*

(2) **Light Energy** When light strikes certain metal surfaces, electrons may be freed from the atoms near the surface. The electrons repel each other through the wire and give their energy to some device. Spacecraft and satellites often use photocells to change light energy into electrical energy. Refer to Figure 16-3(a).

(3) **Heat Energy** Two wires made of different metals may be used to change heat energy into electrical energy. The wires must be joined at the ends and must be at different temperatures. One way of doing this is shown in Figure 16–3(b). Such a device has several applications, including the measurement of very high and very low temperatures.

(4) **Mechanical Energy** Most of our electricity comes from spinning generators that change the mechanical energy of motion into electrical energy. The generators are forced to spin by the use of running water or by steam pressure. Electrons in the wires of the generators gain energy that is then distributed to homes, offices and industries. This source of energy will be discussed in more detail later in the book. [See Figure 16–3(c).]

(5) **Geothermal Energy** Electrical generators can be run by using steam produced by heat inside the earth. See Figure 16–3(d).

Figure 16–3 Other sources of electrical energy

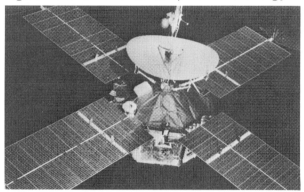

(a) A spacecraft uses thousands of solar cells to generate electricity.

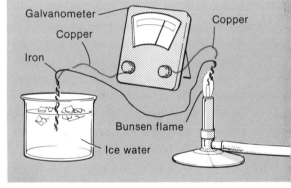

(b) Heat energy

(c) This hydro-electric generating station is located on a river in the interior of British Columbia. Falling water forces turbines to spin. The turbines are connected to electrical generators.

(d) This geyser generating station is located in California, U.S.A.

PRACTICE
2. Name an electric device that receives energy from:
   (a) a primary cell
   (b) a secondary cell
   (c) the sun
   (d) falling water
3. Discuss in class how the following may be used to create electrical energy:
   (a) wind   (b) ocean tides   (c) ocean waves

16.3 Electric Current

In this chapter we will study three important electrical quantities: current, voltage and resistance.

Electric **current** is a measure of how many electrons pass through a wire or other electric device each second. The symbol for current is I. Its unit of measurement is the ampere or amp (A). This unit is named in honour of André-Marie Ampère, a French physicist who lived from 1775 to 1836.

An instrument used to measure electric current is called an **ammeter**. It must be connected directly into the path of the moving electrons. This kind of connection is called a **series** connection. Figure 16–4 shows the correct way of doing this. Follow the path of the electrons (e⁻) in that diagram. They leave the negative terminal of the battery. Then they enter the negative terminal of the ammeter, causing the needle to swing to the right. As they pass through the light bulb they give energy to the bulb. Then they are attracted back to the positive terminal of the source where they gain more energy.

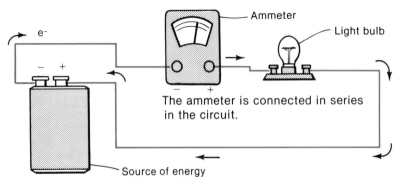

Figure 16–4 Measuring electric current

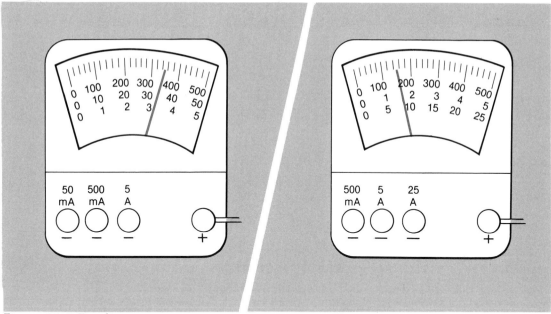

Figure 16-5 Reading an ammeter   Figure 16-6.

You will use an ammeter in experiments later in this chapter. With practice you can learn how to read the scale on an ammeter. One type of ammeter, with three different scales, is shown in Figure 16–5. If you are required to change milliamps (mA) to amps, remember that milli means 1/1000. Thus, 1 A = 1000 mA.

Sample problem 1: What does the ammeter in Figure 16–5 read if the wire is connected to the:
(a) 0–50 mA scale?
(b) 0–500 mA scale?
(c) 0–5 A scale?
Solution: (a) 35 mA or 0.035 A
            (b) 350 mA or 0.35 A
            (c) 3.5 A

PRACTICE
4. Change to amps:
   (a) 600 mA   (b) 80 mA   (c) 4000 mA
5. State the current in amps for the ammeter in Figure 16–6 when the wire is connected to the:
   (a) 0–500 mA scale
   (b) 0–5 A scale
   (c) 0–25 A scale

16.4 Electric Voltage

The purpose of electricity is to take energy from a source and deliver it to a useful device called a **load**. The source of electric current gives energy to the electrons it sends out. Electric **voltage rise** is a measure of the amount of energy given to a group of electrons that leave the source. When the energetic electrons reach a load, they give their energy to it. Electric **voltage drop** is a measure of the amount of energy given by electrons to a load.

Both voltage rise and voltage drop have the symbol V and are measured in **volts** (symbol V). This unit is named after Alessandro Volta, mentioned in Section 16.1.

An instrument used to measure electric voltage is called a **voltmeter**. To measure voltage rise, the voltmeter must be connected across the source. To measure voltage drop, the voltmeter must be connected across the load. This kind of connection is called a **parallel** connection. It is illustrated in Figure 16–7. Compare this parallel connection to the series connection for ammeters in Figure 16–4.

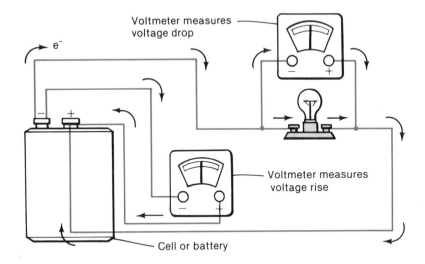

Figure 16–7 Measuring voltage rise and voltage drop

Notice in Figure 16–7 that the electrons leave the negative terminal of the source and enter the negative terminal of each voltmeter. This causes the needle in each voltmeter to swing to the right.

Reading a voltmeter is similar to reading an ammeter, as you will discover in the practice question.

PRACTICE

6. What is the voltage reading on the voltmeter in Figure 16–8 when the wire is connected to the:
   (a) 0–5 V scale?   (b) 0–15 V scale?   (c) 0–300 V scale?

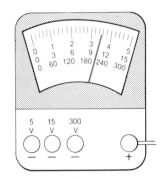

Figure 16–8

16.5 Electric Resistance

Electric conductors and insulators were discussed in Section 15.5. Conductors allow electrons to move more freely than do insulators. Electric **resistance** is a measure of how much a conductor fights the flow of electrons through it. The greater the resistance, the greater the amount of energy the electrons give up as they pass through a conductor.

An ordinary light bulb operates on this idea. Figure 16–9 shows the inside of such a bulb. The thick outside wires have low resistance and gain little energy from electrons. The very thin, coiled wire strung across the top has a high resistance to electron flow. Thus it gains much energy from the electrons, heats up and gives off light. If the thin wire were replaced with straight copper wire, the bulb would not heat up.

The resistance of an electric conductor depends on four factors:

(1) **Length**—The longer the conductor, the greater the resistance.
(2) **Cross-sectional area**—The smaller the area, the greater the resistance.
(3) **Type of material**—Silver, copper and gold have very low resistance; aluminum and tungsten have slightly higher resistance; iron has an even higher resistance.
(4) **Temperature**—At very low temperatures some conductors have almost no resistance.

The symbol for resistance is R. Its unit of measurement is the ohm (symbol Ω, the Greek letter omega). This unit is named in honour of Georg Simon Ohm, 1787-1854, a German physicist. Resistance may be measured using an ohmmeter. Rather than measuring it directly, however, we will be using calculations to find resistance (Experiment 34).

Resistors are devices that have a known resistance. They are often used in electronic devices and science laboratories. Two common materials used to make resistors are wire and granulated carbon. Figure 16–10 illustrates three resistors, a wire-wound resistor of constant or fixed value, a carbon resistor of fixed value and a wire-wound variable resistor called a **rheostat**.

Figure 16–9 Resistance in an electric light bulb

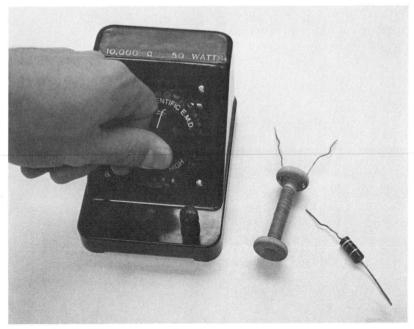

Figure 16-10 Three types of resistors

PRACTICE

7. If you were designing an electric toaster, would you use wire of high resistance or extremely low resistance? Why?
8. What happens to the resistance of an electric conductor when its:
   (a) length is decreased?
   (b) cross-sectional area is decreased?

16.6 Electric Circuits

An electric **circuit** is a path through which electrons may flow. It consists of three main parts:
(1) the source of energy (cell, battery, etc.)
(2) the transporter of energy (wires)
(3) the user of energy or load (light bulb, etc.)

A circuit may also have switches and measuring instruments such as ammeters and voltmeters to determine current, voltage and resistance. Figures 16-4 and 16-7 show examples of circuits.

Figure 16-11 shows another simple but complete electric circuit. The voltmeter connected across the cell measures how much

energy the cell gives to the electrons. The negatively charged electrons (e⁻) leave the cell and pass through an ammeter, which
counts how many electrons travel through the circuit each second.
Then the electrons pass through the light bulb, giving their energy
to it. The voltmeter across the bulb registers how much energy the
electrons give to the bulb. Finally the electrons are attracted back
to the cell where they gain more energy.

So far we have seen circuits with only one load (a single light
bulb). If a circuit has more than one load, the energy from the electrons must be shared. This is presented in more detail later in the
chapter.

The circuit in Figure 16–11 is a **closed circuit**. This means that
electrons have a path they can follow. An **open circuit** results
when the electron path is broken. For example, cutting the wire at
point "a" in Figure 16–11 would cause an open circuit.

PRACTICE
9. In an electric circuit, what is the function of each of the follow-
    ing?
    (a) battery   (c) load   (e) voltmeter
    (b) wires   (d) ammeter

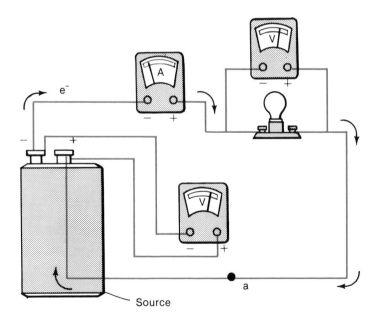

Figure 16–11 A simple electric circuit

16.7 Symbols in Electric Circuits

An electric circuit such as the one in Figure 16–11 is time consuming to draw. Thus, symbols shown in Figure 16–12 are used to draw simpler diagrams of electric circuits. You should learn these symbols.

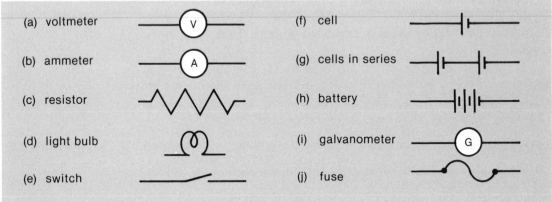

Figure 16–12 Symbols for drawing electric circuits

When cells are connected in series, as in Figure 16–12(g), the total voltage is the sum of the individual voltages. For example, if three 1.5 V cells are connected in series, the total voltage is 4.5 V.

The advantage of using symbols in circuit diagrams is evident in Figure 16–13. Diagram (a) shows an electric circuit and diagram (b) shows the same circuit using symbols.

Figure 16–13 Circuit diagrams

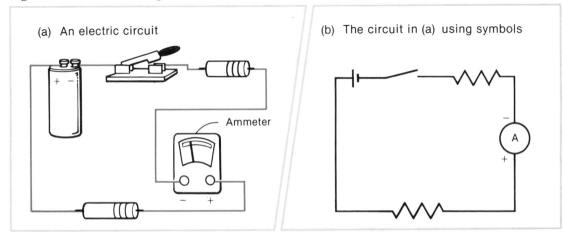

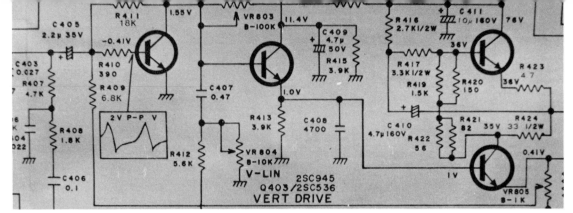

Figure 16–14 Partial circuit diagram of a television set

Of course, circuit diagrams may be much more complex than those studied in this book. Figure 16–14 illustrates an example of a circuit diagram having many parts and symbols.

PRACTICE
10. What is the total voltage for a series connection of these cells:
    (a) 2.0 V, 5.0 V?   (c) 50 V, 100 V, 70 V?
    (b) 2 V, 3 V, 4V?
11. Use symbols to draw a diagram of the circuit shown in:
    (a) Figure 16–4   (b) Figure 16–7   (c) Figure 16–11

16.8 Experiment 34: Ohm's Law

INTRODUCTION
It was the German scientist, Georg Simon Ohm, who discovered one of the most important laws of physics. Ohm's law describes how electric current, voltage and resistance are related. It states that:

> the ratio of the voltage across a resistor to the current through it is constant, if we neglect the effect of temperature

This ratio of voltage to current is what we call resistance, R (Section 16.5). Thus, Ohm's law can be written in equation form.

$$R = \frac{V}{I}$$

Since voltage is measured in volts and current in amps, resistance is measured in volts per amp or ohms (Ω).

To have consistent units, current readings in milliamps should be changed to amps before calculations are made.

PURPOSE: To learn how to set up electric circuits and to verify Ohm's law.

APPARATUS: 3 fixed resistors of different values (e.g., 100 Ω, 50 Ω and 25 Ω); 3 dry cells or a variable power supply; voltmeter; ammeter; connecting wires
Caution: Never leave an electric circuit connected longer than necessary because overheating may result.

PROCEDURE

1. Connect the circuit as shown in Figure 16–15 using the first resistor (R_1). Record the resistance of R_1 in a chart based on Table 16–1.
2. Measure the current through the resistor and the voltage drop across it. Calculate the ratio of V/I and compare your answer to the value of the resistance in ohms. Record your observations and calculations.
3. Repeat # 2 using the same resistor but a higher supply voltage.
4. Repeat # 2 using a third, still higher supply voltage.
5. Repeat # 1 to # 4 using a second resistor (R_2) and finally a third resistor (R_3).

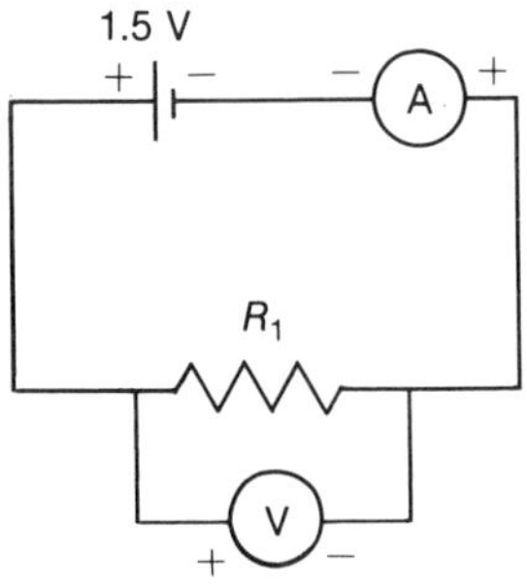

Figure 16–15

Table 16–1 Ohm's Law Experiment

R (ohms)	V (volts)	I (amps)	$\dfrac{V}{I}$ (ohms)
$R_1 =$			

QUESTIONS

1. Calculate the value of the resistance in each case:
   (a) $V = 20$ V, $I = 4$ A (c) $V = 8$ V, $I = 0.5$ A
   (b) $V = 120$ V, $I = 8$ A (d) $V = 10$ V, $I = 400$ mA
2. On a single graph of voltage (vertical axis) versus current (horizontal axis) plot the results of the experiment. There should be three separate, straight lines, one for each resistor. Calculate the slope of each line. Relate the slopes to the values of the resistances.
3. Rearrange the equation $R = \dfrac{V}{I}$ to express:
   (a) V by itself   (b) I by itself
4. The current through a 50 Ω resistor is 2.5 A. What is the voltage across the resistor?
5. The voltage across a 20 Ω resistor is 10 V. What is the current through the resistor?

16.9 Series and Parallel Circuits

There are two different methods of connecting resistors. In a **series circuit** all the electrons that leave the source must follow the same path. Refer to Figure 16–16(a). In a **parallel circuit** the electrons leave the source and come to a junction or dividing point. At that point the electrons must choose which path to take. See Figure 16–16(b). Circuits may have both series and parallel connections, as shown in Figure 16–16(c).

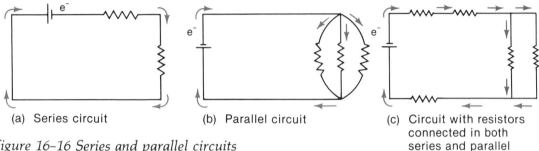

(a) Series circuit

(b) Parallel circuit

(c) Circuit with resistors connected in both series and parallel

Figure 16–16 Series and parallel circuits

Series and parallel circuits have important differences. These will be studied in the next two experiments.

16.10 Experiment 35: Resistors in Series

INTRODUCTION
To follow the instructions in this experiment and the next one, you must understand the meanings of symbols such as I_a and V_{bc}.

The symbol I_a means the current through the wire at point "a". The symbol V_{bc} means the voltage drop (or rise) from point "b" to point "c". These are illustrated in Figure 16–17.

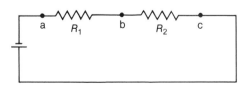

(a) Circuit with points a, b and c

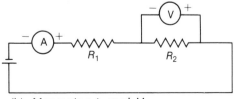

(b) Measuring I_a and V_{bc}

Figure 16–17

PURPOSE: To study current, voltage and resistance for resistors connected in series.

APPARATUS: 2 fixed resistors of different size (e.g., 100 Ω and 50 Ω); dry cells or variable power supply; voltmeter; ammeter; connecting wires

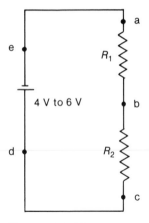

Figure 16–18

PROCEDURE
1. Set up the series circuit shown in Figure 16–18 but place an ammeter at "a" to measure I_a. Record the current in amps.
2. Using the same basic circuit and supply voltage, measure I_b and I_c. Compare these values to I_a.
3. Using the same circuit, measure V_{ed}, V_{ab}, and V_{bc}. Compare the supply voltage (V_{ed}) to the sum of the voltages across the resistors ($V_{ab} + V_{bc}$).
4. Use Ohm's law to calculate the total resistance (R_T) of the circuit

   ($R_T = \frac{V_{ab}}{I}$). Use the same law to calculate $R_1 = \frac{V_{ab}}{I}$ and

   $R_2 = \frac{V_{bc}}{I}$. Compare R_T with the sum of $R_1 + R_2$.

QUESTIONS
1. In a series circuit, what happens if one resistor is suddenly removed?
2. Write three rules for resistors connected in series based on these questions:
   (a) How does the current in one part of a series circuit compare with the current in another part?
   (b) How does the voltage rise (given by the source) compare with the sum of the voltage drops (used by the loads)?
   (c) How does the total resistance in a series circuit compare with the sum of the individual resistances?

16.11 Experiment 36: Resistors in Parallel

INTRODUCTION
When resistors are connected in parallel, the total resistance (R_T) can be found using the following equation.

$$\frac{1}{R_T} = \frac{1}{R_1} + \frac{1}{R_2} + \frac{1}{R_3} + \ldots \text{ where } R_1, R_2, \text{ and } R_3 \text{ are}$$
the individual resistances

Sample problem 2: Calculate the total resistance when two
10 Ω resistors are connected in parallel.

Solution: $\dfrac{1}{R_T} = \dfrac{1}{R_1} + \dfrac{1}{R_2}$

$\dfrac{1}{R_T} = \dfrac{1}{10} + \dfrac{1}{10}$

$\dfrac{1}{R_T} = \dfrac{2}{10}$

Now find the reciprocal of each side of the equation.

$$\therefore R_T = \dfrac{10}{2} = 5\,\Omega$$

Notice in sample problem 2 that the total resistance (5 Ω) is less
than either resistance connected in parallel. This occurs because
the electrons have two paths to take instead of one, so the resistance actually decreases.

PURPOSE: To study current, voltage and resistance for resistors
connected in parallel.

APPARATUS: as in Experiment 35

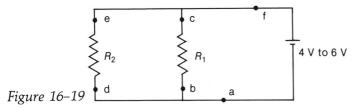

Figure 16–19

PROCEDURE
1. Set up the parallel circuit shown in Figure 16–19, but place an
   ammeter at "a" to measure the total current (I_a) coming from the
   power supply.
2. Alter the circuit to measure I_b and I_d. What does the sum $I_b + I_d$
   represent? (This is the most difficult experimental procedure in
   the chapter. Have your teacher check the results before you
   proceed to # 3.)
3. In the same circuit measure the voltages V_{bc} and V_{de}. What can
   you conclude?
4. Use Ohm's law to calculate the total resistance of the circuit
   ($R_T = \dfrac{V}{I_a}$).

5. Calculate the total resistance using the equation described in the Introduction to the experiment. Compare this answer to the one in #4.

QUESTIONS
1. In a parallel circuit, what happens if one of the resistors is suddenly removed?
2. Write three rules for resistors connected in parallel based on the following questions:
   (a) How does the total current coming from the source compare with the sum of the currents through the individual resistors?
   (b) How does the voltage drop across one resistor compare with the voltage drop across the other resistors?
   (c) How does the total resistance in the parallel circuit compare with the individual resistances?

16.12 Calculations Involving Electric Circuits

In this section we will apply some of the important facts studied in this chapter. Ohm's law indicates how voltage, current and resistance are related ($R = \dfrac{V}{I}$). Table 16–2 summarizes the facts regarding series and parallel circuits. (Appendix F uses these facts to derive the equations for total resistance.)

Table 16–2 Series and Parallel Circuits

Type of circuit	Series	Parallel
Example		
Current	current is constant throughout the entire circuit	individual currents add up to the total current
Voltage	individual voltages add up to the total voltage	voltage is constant across all parts of the circuit
Resistance	$R_T = R_1 + R_2 + \ldots$	$\dfrac{1}{R_T} = \dfrac{1}{R_1} + \dfrac{1}{R_2} + \ldots$

Sample problem 3: Find the total resistance when 5 Ω, 10 Ω and 30 Ω are connected in:
(a) series
(b) parallel
Solution: (a) series $R_T = R_1 + R_2 + R_3$

$$= 5 \ \Omega + 10 \ \Omega + 30 \ \Omega$$

$$= 45 \ \Omega$$

(b) parallel $\dfrac{1}{R_T} = \dfrac{1}{R_1} + \dfrac{1}{R_2} + \dfrac{1}{R_3}$

$$\dfrac{1}{R_T} = \dfrac{1}{5} + \dfrac{1}{10} + \dfrac{1}{30}$$

$$\dfrac{1}{R_T} = \dfrac{6}{30} + \dfrac{3}{30} + \dfrac{1}{30}$$

$$\dfrac{1}{R_T} = \dfrac{10}{30}$$

$$\therefore R_T = \dfrac{30}{10} = 3 \ \Omega$$

Sample problem 4:
Given: $V_{ab} = 20$ V; $R_1 = 10 \ \Omega$;
        $I_a = 2$ A

Find: (a) I_b (b) V_{bc} (c) R_T (d) R_2

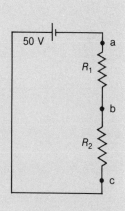

Solution: (a) $I_b = 2$ A because the current is constant in a series circuit.
(b) $V_{bc} = 50$ V – 20 V
        = 30 V because the total voltage is the sum of the individual voltages.
(c) $R_T = \dfrac{V_T}{I}$

$$= \dfrac{50 \text{ V}}{2 \text{ A}}$$

$$= 25 \ \Omega$$
(d) $R_2 = 25 \ \Omega - 10 \ \Omega$
        = 15 Ω because the total resistance is the sum of the individual resistances.

312

Sample problem 5
Given: $V_{ab} = 20$ V; $I_a = 4$ A; $I_c = 1$ A
Find: (a) V_{cd} (b) I_e
Solution: (a) $V_{cd} = 20$ V because the voltage is constant in
a parallel circuit.
(b) $I_e = 4$ A $- 1$ A
$= 3$ A because the total current equals
the sum of the individual currents.

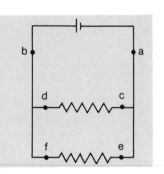

PRACTICE
12. Find the total resistance when the following resistors are connected in series:
(a) $2.2 \, \Omega$, $5.5 \, \Omega$
(b) $80 \, \Omega$, $320 \, \Omega$, $800 \, \Omega$
(c) $1 \, \Omega$, $10 \, \Omega$, $100 \, \Omega$, $1000 \, \Omega$
13. Find the total resistance when the following resistors are connected in parallel:
(a) $8 \, \Omega$, $8 \, \Omega$
(b) $30 \, \Omega$, $30 \, \Omega$, $30 \, \Omega$
(c) $15 \, \Omega$, $60 \, \Omega$, $60 \, \Omega$
14. In the circuit diagram, $V_{ab} = 4$ V,
$I_a = 2$ A, and $R_1 = 2 \, \Omega$
Find (a) I_c (b) V_{bc} (c) R_T (d) R_2

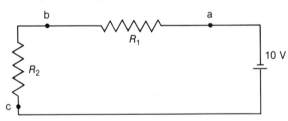

15. In the circuit diagram, $I_b = 2$ A
and $I_c = 3$ A.
Find: (a) V_{bd} (b) I_a (c) R_T

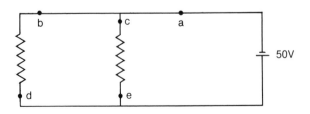

16.13 Review Assignment

1. Describe the difference between static electricity and current electricity.   (16.1)
2. Name a common device that changes chemical energy into electrical energy.   (16.2)
3. State one advantage of a:
   (a) primary cell
   (b) secondary cell   (16.2)
4. What is our most common source of electrical energy? (16.2)
5. Write the symbol for each of the following:
   (a) current   (e) volts
   (b) voltage   (f) ohms
   (c) resistanc   (g) milliamps
   (d) amps   (h) kilovolts   (16.3–16.5)
6. Name the instrument used to measure:
   (a) electric current
   (b) electric voltage   (16.3, 16.4)
7. Change the following to amps or volts:
   (a) 350 mA   (c) 905 mV
   (b) 42 mA   (d) 1050 mV   (16.3, 16.4)
8. State the kind of connection (series or parallel) for:
   (a) an ammeter
   (b) a voltmeter   (16.3, 16.4)
9. What is the difference between voltage rise and voltage drop?   (16.4)
10. State four factors that affect the resistance of an electric conductor.   (16.5)
11. Why is copper a good choice for electric wiring?   (16.5)
12. Describe the difference between a closed circuit and an open circuit.   (16.6)
13. Use proper symbols to draw a diagram of the following series circuit. Two cells are connected to a switch, a light bulb, an ammeter and two resistors. Include a voltmeter to measure the voltage drop across the light bulb.   (16.7)
14. What is the total voltage when four 2.5 V cells are connected in series?   (16.7)
15. State Ohm's law in equation form.   (16.8)
16. Calculate the resistance of each of the following appliances:
    (a) an electric toaster draws 10 A of current from a 120 V circuit

(b) an electric razor operates at 0.25 A when the voltage is 5.0 V

(c) a clothes dryer, operating at 240 V, draws 12 A of current   (16.8)

(17.) Calculate the unknown quantities: (16.8)

	R (Ω)	V (V)	I (A)
(a)	?	120	15
(b)	70	?	0.3
(c)	24	12	?

(18.) If 0.8 A of current flow through a 75 Ω resistor, what is the voltage drop across the resistor?   (16.8)

(19.) A certain fuse used in a 120 V circuit has a resistance of 8 Ω. What is the current rating of the fuse?   (16.8)

20. Redraw the circuit shown here in your notebook.
    (a) Indicate the direction of the flow of electrons in all parts of the circuit.
    (b) State the purpose of the cell.   (16.6, 16.9)

21. Calculate the total resistance when 100 Ω, 200 Ω and 600 Ω are connected in:
    (a) series
    (b) parallel   (16.10–16.12)

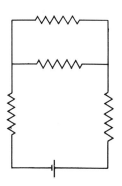

22. In the circuit shown, R_1 = 20 Ω.
    The voltage drop across R_1 is 10 V and across R_2 is 20 V.

(a) What type of circuit is shown?
    (b) What is the total voltage drop across both resistors?
    (c) What is the voltage rise given by the source?
    (d) Calculate the total resistance in the circuit if the current is 0.5 A.
    (e) Find the value of R_2.   (16.10, 16.12)

23. In the circuit shown, I_a = 2 A and I_c = 0.5 A. The voltage rise given by the source is 3 V.
    (a) What type of circuit is shown?
    (b) What is the voltage drop across R_1?
    (c) What is the voltage drop across R_2?
    (d) What is the value of I_b?
    (e) Calculate the total resistance of the circuit.   (16.11, 16.12)

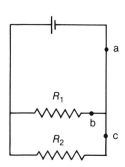

16.14 Answers to Selected Problems

PRACTICE QUESTIONS

4. (a) 0.6 A
   (b) 0.08 A
   (c) 4.0 A
5. (a) 150 mA or 0.15 A
   (b) 1.5 A
   (c) 7.5 A
6. (a) 3.5 V
   (b) 10.5 V
   (c) 210 V
10. (a) 7.0 V
    (b) 9 V
    (c) 220 V

12. (a) 7.7 Ω
    (b) 1200 Ω
    (c) 1111 Ω
13. (a) 4 Ω
    (b) 10 Ω
    (c) 10 Ω
14. (a) 2 A
    (b) 6 V
    (c) 5 Ω
    (d) 3 Ω
15. (a) 50 V
    (b) 5 A
    (c) 10 Ω

REVIEW ASSIGNMENT

7. (a) 0.35 A
   (b) 0.042 A
   (c) 0.905 V
   (d) 1.05 V
14. 10 V
16. (a) 12 Ω
    (b) 20 Ω
    (c) 20 Ω
17. (a) 8.0 Ω
    (b) 21 V
    (c) 0.5 A

18. 60 V
19. 15 A
21. (a) 900 Ω
    (b) 60 Ω
22. (b) 30 V
    (c) 30 V
    (d) 60 Ω
    (e) 40 Ω
23. (b) 3 V
    (c) 3 V
    (d) 1.5 A
    (e) 1.5 Ω

17

Using Electrical Energy

GOALS: After completing this chapter you should be able to:
1. State the difference between direct current and alternating current.
2. Name ways of changing electrical energy into other types of energy.
3. Describe how a three-wire system can create 240 V and 120 V circuits in a home.
4. State the advantage of using parallel connections for electric circuits in homes and commercial buildings.
5. Given the voltage across an electric device and the current through it, find its electric-power rating ($P = VI$) in watts.
6. Given the power of an electric device and the amount of time it is used, calculate the energy consumed ($E = Pt$) in joules, megajoules or kilowatt hours.
7. Given the cost rate of electricity and the energy consumed, calculate the total cost of the electrical energy (cost = rate × energy).
8. Explain the need for safety when dealing with current electricity.
9. Describe the function of fuses and circuit breakers.
10. Give reasons for the need for conservation of electrical energy.

Knowing the information in this chapter will be especially useful if you plan a career in:

• electricity
• telecommunications
• laboratory technician
• power distribution (linemen, etc.)
• electronics
• computers
• astronomy

17.1 Direct and Alternating Currents

Current electricity is a flow of electrons. If the electrons flow continuously in a path without reversing direction, the current is called **direct current** or **DC**. Battery-operated devices, such as flashlights, electrical systems in cars and portable radios, use direct current. The experiments performed in Chapter 16 also used direct-current circuits. (See Figure 17–1.)

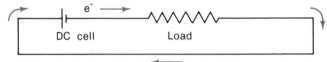

Electrons repel each other through the circuit without reversing direction.
Figure 17–1 A DC circuit

If the electrons in a circuit are forced to reverse their direction periodically, the current is called **alternating current** or **AC**. This type of current is created by our huge electrical generators. The generators force nearby electrons back and forth in the wires. Those electrons repel other electrons almost immediately back and forth through the entire circuit. Most of our industries and most of our appliances use, alternating current. See Figure 17–2. (Electric generators are discussed in greater detail in Chapter 19.)

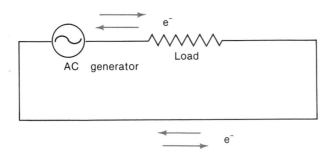

e⁻

Load

AC   generator

Electrons are repelled back and forth in the circuit. (Notice the new symbol introduced in the circuit diagram.)

e⁻

Figure 17–2 An AC circuit

The AC generators in North America force the electrons to repeat their back-and-forth motion sixty times per second. Thus, our AC is rated at a frequency of 60 Hz. (Some countries, such as those in Europe, use a frequency of 50 Hz.) Our eyes cannot tell that lights are flashing on and off sixty times each second. At lower frequencies, say around 20 Hz to 25 Hz, we can see a flickering.

The reason AC is used in our electrical systems is that, with our present technology, AC is easier to transmit than DC. Electrical energy produced by generators can be transmitted large distances with little energy loss by great networks of power lines and transformers. (Transformers are discussed in Chapter 19.)

The electricity available in most outlets in homes and schools is 120 V (AC). That voltage can be changed into some other voltage using a device called an adaptor. Figure 17–3 shows an adaptor that can operate an instrument such as a calculator. The adaptor changes alternating current (120 V, 60 Hz) into direct current (9 V, 100 mA). In this case the instrument can also be operated with a 9 V (DC) battery.

Figure 17–3 An adaptor can change AC into DC

PRACTICE

1. Describe how electrons move in:
   (a) a direct-current circuit
   (b) an alternating-current circuit

2. State which type of current (AC or DC) is used to operate:
   (a) a refrigerator   (c) an automobile cigarette lighter
   (b) a portable radio   (d) an ordinary light bulb

17.2 Uses of Current Electricity

Anyone who has experienced an electrical "blackout" can appreciate how much we rely on electricity. Electricity provides comfort, convenience and entertainment, and does a great amount of work for us.

Section 16.2 described how chemical, heat, light and mechanical energies can be changed into electrical energy. Once the electrical energy is distributed to the user, it can be changed back into other types of energy.

Electrical energy is easily changed into light and heat. Light energy is one of the most common uses of electricity. Heat energy is used to cook food, iron clothes and heat car engines on winter nights in cold regions.

Electrical energy can be changed into chemical energy, creating special chemical reactions. One process uses electrical energy to separate certain metals, such as aluminum, from their ores. Another process uses electrical energy to electroplate one metal onto another. For example, steel bumpers on cars are covered with thin layers of copper, nickel and chromium.

Electrical energy is often turned into mechanical energy or energy of motion. This occurs in devices such as typewriters, egg beaters, drills, lathes and many other tools.

Because the human body has its own electrical system, electrical energy also has many medical uses. Injuries, especially bone fractures, heal faster with increased electrical activity. If this increase is not provided from inside the body, it can be supplied from outside the body. As another example, some people have abnormally low heartbeats, so low that they could not live without an aid. That aid is given by a small electric device called a pacemaker. A patient using a pacemaker can lead a normal life.

Of course, there are hundreds of other uses of current electricity, too numerous to mention here.

PRACTICE
3. Name two devices, other than those mentioned in this section, that change electrical energy into:
   (a) heat energy (b) mechanical energy

Figure 17-4 Uses of current electricity

4. Sound energy was not mentioned in this section. Name two
   devices that change electrical energy into sound energy.

17.3 Current Electricity in the Home

Figure 17-5 shows the basic electrical wiring entering an ordinary
household. Electrical energy is delivered to homes either through
underground cables or from utility poles. Three wires are fed into
a meter and then into a fuse box or circuit-breaker box. Two
wires, covered with red insulation (R) and black insulation (B), are
called "hot". The third wire, either bare or covered with white in-
sultion (W), is called neutral.

Several other things can be seen in Figure 17-5. The neutral
wires are connected to the neutral bar. They have a direct path to

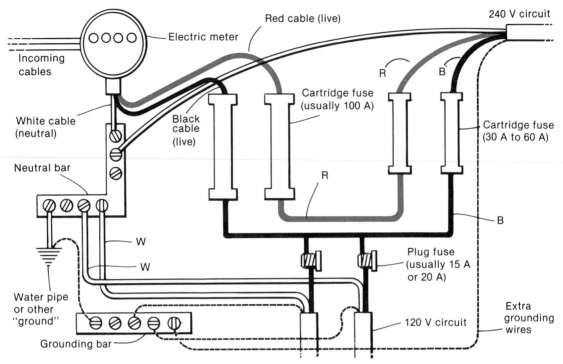

Figure 17-5 Electricity delivered to the home

the ground because they are not connected to any fuses. The black and red wires pass through large cartridge fuses. Then they are directed, again through cartridge fuses, to a circuit for large appliances. The voltage from the black wire to the red wire is 240 V. This is the voltage needed for major appliances such as a clothes dryer, stove, water heater and air conditioner.

The black wire in Figure 17-5 is also shown connected through plug fuses to two 120 V circuits. The voltage between B and W is 120 V. This voltage is used for most lights, outlets and small appliances. The 120 V circuits may also be connected between R and W, although this is not shown in the diagram.

The broken lines in Figure 17-5 are grounding wires, which are included for extra safety.

The fuses are used as protection against overheating. Each fuse has a wire that burns if the current exceeds a certain limit, such as 15 A. This causes an open circuit and the electrons no longer flow. Then the fuse must be replaced.

Circuit breakers have the same function as fuses, but are more convenient. Each circuit breaker has a bar made of two strips of different metals. The metals expand at different rates when heated, causing the bar to bend. If a circuit overheats, the bar bends far enough to trip a switch. The switch must then be reset. (The compound bar was discussed in Section 10.11.)

In each household there are several 120 V circuits connected either to B-W or R-W. A simple 120 V circuit is shown in Figure 17–6(a). It consists of two lights, two single outlets and a plug fuse to the black or red wire. Diagram (b) shows the same circuit with grounding added. Diagram (c) shows a three-prong plug and receptacle.

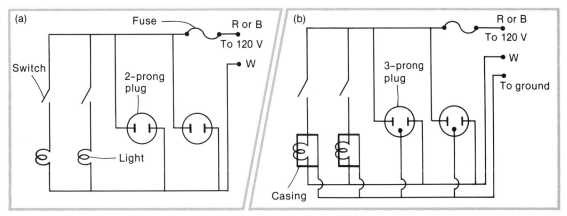

Figure 17–6 A simple 120 V circuit

(c)

In Figure 17–6 the loads are connected in parallel. As you learned in Experiments 35 and 36, this type of connection has an advantage over a series connection. In a series connection, if one load is burned out or removed, the other loads do not work. In a parallel connection each load can work without the other loads.

PRACTICE
5. What voltages are possible in household circuits?
6. In a household electrical circuit, what is the function of each of the following?
   (a) the electric meter   (c) grounding wires
   (b) a fuse or circuit breaker   (d) a switch

17.4 Electric Power

Power is the rate of using energy. (See Section 8.7.) Its equation is:

$$\textbf{power} = \frac{\textbf{energy}}{\textbf{time}} \text{ or } P = \frac{E}{t}$$

Energy is measured in joules (J), time in seconds (s), and power in joules per second or watts (W).

The equation $P = \dfrac{E}{t}$ could be used to derive an equation for electric power in terms of voltage and current. That equation, derived in Appendix F, is:

$$\textbf{electric power} = \textbf{voltage} \times \textbf{current or } P = VI$$

Voltage is measured in volts (V), currents in amps (A) and power in watts (W). A watt is a small unit, so power is often stated in kilowatts (1 kW $= 10^3$ W), or megawatts (1 MW $= 10^6$ W).

Sample problem 1: A small colour television, connected to a 120 V outlet, draws 1.5 A of current. Calculate its power rating.
Solution:
$$P = VI$$
$$= 120 \text{ V} \times 1.5 \text{ A}$$
$$= 180 \text{ W}$$

PRACTICE
7. Calculate the power rating of each appliance:
    (a) a 120 V electric sander draws 3.0 A of current
    (b) an electric can opener, used in a 120 V circuit, operates at 2.2 A
    (c) an electric handsaw operates at 9.5 A when connected to 120 V
    (d) a portable radio, using four 1.5 V cells in series, draws a current of 0.6 A
8. Rearrange the equation $P = VI$ to express
    (a) V by itself
    (b) I by itself
9. Calculate the voltage of a 0.9 W calculator that draws a current of 0.1 A.
10. What is the current used by a 1200 W electric kettle in a 120 V household circuit?

17.5 The Cost of Current Electricity

Whether we use electrical or other forms of energy, the cost is important in our daily lives. To learn how power companies charge customers for energy, we shall begin with the equation for power in terms of energy and time, $P = \dfrac{E}{t}$. If that equation is rearranged to express E by itself, we have:

energy = power × time or $E = Pt$

Thus, if we know the power rating (in watts) of an appliance and the time (in seconds) it is used, we can find the energy consumed (in joules).

After the energy has been calculated, it can be used in another equation to find the cost of the electricity. That equation is:

cost = rate × energy

The rate is expressed in some convenient unit, as you will learn in the sample problems that follow.

In sample problem 2, assume that a power company charges 1.2¢ for each one million joules (1.0 MJ) of energy consumed. In other words, the rate is 1.2¢/MJ.

Sample problem 2: A 500 W hair dryer is used for 5 min (300 s). Calculate the:
(a) energy consumed (in joules and megajoules)
(b) cost of the energy
Solution: (a) $E = Pt$

$\quad\quad = 500\ W \times 300\ s \quad (W = \dfrac{J}{s})$

$\quad\quad = 150\ 000\ J$ or $0.15\ MJ$ (Move the decimal six places.)

(b) cost = rate × energy

$\quad\quad = 1.2\ \dfrac{\text{¢}}{MJ} \times 0.15\ MJ$

(Note the division of units.)

$\quad\quad = 0.18$¢

In many regions, power companies calculate energy in units other than megajoules. Thus, in the equation $E = Pt$, if power is measured in kilowatts (kW) and time in hours (h), then energy is stated in kilowatt hours (kW·h).

In sample problem 3, assume that the rate charged by a power company is 5¢/(kW·h).

Sample problem 3: A 100 W (0.1 kW) light bulb is turned on for 4 h. Calculate the:
(a) energy consumed (in kilowatt hours)
(b) cost of the energy

Solution: (a) $E = Pt$
$$= 0.1 \text{ kW} \times 4 \text{ h}$$
$$= 0.4 \text{ kW·h}$$

(b) cost = rate × energy
$$= 5 \frac{\text{¢}}{\text{kW·h}} \times 0.4 \text{ kW·h}$$
(Note the division of units.)
$$= 2.0\text{¢}$$

Figure 17-7 Reading an electric meter

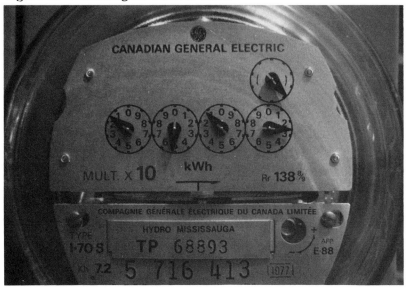

(a) A typical electric meter

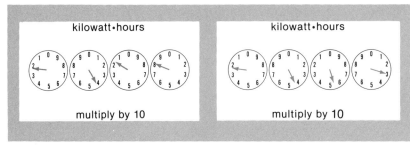

(b) March 1 reading　　　　(c) April 1 reading

An electric meter is used to determine how much energy (in megajoules or kilowatt hours) is consumed by a household each month. Figure 17-7(a) shows a typical electric meter. The dials are read from left to right and the reading is multiplied by 10. In Figure 17-7(b) the reading is 2418 × 10 = 24 180 kW·h.

Sample problem 4:
(a) What is the meter reading in Figure 17-7(c)?
(b) Assume that one month elapsed between the readings in (b) and (c), Figure 17-7. How much energy was used in the month?
(c) At a rate of 4¢/(kW·h), how much did the electrical energy cost?

Solution: (a) 24 530 kW·h
(b) 24 530 kW·h – 24 180 kW·h = 350 kW·h
(c) cost = rate × energy

$$= 4 \ \frac{¢}{kW·h} \times 350 \ kW·h$$

$$= 1400¢ \ or \ \$14.00$$

PRACTICE

11. For each appliance listed calculate the **energy** consumed (in megajoules) and the **cost** of the energy. Assume a rate of 2¢/MJ.
(a) a 50 W stereo is operated for 15 h (54 000 s)
(b) a 200 W air conditioner is run for 10 h (36 000 s)
(c) a 1500 W coffee-making machine is left on for 2 h (7200 s)
12. For each appliance listed calculate the **energy** consumed (in kilowatt hours) and the **cost** of the energy. Assume a rate of 6¢/(kW·h).
(a) a 300 W drill (0.3 kW) is used for 0.5 h
(b) a 1500 W oven (1.5 kW) is operated for 2 h
(c) an 800 W block heater (0.8 kW) is used for 8.0 h
13. The diagrams show meter readings two months apart. The electricity costs 5¢/(kW·h).
(a) How much energy was used during the two months?
(b) Calculate the cost of the energy.

 May 1 × 10

 July 1 × 10

14. Read the electric meter in your house, apartment or school once or twice daily for at least two weeks. Tabulate your readings and discuss them in class. You may wish to discuss factors that affect the use of electricity (e.g., the outside temperature, the difference between daytime use and night-time use, the difference between weekday use and weekend use and other special circumstances).

17.6 Electrical Safety

Each year many lives are lost and properties damaged due to the careless use of electricity. Electricity can be a hazard and must be treated wisely.

If a fuse blows, it should be replaced with a fuse of the correct size. Overheating causes fuses to blow, so the reason for overheating should be found and corrected. The reason may be too many appliances connected to one circuit, a frayed cord or a faulty appliance. Similar precautions apply to the use of circuit breakers.

Wherever there is water or moisture, electricity should be avoided. For example, a telephone or electric radio should not be used near a bathtub. Impure water is a good conductor of electric-ity, and the electricity is powerful enough to kill a person.

Extension cords can be a fire hazard. Often they can carry no more than 7 A of current safely. Thus, only low-power appliances should be connected to ordinary extension cords. For example, the maximum power for a 7 A cord is $P = VI = 120$ V $\times$ 7 A or 840 W. If a 1200 W kettle is plugged into the cord, overheating will result. A 15 A fuse will not provide safety in such a cir-cumstance.

PRACTICE
15. Why would it be dangerous to replace a burned-out fuse with a copper penny?

17.7 Conserving Electrical Energy

North Americans are among the greatest users of energy in the world. We have more electrical appliances and gadgets than even science-fiction writers could have imagined just thirty years ago.

Electrical energy has brought us comfort, convenience and much leisure time. It has helped increase our store of knowledge and ability to communicate. If we wish to continue enjoying these benefits, we must take care not to waste our resources. Everyone must do his or her part to conserve electrical energy.

17.8 Review Assignment

1. Describe the difference between alternating current and direct current.   (17.1)
2. What is the frequency of AC in North America?   (17.1)
3. Name the type of energy created by electrical energy in each of the following:
   (a) an oven   (b) a doorbell
   (c) an electric sander   (17.2)
4. State the voltage between these pairs of wires in a household circuit:
   (a) black and white   (b) black and red
   (c) red and white   (17.3)
5. What is the advantage of connecting appliances in parallel in a household circuit?   (17.3)
6. Calculate the power rating of each device:
   (a) an electronic toy, using a 9 V battery, draws 0.2 A of current
   (b) a 240 V water heater draws 20 A of current
   (c) an electric typewriter operates at 0.9 A and 120 V
   (17.4)
7. Calculate the unknown quantities:

	P(W)	V(V)	I(A)
(a)	?	120	5.0
(b)	6.0	?	0.5
(c)	40	8.0	?

8. An electric clock uses 2.4 W of power in a 120 V circuit. What is the current through the clock?   (17.4)
9. A 30 W stereo tape-deck uses 2.5 A of current. What is the voltage across the tape-deck?   (17.4)
10. A 2000 W stove burner is used for 30 min (1800 s). The rate of electricity is 1.5¢/MJ. Calculate the:
    (a) energy consumed (in megajoules)
    (b) cost of the energy   (17.5)

11. Many light bulbs are planned to burn out after about 3000 h of operation. Assume a 100 W (0.1 kW) light bulb is left on for 3000 h (about four months), and that the rate of electricity is 5¢/(kW·h). Find the:

(a) energy consumed (in kilowatt hours)

(b) cost of the energy consumed   (17.5)

12. Read the meters in the diagram and find the cost of the electricity assuming the rate is 6¢/(kW·h).   (17.5)

Sept 15  × 10   Oct 15  × 10

13. What is the purpose of a fuse or circuit breaker?   (17.6)

14. Calculate the maximum power available to each circuit:

(a) a 15 A fuse is used in a 120 V circuit

(b) a 60 A fuse is used in a 240 V circuit

(c) a 6 A extension cord is used in a 120 V circuit   (17.4 and 17.6)

15. List three ways that could be used to conserve electrical energy in your:

(a) school   (b) home

17.9 Answers to Selected Problems

PRACTICE QUESTIONS

7. (a) 360 W

(b) 264 W

(c) 1140 W

(d) 3.6 W

8. (a) $V = \dfrac{P}{I}$

(b) $I = \dfrac{P}{V}$

9. 9 V

10. 10 A

11. (a) 2.7 MJ and 5.4¢

(b) 7.2 MJ and 14.4¢

(c) 10.8 MJ and 21.6¢

12. (a) 0.15 kW·h and 0.9¢

(b) 3.0 kW·h and 18¢

(c) 6.4 kW·h and 38.4¢

13. (a) 580 kW·h

(b) $29.00

REVIEW ASSIGNMENT

4. (a) 120 V

(b) 240 V

(c) 120 V

6. (a) 1.8 W

(b) 4800 W

(c) 108 W

7. (a) 600 W

(b) 12 V

(c) 5.0 A

8. 0.02 A or 20 mA

9. 12 V

10. (a) 3.6 MJ

(b) 5.4¢

11. (a) 300 kW·h

(b) $15.00

12. $27.60

14. (a) 1800 W

(b) 14 400 W

(c) 720 W

18

Electromagnetism

Knowing the information in this chapter will be especially useful if you plan a career in:

- electricity
- telecommunications
- laboratory technician
- power distribution (linemen, etc.)
- electronics
- computers
- astronomy

18.1 The Force of Magnetism

Every student probably knows something about magnets and compasses. The force that attracts nails to a magnet or causes a compass to point north is called the **magnetic force** or **force of magnetism**.

The early Greeks knew about the force of magnetism. They discovered a mysterious metal they named lodestone near the city of Magnesia. This mineral attracted iron and some other substances. The force of attraction was called magnetism (after Magnesia).

You have likely experimented with magnetism in previous science classes. The first three sections of this chapter will review activities involving magnetism.

Figure 18–1 shows three suspended objects: a chunk of lodestone, a magnet and the needle of a compass. All three have come to rest in a north-south direction. They are influenced by the earth's magnetic force. Even the Greeks realized that one end of a suspended piece of lodestone always faces toward the North Star in the sky. That star was called the leading or "lode" star, thus the name "lodestone".

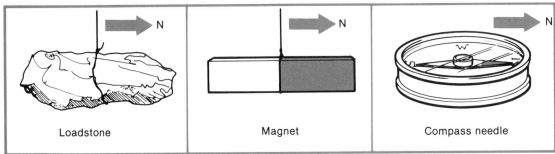

| Loadstone | Magnet | Compass needle |

Figure 18–1 Objects influenced by the earth's force of magnetism

In Figure 18–1 the end of each object that faces north is called the **north-seeking pole** and has the symbol N. The opposite end is called the **south-seeking pole** and has the symbol S. Every magnet has an N-pole and an S-pole.

If the N-pole of one magnet is brought close to the N-pole of another magnet, they repel each other. Also, the S-pole of one magnet repels the S-pole of another magnet. However, if N and S are brought close together, they attract each other. (See Figure 18–2.) These facts are summarized by the **laws of magnetism**:

(a) **like poles repel**
(b) **unlike poles attract**

PRACTICE

1. Compare the laws of magnetism to the laws of static electricity (Experiment 32, Section 15.3).
2. Is the south magnetic pole of the earth south-seeking or north-seeking? Explain your answer.

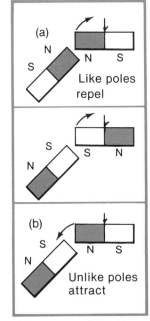

Figure 18-2 The Laws of magnetism

18.2 Explaining Magnetism

If a magnet is brought close to paper, there is no magnetic action. If a magnet is brought close to an iron nail, the nail is attracted to the magnet. Iron is a substance that can be influenced by the force of magnetism. Such a substance is called **magnetic**. Iron, nickel and cobalt are important magnetic substances. Most other substances, such as paper, are **non-magnetic**.

Scientists believe that magnetic substances have sets of atoms that are lined up like little magnets. This is illustrated in Figure 18–3(a). If the sets of atoms are arranged so that many of them face the same direction, a weak magnet results, as shown in diagram (b). If all the sets of atoms face the same direction, a strong magnet results, as in diagram (c). If this arrangement lasts for a long time, the magnet is called a **permanent magnet**.

One way to make a magnetic substance into a magnet is to stroke it with another magnet. For example, an iron needle becomes a magnet after being stroked several times with a strong magnet. See Figure 18–4(a). If the needle is suspended, as in diagram (b), it will line up in the north-south direction.

Set of atoms
(a) A magnetic substance

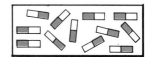
Several sets face the same direction
(b) A weak magnet

All sets face the same direction
(c) A strong magnet

Figure 18–3 A magnetic substance can become a magnet.

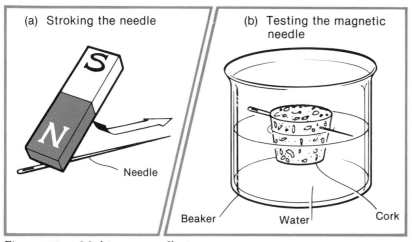

(a) Stroking the needle

S

N

Needle

(b) Testing the magnetic needle

Beaker   Water   Cork

Figure 18–4 Making a needle into a magnet

Another way to make a magnetic substance into a magnet is to hold it close to a permanent magnet. An iron nail becomes a magnet when held close to a magnet, as shown in Figure 18–5. This procedure is called **induction**, because touching is not necessary. In this case the nail is a **temporary magnet**. When the nail is moved away from the magnet, it loses its magnetism.

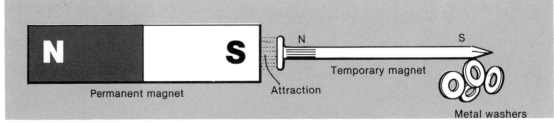

Figure 18-5 Making a temporary magnet by induction

If a permanent magnet is broken in two, each part of the original magnet becomes a new magnet with its own N- and S-poles, as shown in Figure 18-6(a). The breaking process can continue with similar results, as shown in diagram (b). This helps prove the scientists' view that a magnet has many sets of atoms that face the same direction.

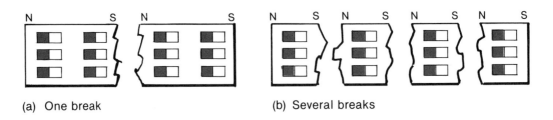

(a) One break   (b) Several breaks

Figure 18-6 Breaking a permanent magnet

PRACTICE
3. Describe how you could use a compass to determine which end of the needle in Figure 18-4 is the N-pole.
4. In the diagram, which end of the needle becomes the north-seeking pole?

Needle

18.3 Magnetic Fields

A baseball field is a space in which a baseball game is played. A **magnetic field** is a space in which a magnetic force exists.

In science laboratories there are two common ways to detect a magnetic field. One is with a compass and the other is with small particles of a magnetic substance, such as iron filings. Figure

(a) Using compasses

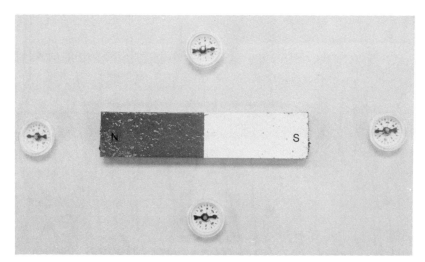

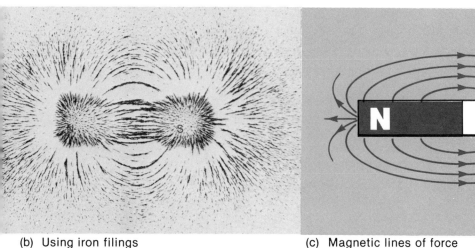

(b) Using iron filings   (c) Magnetic lines of force

Figure 18-7 The magnetic field around a bar magnet

18-7(a) shows several compasses in the magnetic field around a permanent bar magnet. Diagram (b) shows iron filings sprinkled on paper over the same magnet. Diagram (c) illustrates how we draw lines representing the forces in the magnetic field. These lines are called **magnetic lines of force** and are simply a way of drawing the field.

By referring to Figure 18-7, we can discover three important **characteristics of lines of force** in a magnetic field.

(1) The lines of force appear to repel each other. To illustrate this, we draw the lines of force so they do not cross each other.

(2) The lines of force are concentrated at the poles.

(3) The lines of force leave N and proceed to S outside the magnet, as shown when a compass is used to test the direction.

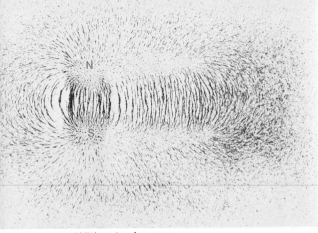

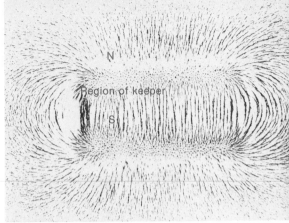

(a) Without a keeper   (b) With a keeper

Figure 18–8 A horseshoe magnet

It is interesting to view the shape of the magnetic field near other types of magnets. For example, Figure 18–8(a) uses iron filings to show the magnetic field near a horseshoe magnet. Diagram (b) shows what happens to that field when an iron bar, called a **keeper**, joins the poles. The keeper sustains the magnetism when the magnet is not in use.

Iron filings are also useful to illustrate the magnetic fields near pairs of magnets. In Figure 18–9(a) two like poles are repelling each other. In diagram (b) two unlike poles are attracting each other.

Figure 18–9 Magnetic fields near pairs of magnets

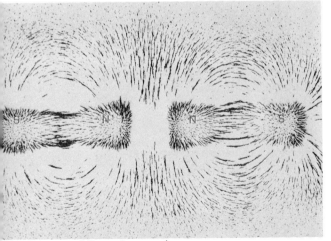

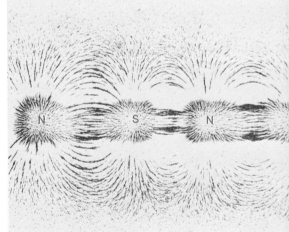

(a) Like poles repel   (b) Unlike poles attract

An orienteering compass works because the earth has a magnetic field around it. In that field a compass needle points toward the magnetic north pole (actually a south-seeking or

S-pole) of the earth. The magnetic north pole of the earth is not located at the geographic North Pole. Thus, for most locations the compass needle does not point toward true north. See Figure 18–10.

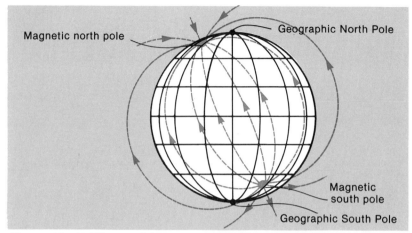

Figure 18–10 The earth's magnetic field

PRACTICE

5. In the diagrams below each circle represents a compass. Draw the diagrams and show the direction of the needle in each compass.
6. Draw a diagram, similar to Figure 18–7(c), showing the lines of force near a horseshoe magnet. Include the directions of the lines of force.
7. Are the north magnetic pole of the earth and the N-pole of a magnet the same? Explain your answer.

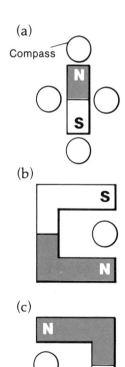

18.4 Electricity and Magnetism

One of the most important discoveries ever made in science is that electricity and magnetism cannot be separated. The study of how electricity and magnetism interact is called **electromagnetism**, which is the title of this chapter.

Current electricity exerts a magnetic force. The nature of that force and its many uses will be discussed in the remainder of this chapter. In the next chapter you will learn how magnetism can be used to create electricity.

To appreciate the amazing ways that electromagnetism acts, you should perform all the experiments and try to understand all of the diagrams in the rest of this chapter.

18.5 Experiment 37: Magnetic Forces around a Straight Conductor

INTRODUCTION

In 1819 a Danish scientist named Hans Christian Oersted (1777-1851) discovered by accident what you will also find out in this experiment. He was trying to prove that electricity does not create magnetism. Instead he proved the opposite.

To help explain what happens in this experiment and others, diagrams with special symbols will be used. The following symbols make diagrams much easier to draw:

(a) $\otimes$ This represents a wire with electrons going away from you, into the page. (Imagine the tail of an arrow travelling away from you.)

(b) $\odot$ This represents a wire with electrons coming toward you, out of the page. (Imagine the tip of an arrow travelling toward you.)

A magnetic compass will be used in the experiment to find the direction of the magnetic lines of force around the conductor. Before you use a compass, always check to be sure that the N-pole faces north, not south.

Iron filings may also be used in this experiment, especially for demonstration purposes.

PURPOSE: To study magnetic forces around straight conductors.

APPARATUS: dry cell or DC power supply; magnetic compass; switch; connecting wires

Caution:

(a) To prevent overheating, do not leave the switch closed very long at any one time.

(b) If a DC power supply is used, a 5 Ω or 10 Ω resistor should be connected in series in the circuit.

Figure 18–11

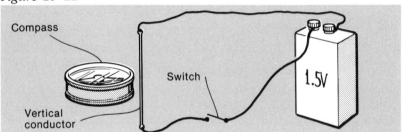

Compass

Switch

Vertical conductor

1.5V

PROCEDURE
1. Set up the apparatus as shown in Figure 18–11. The straight conductor should be vertical and the compass should be held close to it. Determine whether the electrons will be flowing down or up the wire when the switch is closed.
2. Close the switch for a moment and determine the **final direction** the compass needle faces. (Do not worry about which way the needle spins.) Move the compass around the wire and find the direction of the lines of force.
3. Reverse the terminals so the electrons flow in the opposite direction. Repeat #2.
4. Relate the observations in #2 and #3 to the following **left-hand rule (LHR) for straight conductors:** Point the thumb of your left hand in the direction of the flow of electrons. Then the fingers wrapped around the conductor point in the direction of the magnetic lines of force. (Check to see if the compass needle points in the same direction as your fingers.)
5. Draw diagrams of the results in #2 and #3 using the symbols described in the Introduction.
6. Predict what happens to the strength of the magnetic field around a straight conductor when you:
   (a) increase the current (by adding another cell in series)
   (b) increase the distance between the compass and the conductor
   If possible, check your predictions.

QUESTIONS
1. Each empty circle represents a compass. Show the direction of the needle of each compass.
   (a)   (b)

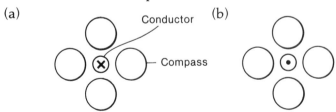

2. Each empty circle represents a conductor with a magnetic field around it. Place either a dot or an × in each empty circle.
   (a)   (b)

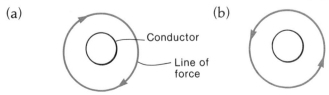

18.6 Experiment 38: Magnetic Forces in a Coiled Conductor

INTRODUCTION

The magnetic force around a straight conductor is rather weak. To make the force stronger, the conductor can be wound around and around to create a coil.

A **galvanoscope**, suggested for this experiment, is a device with two or three sets of coils. Each set has a different number of windings. Refer to Figure 18–12(a).

If a commercial galvanoscope is not available, you can make your own. Simply wind about 50 turns of fine, insulated copper wire around a beaker. See Figure 18–12(b).

Symbols can make diagrams easier to draw. Figure 18–13 shows two common ways of illustrating a coiled conductor with electrons travelling through it.

PURPOSE: To study magnetic forces in a coiled conductor.

APPARATUS: dry cell or DC power supply; magnetic compass; galvanoscope; switch; connecting wires

PROCEDURE

1. Set up the apparatus as shown in Figure 18–14. (Remember to add a small resistor in series if a DC power supply is used.) Be sure that the compass needle points in the same direction as the wires of the coil. Determine which direction the electrons will be flowing in the coil when the switch is closed.
2. Close the switch for an instant and determine the **final direction** of the compass needle.
3. Reverse the terminals so the electrons flow in the opposite direction. Repeat #2.
4. Relate the observations in #2 and #3 to the following **LHR for coiled conductors:** Wrap the fingers of your left hand around the coil in the direction of the flow of electrons. Then the thumb points in the direction of the magnetic force **inside** the coil. The compass needle should also point in this direction. Thus, your thumb shows where the N-pole of the compass should be. Check to see if this is the case.
5. Predict what happens to the strength of the magnetic field inside the coiled conductor when you:

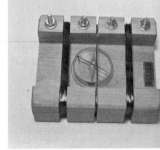

(a) A commercial galvanoscope

(b) A student-made galvanoscope

Figure 18–12 Galvanoscopes

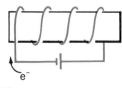

(a)

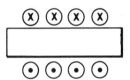

(b)

Figure 18–13 Symbols for coiled conductors

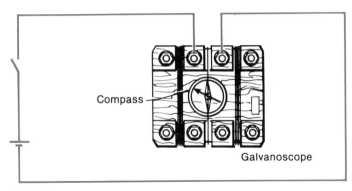

Figure 18–14

(a) increase the current (by adding another cell in series)
(b) increase the number of windings
If possible, check your predictions.

QUESTION

1. Each empty circle represents a compass near one end of a coiled conductor.
   (a) Label the ends of the coils as N or S.
   (b) Show the direction of the needle of each compass.

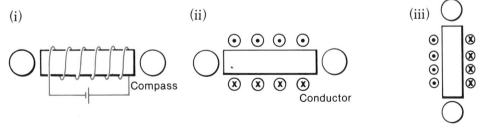

18.7 Experiment 39: Electromagnets

INTRODUCTION

Both permanent magnets and electron-carrying conductors exert a magnetic force. A permanent magnet attracts magnetic objects, such as iron washers and nails. An **electromagnet** is an electron-carrying conductor that attracts magnetic objects.

The straight conductor and coiled conductor used in the previous two experiments had weak magnetic forces. To make a strong electromagnet, an iron core is placed inside a coiled conductor. This concentrates the magnetic lines of force.

If a manufactured coil and core set is not available, you can make your own by wrapping a fine insulated wire many times around an iron nail.

PURPOSE: To make an electromagnet and to determine how its strength is affected by the current and number of windings.

APPARATUS: 2 dry cells or a DC power supply; switch; coiled conductors with iron core; metal objects (washers, nails or thumb-tacks); compass; connecting wires

Figure 18-15

PROCEDURE

1. Set up the apparatus as shown in Figure 18-15. Find out how many metal objects the electromagnet can pick up when the switch is closed. Do not leave the switch closed for more than about 10 s. What happens when you open the switch?
2. Use the LHR for coiled conductors to predict which end of the electromagnet is N. With the electromagnet horizontal, check your prediction with a compass.
3. Predict the effect on the strength of the electromagnet when the current is increased. Try it by adding a second cell in series.
4. Predict the effect on the strength of the electromagnet if the number of windings is changed. If possible, check your prediction.

QUESTION

1. The diagram shows one way of making a U-shaped electromagnet. Use the LHR for coiled conductors to determine which pole is N and which is S.

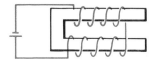

18.8 Applications of Electromagnets

Several devices make use of the force created by an electromagnet. The electromagnetic lift, the doorbell and the tape recorder will be described here. Other applications include the electromagnetic relay, the telephone earpiece and the recording timer used in experiments in Chapters 5 and 6.

The electromagnetic lift

Huge electromagnets are used to lift or move scrap iron or iron sheets as shown in Figure 18–16(a). Some electromagnets can lift more than 20 000 kg of metal.

One way of making an electromagnetic lift is shown in Figure 18–16(b). When the switch is closed, the U-shaped device becomes a powerful magnet. The poles of the magnet are labelled N and S according to the LHR for coiled conductors. When the switch is opened, the magnetism no longer exists and the load drops.

The electric doorbell

Figure 18–17 illustrates the design of a doorbell that uses an electromagnet. When the switch is closed the electrons flow, making the U-shaped part into an electromagnet. The electromagnet attracts the iron bar away from point P. This opens the circuit briefly so the electromagnet does not work. Then the spring pulls the iron bar back to P and the process starts all over again. As long as the switch is closed, the bar moves back and forth. A hammer attached to the bar strikes a gong, creating the sound of the doorbell.

The tape recorder

The tape used in a tape recorder has millions of magnetic particles made of iron oxide that can be lined up using an electromagnet. The process works as follows.

(a) Moving scrap iron

(b) Basic design of the lift

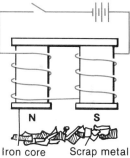

Iron core   Scrap metal

Figure 18–16 The electromagnetic lift

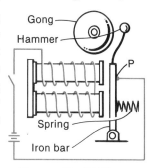

Figure 18–17 The design of an electric doorbell

342

A microphone receives sound signals of various frequencies and amplitudes. Those signals are changed into electrical signals, which are sent to an electromagnet called the recording head. The magnetic field of the electromagnet changes according to the electric signals. As a tape is pulled past the electromagnet, the magnetic field causes the iron oxide particles to form certain patterns. Those patterns will remain on the tape.

To produce the recorded sound, the tape, now having its own magnetic field, must be pulled past the electromagnet. Electric signals are then changed back into sound signals.

PRACTICE

8. Use the LHR for coiled conductors to determine which pole is N and which is S in the electromagnet in Figure 18–17.
9. It is better to use plastic scissors than steel scissors when repairing a broken magnetic tape from a tape recorder. Why?

18.9 The Interaction of Magnetic Forces

After Hans Christian Oersted discovered in 1819 that electricity can cause magnetism, many other scientists experimented with electricity and magnetism. For example, an English scientist named Michael Faraday (1791-1867) discovered that the magnetic force from a permanent magnet can interact with the magnetic force from an electron-carrying conductor. Faraday's discovery is called the **motor principle** because it is the basis of the operation of an electric motor.

Consider Figure 18–18, which explains how the motor principle works. Diagram (a) shows the magnetic lines of force around a permanent U-shaped magnet. Diagram (b) shows the magnetic lines of force around a conductor with electrons flowing away from you. (Remember the LHR for straight conductors.) Diagram

Michael Faraday (1791-1867)

(a) (b) (c) Cancellation of forces (d) Conductor is forced to the right

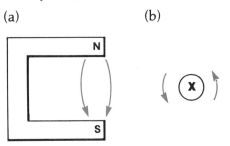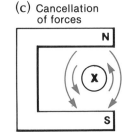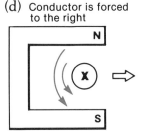

Figure 18–18 The motor principle

(c) shows the two sets of forces together. To the right of the conductor the forces are opposite in direction, so they tend to cancel. The lines of force are concentrated to the left of the conductor and reinforce each other, pushing the conductor to the right, as seen in diagram (d).

An interesting example of the interaction of electricity and magnetism can be seen if a showcase bulb is available. Hold a permanent magnet close to the bulb without the light on. Notice the effect. Now turn the light on and again hold the magnet close to it. (See Figure 18–19.)

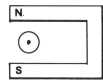

PRACTICE

10. Redraw this diagram, then:
    (a) Draw the lines of force for the magnet and the conductor.
    (b) Show where the cancellation of forces occurs.
    (c) Determine the direction of motion of the conductor.

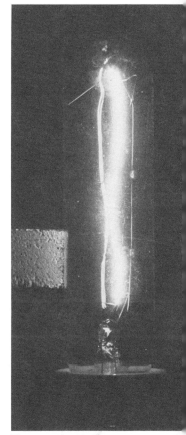

Figure 18–19
Demonstrating the interaction of electricity and magnetism

18.10 Experiment 40: The Motor Principle

INTRODUCTION

This experiment is fairly easy to perform. However, the explanation is difficult. Be sure you understand Section 18.9 before you begin the procedure.

PURPOSE: To observe and learn to predict the effect of the interaction of two sets of magnetic forces.

APPARATUS: U-shaped magnet; dry cell; switch; connecting wires

PROCEDURE

1. Set up the apparatus as shown in Figure 18–20 with the S-pole resting on the bench top. Leave the switch open while you determine the direction of motion of the electrons. Use the theory from Section 18.9 to predict which way the conductor will be pushed.
2. Check your prediction by closing the switch for a few seconds.

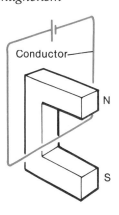

Figure 18–20

If your prediction was correct, proceed to # 3. If your prediction was wrong, find out why and try again.

3. Reverse the magnet so the N-pole is resting on the bench. Repeat #1 and #2.
4. Reverse the wires on the cell so the electrons flow in the opposite direction. Repeat #1, #2 and #3.

QUESTIONS

1. What do you think happens to the force pushing the conductor if the:
   (a) magnet is made stronger?
   (b) current through the conductor is increased?
2. Determine the direction of motion of the conductor in each case. Use a diagram to explain each answer.

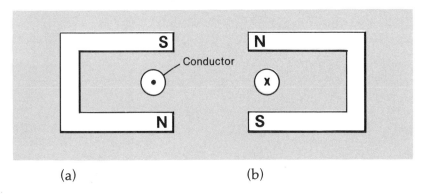

(a)   (b)

18.11 Designing Electric Motors

The motor principle is used in the design of several devices, including ammeters, voltmeters, galvanometers and electric motors. Only the electric motor will be discussed in detail here.

In Experiment 40, the conductor simply swings back and forth. If the conductor can somehow be made to rotate around and around, the result is called an electric motor.

An **electric motor** is a device that changes electrical energy into mechanical energy. It consists of two main parts: a magnet and an electron-carrying conductor. The easiest way to see how a motor works is to follow a set of diagrams step by step.

Figure 18–21 shows how we can apply the motor principle to a conducting loop. Diagram (a) has a conductor with electrons coming toward you. Use the motor principle to prove to yourself that the force on the conductor is downwards. Diagram (b) shows a conductor with the force upwards. Diagrams (a) and (b) are com-

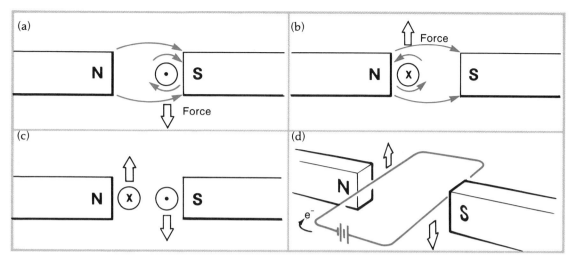

Figure 18–21 Applying the motor principle to a loop of wire

bined to obtain (c). Then in diagram (d) we have a three-dimensional view of the loop of wire. Prove to yourself that (d) is the same as (c) in Figure 18–21.

In Figure 18–21(d) the loop of wire is experiencing a clockwise force. After making half a turn, the loop would be forced counter-clockwise. This would not allow a constant rotation of the motor. This problem is solved by an important invention called a com-mutator.

A **commutator** is a metal ring split into two parts. It allows a loop in an electric motor to continue rotating in one direction. Its operation is shown in Figure 18–22. Diagram (a) shows a com-mutator made of two parts, R_1 and R_2. The parts touching the split ring are called **brushes**, B_1 and B_2. They are connected to a DC bat-tery.

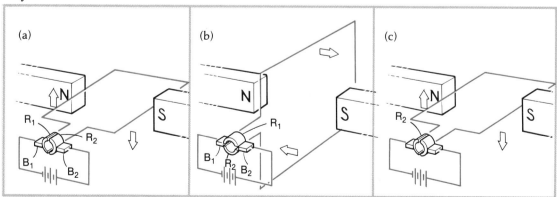

Figure 18–22 Creating a force in a clockwise direction

Figure 18–22(a) shows the same situation as Figure 18–21(d). The force on the loop is clockwise. In Figure 18–22(b) the loop has swung to its vertical position. Here the electrons stop flowing for an instant because the brushes do not touch the split ring. However, the loop keeps moving because it is already in motion. In diagram (c), R_2 is contacting B_1. This situation is similar to diagram (a) in which R_1 is touching B_1. Thus, the force on the loop is again clockwise. This process continues as the electrons reverse directions in the loop every half turn of the loop. Try to understand Figure 18–22 before you proceed to the next section.

PRACTICE

11. The conductors represent a loop in a magnetic field. Determine whether the force on the loop is clockwise or counter-clockwise.

12. For the instant shown in the diagram, is the force on the loop clockwise or counter-clockwise?

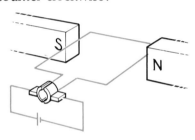

18.12 Constructing and Using Electric Motors

Although Section 18.11 describes how an electric motor operates, it does not show the actual construction. To make the motor strong enough to be useful, some additions must be made. First, instead of using a single loop, a set of several loops can be added. In a motor this set of loops is called the **armature**. An iron core is also added inside the armature. This helps increase the force, as you learned in the electromagnet experiment. The result is a motor strong enough to operate some device.

To help students understand the parts of a motor, special motors are available. One such motor, called a St. Louis motor, is shown in Figure 18–23(a). You can see the armature, commutator and metal brushes. The permanent magnets are called **field magnets** because they create the magnetic field.

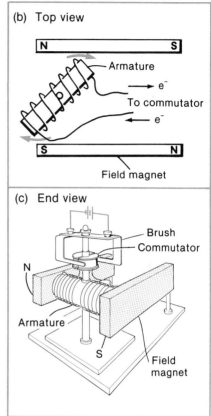

(b) Top view

Armature

e^-

To commutator

e^-

Field magnet

(c) End view

Brush

Commutator

N

Armature

S

Field magnet

(a) General construction of the motor

Figure 18–23 The St. Louis motor

In Figure 18–23(b) and (c) follow the flow of electrons from the source and through the coil. Use the LHR for coiled conductors to see why one end of the armature is N and the other S. The force on the armature is clockwise because N repels N and S repels S. If a St. Louis or similar motor is available, use it to learn first-hand how an electric motor operates.

The field magnets in the motors described have been permanent magnets. It is also possible to use electromagnets as field magnets. This will be the case when you build your own motor in the next experiment.

Up to this point, the description of how a motor works has been limited to DC motors. AC motors are actually more common. They work in much the same manner as DC motors. The main difference between an AC motor and a DC motor is in the commutator arrangement.

Electric motors range in size from tiny to huge. Small motors are used in toys and ripple tanks. Larger motors are used for electric tools and machines. Very large motors are used in subway trains and electric locomotives.

PRACTICE

13. State the function of each of the following parts of a DC motor:
   (a) commutator   (c) armature
   (b) brushes   (d) field magnets

18.13 Experiment 41: Making a DC Electric Motor

INTRODUCTION

Be sure you understand Sections 18.11 and 18.12 before you begin this experiment.

There are several ways to build an electric motor, depending on the materials available. Instructions could fill many pages. To prevent wordy instructions, diagrams and suggestions will be used. You will have to decide on the details.

The motor you build will have electromagnets for the field. Only one DC circuit is needed to operate both the field magnets and the armature. A well-built motor will be self-starting using only 1.5 V as the source. A poorly built motor will need a push to get started and may need up to 6 V. A useless motor will not spin at all.

Good luck!

PURPOSE: To build a working DC electric motor and explain its operation.

APPARATUS

(a) **Materials required to build one motor:**
   1 pine or plywood board, about 12 cm × 6 cm × 2 cm;
   5 common nails, about 8 to 10 cm long;
   1 piece of glass tubing, about 5 to 6 cm long and large enough in diameter to fit over a nail;
   12 m of #18 insulated copper magnet wire;
   2 pieces of thin copper sheeting, about 2.5 cm × 1.5 cm;
   2 thumbtacks;
   1 DC cell, 1.5 V

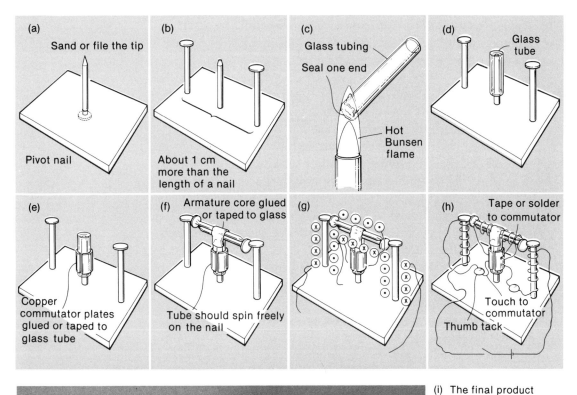

(a) Sand or file the tip

Pivot nail

(b) About 1 cm more than the length of a nail

(c) Glass tubing
Seal one end
Hot Bunsen flame

(d) Glass tube

(e) Copper commutator plates glued or taped to glass tube

(f) Armature core glued or taped to glass
Tube should spin freely on the nail

(g)

(h) Tape or solder to commutator
Touch to commutator
Thumb tack

(i) The final product

Figure 18–24 Steps in making a DC electric motor

(b) **Apparatus needed to help construct the motor:**
hammer; knife; solder; soldering iron; sandpaper; file; masking tape; glue; Bunsen burner; heat-insulating board

PROCEDURE
1. Read through all the instructions and look at all the diagrams
   before you start building your motor.
2. Put together the parts of the motor using the steps shown in
   Figure 18–24. The following suggestions may help:
   (a) Sand or file down the tip of the nail that acts as a pivot.
   (b) When cooling the glass tubing, press the sealed end against a
       heat-insulating board to flatten the end.
   (c) From the 12 m of wire you need three pieces. You must
       decide on the length of each piece.
   (d) When you make the electromagnets or the armature, wrap
       the wire in only one direction. It might be interesting to
       count the number of windings in each case.
   (e) Wherever an electrical contact is needed, the insulation
       must be sanded off the wire.
   (f) Be sure the field magnets are wound so that one is N and the
       other S. (You can check the poles with a compass.)
3. Try your motor. Be able to explain its operation.
4. If you wish, try to design changes to improve your motor.

QUESTIONS
1. In Figure 18–24(h) which field magnet is N and which S?
2. If you were to start the project over again, what changes would
   you make?

18.14 Review Assignment

1. State the laws of magnetism.   (18.1)
2. Describe the difference in the structure of magnetic and non-
   magnetic substances.   (18.2)
3. What is the difference between a magnet and a magnetic
   substance?   (18.2)
4. An iron bar is held close to a magnet, as shown in the diagram.
   (a) Which end of the bar becomes N?
   (b) What is the name of the process that makes the bar a tem-
       porary magnet?   (18.2)
5. List three characteristics of magnetic lines of force.   (18.3)

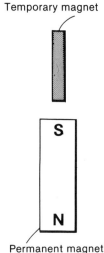

Temporary magnet

S

N

Permanent magnet

6. Draw the shape of the magnetic field in each case. Include the direction of the lines of force.   (18.3)

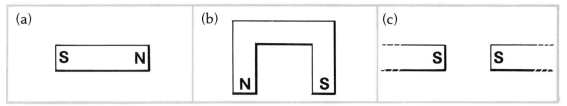

(a)   (b)   (c)

7. In each case determine which end of the permanent magnet is N and which is S.   (18.3)

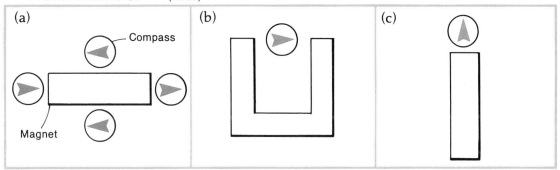

(a)   (b)   (c)

8. Define electromagnetism.   (18.4)
9. State the left-hand rule for:
   (a) straight conductors   (b) coiled conductors   (18.5, 18.6)
10. State whether the magnetic lines of force are clockwise or counter-clockwise when the electrons in a conductor are travelling:
   (a) toward you   (b) away from you   (18.5)
11. Each empty circle represents a compass near one end of a coiled conductor.
   (a) Label the ends of the coils as N or S.
   (b) Show the direction of the needle of each compass.
       (18.6)

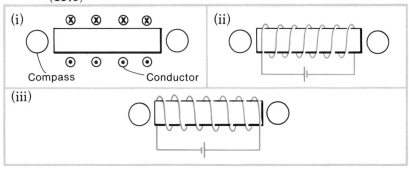

(i)   (ii)

(iii)

12. The small circles represent conductors around a coil. Determine which circles should have a "·" and which an "×". (18.6)

(a)

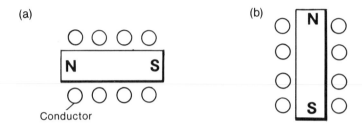

Conductor

(b)

13. If you were making an electromagnet, what would you do to improve its strength? (18.7)

14. The diagrams show electromagnets. Determine which poles are N and which are S. (18.7, 18.8)

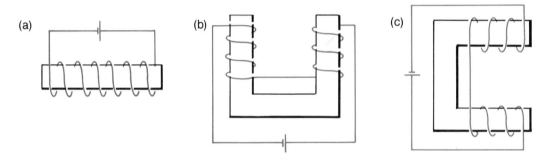

(a)

(b)

(c)

15. In the diagram a current-carrying conductor is in the magnetic field of a U-shaped magnet. With the aid of a diagram, explain why the conductor experiences an outward force. (18.9)

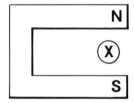

16. In each case determine the direction of the force on the conductor. (18.9, 18.10)

(a)

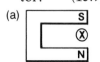

(b)

(c)

17. The diagram represents a single loop in an electric motor. Determine whether the force on the loop is clockwise or counter-clockwise.   (18.11)

18. Refer to the diagram of the St. Louis motor.
    (a) List the names of the parts of the motor labelled A, B, C and D.
    (b) Determine which end of the coil is N.
    (c) In which direction will the armature spin?   (18.12)

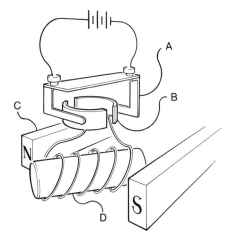

18.15 Answers to Selected Problems

PRACTICE QUESTIONS
  8. The top pole is S and the bottom pole is N.
 10. (c) outwards, away from the magnet
 11. counter-clockwise
 12. clockwise

REVIEW ASSIGNMENT
 10. (a) clockwise
     (b) counter-clockwise
 11. (i) left end is N
      (ii) right end is N
    , (iii) both ends are S

 16. (a) inwards
     (b) inwards
     (c) inwards
 17. clockwise
 18. (b) left end is N
     (c) counter-clockwise when viewed from above

19

Electromagnetic Induction

GOALS: After completing this chapter you should be able to:
1. State the conditions needed for magnetism to create electricity.
2. Define electromagnetic induction.
3. State the factors that affect the size of a current induced in a conductor by a changing magnetic field.
4. Apply Lenz's law to find the direction of an induced current in a coiled conductor.
5. Describe how an electric generator creates electrical energy.
6. Describe how electricity in one part of a transformer creates electricity in another part of the transformer.
7. Define step-up transformer and step-down transformer.
8. Given the number of windings in the primary and secondary coils of a transformer and the voltage of the primary coil, find the voltage of the secondary coil ($V_s = \dfrac{V_p N_s}{N_p}$).
9. Explain why transformers are needed to transmit electrical energy over long distances.
10. State the origin of 60 Hz electromagnetic waves.

Knowing the information in this chapter will be especially useful if you plan a career in:

- electricity
- telecommunications
- laboratory technician
- power distribution (linemen, etc.)
- electronics
- computers
- astronomy

19.1 Using Magnetism to Create Electricity

In Chapter 18 it was shown that current electricity causes magnetism. In this chapter you will study how magnetism can create electric current. Magnetism and electricity are closely related. Thus, as you study this chapter try to recall facts from the previous chapter.

In order for a magnetic field to produce electricity, some kind of change in the field must occur. This is shown by experiment using the setup in Figure 19–1(a). A conductor is held between the poles of a U-shaped magnet. A galvanometer is connected to the con-

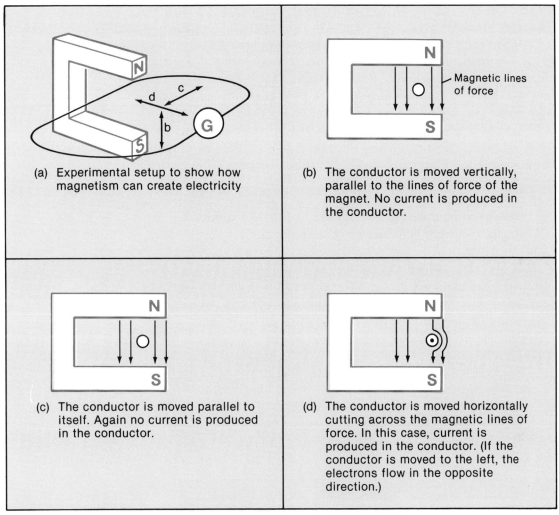

(a) Experimental setup to show how magnetism can create electricity

(b) The conductor is moved vertically, parallel to the lines of force of the magnet. No current is produced in the conductor.

(c) The conductor is moved parallel to itself. Again no current is produced in the conductor.

(d) The conductor is moved horizontally cutting across the magnetic lines of force. In this case, current is produced in the conductor. (If the conductor is moved to the left, the electrons flow in the opposite direction.)

Figure 19-1 One way to create electricity using a magnetic field

ductor to indicate current. The three arrows, b, c and d, show three ways the conductor can be moved in the magnetic field. Diagrams (b), (c) and (d) indicate what happens for each motion.

Notice in Figure 19-1(b) and (c) that the conductor does not cross any lines of force in the magnetic field. In other words, as far as the conductor is concerned, the magnetic field is not changing. No current flows in the conductor.

In diagram (d), however, the conductor cuts across the lines of force. Thus, as far as the conductor is concerned, the magnetic field is changing and current flows in the conductor. We say the current has been **induced**. Therefore, **electromagnetic induction** means the creation of an electric current caused by a changing magnetic field.

In the example given the conductor was moved to cause the magnetic field around it to change. A second way to create a changing magnetic field is to hold the conductor still and move the magnet in the appropriate direction. A third way is to change the size of the magnetic field around a stationary conductor.

It was Michael Faraday (Section 18.9) who discovered in 1831 that a magnetic field can cause electricity. In fact, the **law of electromagnetic induction** is named after him. Faraday's law states that:

> **whenever the magnetic field in the region of a conductor changes, current is induced in the conductor**

Faraday's law helped change the industrial world. Most of the electricity we use is created and transmitted using electromagnetic induction. How we create and transmit electricity is the topic of the remainder of this chapter.

PRACTICE
1. What condition is necessary in order that a magnetic field induces a current in a conductor?
2. In each case state which way(s) the conductor must move in order to have a current induced in it.

(a)   (b)   (c)

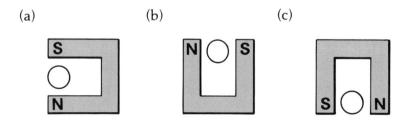

19.2 Experiment 42: Inducing Current in a Coiled Conductor

INTRODUCTION
The examples of electromagnetic induction given in Section 19.1 used a single conductor. In that case the current is very small. Therefore, in this experiment a coiled conductor is used to create a larger current.

PURPOSE

(a) To study how current can be induced in a coiled conductor.
(b) To study factors that affect the size of the induced current.

APPARATUS: galvanometer; set of 2 coils with different numbers of windings (say 50 and 100); 2 bar magnets (one stronger than the other); connecting wires

PROCEDURE

1. Connect the coil with the smaller number of windings to the galvanometer, as shown in Figure 19–2. Move one end of the magnet into the coil. Describe what happens.
2. Determine what happens to the current when the magnet is held still in the coil.
3. Move the magnet out of the coil. Describe what happens.
4. Hold the magnet still and move the coil in various directions. Describe the effects.
5. Determine how the factors listed below affect the size of the current induced in a conductor by a changing magnetic field:
   (a) number of turns of the coil (use two different coils)
   (b) speed of motion of the magnet (use fast, medium and very slow speeds)
   (c) strength of the magnetic field (use a strong magnet and a weak magnet)

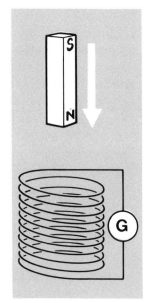

Figure 19–2

QUESTION

1. Assume you are asked to make a "generator" of electricity. Based on this experiment, what conditions would you use to obtain a large current?

19.3 Experiment 43: The Direction of Induced Current: Lenz's Law

INTRODUCTION

In Experiment 42 you likely noticed that the electrons flowed first in one direction and then in the other. It was Heinrich Lenz (1804-1865) from Russia who first discovered how to determine the direction of the flow of electrons in an induced current.

What Lenz proved is that whenever we exert a force, nature opposes us. Let us see how this applies to a magnet and a coil.

Figure 19–3(a) shows a magnet being pushed into a coil. As the
N-pole approaches the coil, electrons start moving in the coil.
When this happens, the coil becomes magnetized with an N-pole
and an S-pole. Because nature opposes the force of the magnet, the
top of the coil must become N, as shown in diagram (b). Then the
left-hand rule for coils (stated backwards) is used to find the direction of the electron flow. The rule states that if the thumb of the
left hand points to the N-pole of the coil, the fingers wrapped
around the coil indicate the direction of the electrons. The
galvanometer proves that our reasoning is correct.

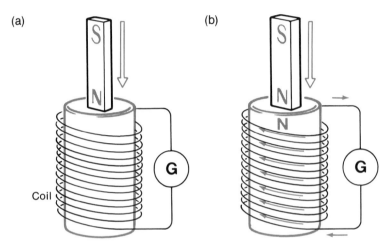

Figure 19–3 Pushing an N-pole into a coil

In Figure 19–4, the N-pole of the magnet is being pulled out of
the coil. In this case the top of the coil becomes an S-pole to try to
prevent the magnet from leaving. Thus, the electrons flow in a
direction opposite to that in Figure 19–3(b).

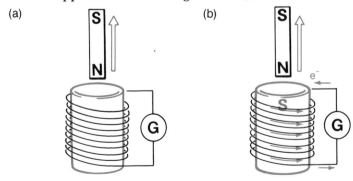

Figure 19–4 Pulling an N-pole out of a coil

Heinrich Lenz summarized these ideas in a statement called **Lenz's law**. It states that:

> for a current induced in a conductor by a changing magnetic field, the direction of the current is in such a direction that its own magnetic field opposes the change that produced it

In this experiment a galvanometer is used to check the direction of the current. If the galvanometer needle swings to the right, the electrons must have entered at the negative terminal.

PURPOSE: To learn to use Lenz's law to predict the direction of the induced current.

APPARATUS: galvanometer; coiled conductor; magnet; connecting wires

PROCEDURE
1. Connect the coil to the galvanometer. Examine the coil to see which way the wires are wound.
2. For each situation listed below, predict which end of the coil will become N. Then use the LHR for coils to predict the direction of the current. Check your prediction.
   (a) N of the magnet into the coil
   (b) N of the magnet out of the coil
   (c) S of the magnet into the coil
   (d) S of the magnet out of the coil

QUESTION
1. In each case state which end of the coil is N and find the direction of the current:

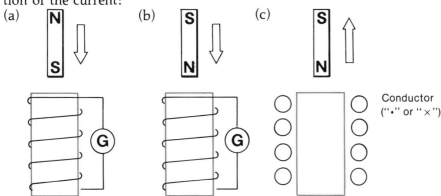

19.4 Electric Generators

In performing Experiments 42 and 43, you were acting as a generator of electricity. You exerted mechanical energy (energy of motion) to create electricity. A device that changes some form of energy into electrical energy is called an **electric generator.**

Manufactured generators are almost identical to motors. To understand how a generator works, all you have to do is relate Lenz's law to motors.

Consider Figure 19–5. It shows a St. Louis generator, which resembles a St. Louis motor. The motor was connected to a DC source, but the generator is connected to a galvanometer. A mechanical push on the armature causes electrons to flow.

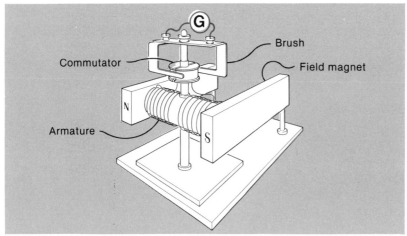

Figure 19–5 A St. Louis generator

To discover which way the electrons flow in the DC generator, we use the same procedure used in the previous experiment. In Figure 19–6 the left end of the armature coil is being pushed toward an N-pole. Thus the left end of the armature becomes N. (Remember from Lenz's law that the forces must oppose each other.)

Now applying the LHR for coils, we can find the direction of the electron flow. This is shown in Figure 19–6(b).

If a St. Louis generator is available, connect it to a galvanometer and check your understanding of these ideas.

Commercial generators are more complex than St. Louis generators. Figure 19–7 shows part of a demonstration generator that illustrates the basic design of a commercial generator. Several sets

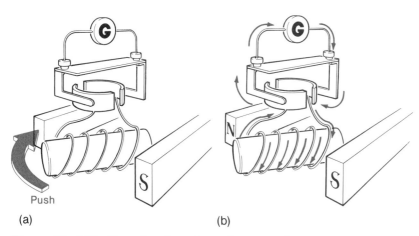

Push

(a)   (b)

Figure 19–6 Finding the direction of current in a generator

of armature windings help create a current that has a nearly constant value.

Most electric generators create AC electricity rather than DC. As is the case with motors, the main difference between an AC generator and a DC generator is in the armature arrangement.

Huge AC generators create the electricity that we use in our homes, buildings and industries. These generators are forced to spin either by falling water or the pressure of steam. Figure 19–8 shows how the kinetic energy of falling water is changed into electrical energy. Figure 19–9 shows how heat energy (or thermal energy) from burning coal is changed into electrical energy. The heat energy to boil water in this type of generator can also be obtained from natural gas, oil or nuclear reactions. (Nuclear generating stations are discussed in Section 25.3.)

Figure 19–7 Design of armature windings for a commercial generator

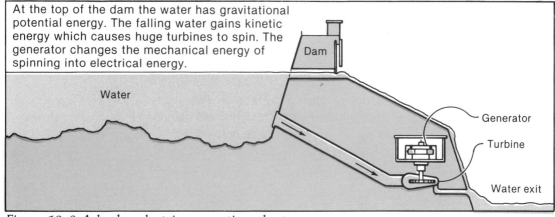

At the top of the dam the water has gravitational potential energy. The falling water gains kinetic energy which causes huge turbines to spin. The generator changes the mechanical energy of spinning into electrical energy.

Dam

Water

Generator

Turbine

Water exit

Figure 19–8 A hydro-electric generating plant

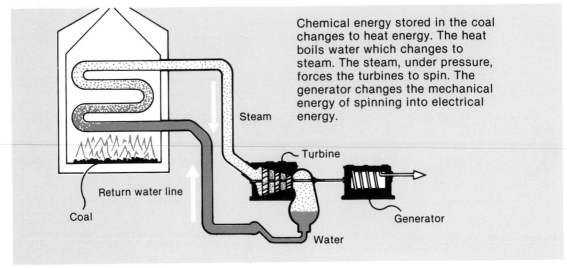

Chemical energy stored in the coal changes to heat energy. The heat boils water which changes to steam. The steam, under pressure, forces the turbines to spin. The generator changes the mechanical energy of spinning into electrical energy.

Figure 19–9 A thermal-electric generating plant

PRACTICE

3. The diagram shows a DC St. Louis generator. A push is exerted on the armature as indicated.
   (a) Determine which end of the armature becomes N.
   (b) Find the direction of the flow of electrons in the circuit.
4. Natural gas can be used in a generating station to create electricity. Describe how the chemical energy stored in natural gas is changed into electrical energy.

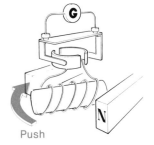

19.5 Using Electricity to Create Electricity

From Section 19.1 we know that a changing magnetic field near a conductor will induce a current in the conductor. This fact has been illustrated several times so far in this chapter. We will illustrate it once more using a device with a different shape.

Refer to Figure 19–10, which shows a solid iron ring with a conductor coiled around part of the ring. When a magnet is suddenly touched to point A, the magnetic field in the ring changes. This induces a current in the coil for an instant. The current is indicated by the galvanometer.

When the magnet is held still at A, the magnetic field does not change. Thus, no current flows in the coil.

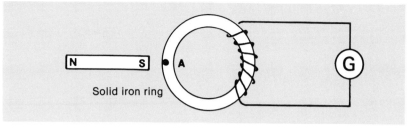

Figure 19-10 Another example of electromagnetic induction

Then when the magnet is suddenly pulled away from A, the magnetic field in the ring changes. Current is again induced in the coil for an instant.

Instead of using a magnet touched to point A, we could use current electricity. This was first discovered by Michael Faraday who designed a device now called Faraday's ring. This device is illustrated in Figure 19-11. There are two parts attached to the ring, a primary circuit and a secondary circuit. When the switch in the primary circuit is suddenly closed, the magnetic field in the ring changes. A current is induced in the secondary circuit for an instant.

As long as the switch remains closed, the magnetic field does not change, so no current flows in the secondary circuit.

Then when the switch is suddenly opened, the magnetic field in the ring changes. Again a current is induced in the secondary circuit for an instant.

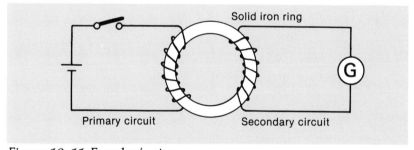

Figure 19-11 Faraday's ring

In Faraday's ring it is not easy to open and close the switch every time current is needed in the secondary circuit. A better way is to use AC electricity. That way, every time the electrons in the primary circuit change direction, the magnetic field in the ring changes. Then a current is induced in the secondary circuit. Thus, 60 Hz (AC) electricity in the primary circuit creates 60 Hz (AC) electricity in the secondary circuit.

In the next two sections you will learn why these ideas are important.

5. Choose the statement that best summarizes the operation of Faraday's ring:
   (a) a magnetic field induces current in a conductor
   (b) electricity creates magnetism
   (c) a changing magnetic field induces current in a conductor
   (d) magnetism creates electricity

19.6 Transformers

The idea of Faraday's ring has been applied to a very useful and important electrical device—the transformer. A **transformer** is a device that changes electricity at one voltage to electricity at a different voltage.

A transformer has two circuits, a **primary** and a **secondary**. The primary circuit has the source of electricity. The secondary circuit has the load that uses the electrical energy. Each circuit has a different number of windings. This allows the voltage in one circuit to be different from the voltage in the other circuit.

Consider Figure 19–12. It shows that a transformer is similar to a Faraday's ring. In this example the primary coil has three turns and the secondary coil has nine turns. If the voltage in the primary is 10 V then the voltage in the secondary is 30 V. The secondary coil has three times as many windings as the primary, so it steps up the voltage three times. This type of transformer is called a **step-up transformer.**

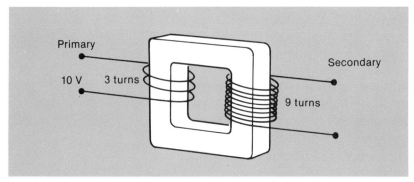

Figure 19–12 A step-up transformer

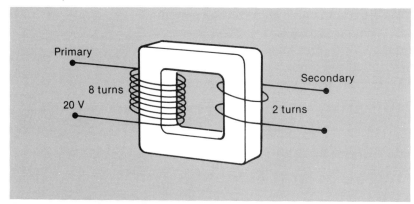

Figure 19–13 A step-down transformer

If the primary circuit has more turns than the secondary circuit, then the transformer is a **step-down transformer**. This is illustrated in Figure 19–13.

In Figure 19–13 there are four times as many windings in the primary as there are in the secondary. Thus the primary voltage is four times as much as the secondary voltage. Therefore the voltage in the secondary is 5 V.

In both these examples the ratio of the secondary voltage to the primary voltage equals the ratio of the secondary windings to the primary windings. This can be written in equation form:

$$\frac{\text{secondary voltage}}{\text{primary voltage}} = \frac{\text{secondary windings}}{\text{primary windings}} \quad \text{or} \quad \frac{V_s}{V_p} = \frac{N_s}{N_p}$$

where V means voltage, N means the number of windings, s means secondary and p means primary.

This equation can be used to find any one quantity if the other three quantities are shown. For example, to find the secondary voltage:

$$V_s = \frac{V_p \, N_s}{N_p}$$

Sample problem 1: Prove that the secondary voltage in the transformer in Figure 19–12 is 30 V.

Solution:

$$V_s = \frac{V_p \, N_s}{N_p}$$

$$= \frac{10 \text{ V} \times 9 \text{ turns}}{3 \text{ turns}}$$

$$= 30 \text{ V}$$

In sample problem 1, 10 V from the source in the primary circuit gives 30 V to the secondary circuit. It appears as if we have received something for nothing. But nature does not allow this to happen. Actually, the current in the secondary circuit is only ⅓ of that in the primary. In other words, in a transformer, if the voltage increases, the current decreases.

If the proper equipment is available, a demonstration of an AC transformer can be shown. Figure 19–14 shows a setup to prove that the ratio of $N_s : N_p$ is the same as the ratio of $V_s : V_p$.

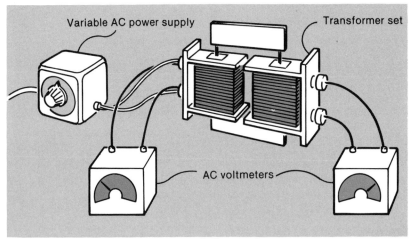

Figure 19–14 A demonstration AC transformer

PRACTICE
6. For each transformer shown:
    (a) state whether it is a step-up or step-down transformer
    (b) calculate the secondary voltage

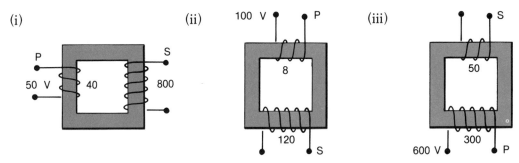

7. Alter the equation $\dfrac{V_s}{V_p} = \dfrac{N_s}{N_p}$ to express:

    (a) N_s by itself   (b) N_p by itself   (c) V_p by itself

⑧ Calculate the unknown quantities:

	N_p	N_s	V_p	V_s
(a)	500	25	120	?
(b)	250	1000	?	960
(c)	50	?	12	3600
(d)	?	80	240	4

19.7 Using Transformers to Distribute Electrical Energy

Electric generating stations are almost always built far from where most of the electricity is used. Thus the electrical energy must be transmitted over long distances.

Generating stations produce AC electricity with voltages of about 20 kV (20 000 V) as well as high currents. If this electricity is transmitted over long distances, much energy is lost due to heating of the wires.

It has been found that less energy is lost if the voltage is increased and the current decreased. The device that increases voltage and at the same time decreases current is the transformer. Therefore, to transmit electricity long distances, we must use transformers connected to high-voltage, low-current lines. This keeps heat loss to a minimum.

Figure 19–15 shows one possible transformer arrangement used to distribute electrical energy. First the voltage is stepped up from the generating station to the main transmission line. Then the voltage is stepped down in stages and finally used in the home. Notice that the electrons in a home circuit do not come from the generating station. They just move back and forth in the very last circuit. Refer also to Figures 19–16 and 19–17.

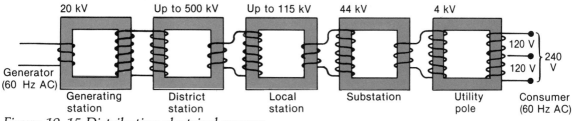

Figure 19–15 Distributing electrical energy

Figure 19–16 A local transformer station

Figure 19–17
A utility-pole
transformer

19.8 Electromagnetic Waves

We have studied how charged particles, namely electrons, move back and forth in an alternating current. The electrons gain energy and speed up. Then they slow down, stop and change direction. Each time they slow down they lose energy. Some of that energy is given off by the electrons in the form of a wave. Such a wave is called an **electromagnetic wave**.

There is an entire set of electromagnetic waves. Those caused by 60 Hz (AC) have a frequency of 60 Hz. You may have noticed the interference effects of such waves on your car radio as you drive under a high-voltage transmission line.

Waves such as radio waves and radar waves are also electromagnetic. They have frequencies between 10^4 Hz and 10^7 Hz. Another type of electromagnetic wave is called light. It is the topic of the next four chapters.

19.9 Review Assignment

1. Suppose you were given a straight conductor and a U-shaped permanent magnet. Describe how you could induce a current in the conductor when the:
   (a) magnet is held still
   (b) conductor is held still   (19.1)
2. Define electromagnetic induction.   (19.1)
3. In each diagram an arrow shows the direction of motion of a conductor. State whether or not a current is induced in the conductor and explain why.   (19.1)

   (a)   (b)   (c)

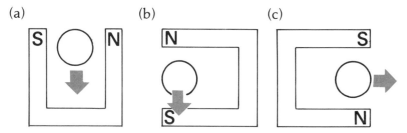

4. Assume you are using a bar magnet to induce a current in a coiled conductor. What happens to the current when you:
   (a) increase the number of windings of the coil?
   (b) increase the speed of motion of the magnet?
   (c) use a stronger magnet?
   (d) stop the magnet?   (19.2)
5. Determine the direction of the current in each coiled conductor.   (19.3)

   (a)   (b)   (c)

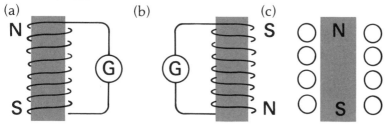

6. Use Lenz's law to decide which end of each coil is N. Then determine the direction of motion of the electrons in the coil.   (19.3)

   (a)   (b)   (c)

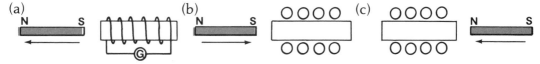

7. What is the purpose of an electric generator?   (19.4)
8. In the DC generator in the diagram, the armature is pushed as shown.
   (a) Which end of the coil is N?
   (b) Which way do the electrons flow?   (19.4)

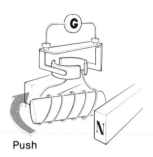

Push

9. Assume you live in a log cabin on a hill far away from any source of electricity. How could you use the wind to generate electricity to operate lights inside the cabin?   (19.4)
10. In a Faraday's ring apparatus, a DC source and a switch are connected in the primary circuit. Under what conditions is a current induced in the secondary circuit?   (19.5)
11. In a transformer, why is it better to use AC electricity then DC electricity?   (19.6)
12. Describe the construction of:
    (a) a step-up transformer
    (b) a step-down transformer   (19.6)
13. A transformer is needed to operate a fluorescent light fixture. The primary coil has 240 windings and the secondary 60. If the primary coil is connected to a 120 V household circuit, what is the voltage that operates the light?   (19.6)
14. A model electric train requires 8 V to operate. A transformer with a primary coil of 900 windings is plugged into a 120 V wall circuit. How many windings does the secondary coil have?   (19.6)
15. An induction coil in an automobile is actually a transformer. It changes 12 V from the primary into perhaps 24 kV (24 000 V) in the secondary. If the number of windings in the secondary is 4.0×10^5 (400 000), find the number of windings in the primary. (The 12 V circuit in a car works on DC. In order to make the induction coil work, the DC must be turned off and on quickly. This is controlled by the car's distributor.)   (19.6)
16. An electric doorbell uses a transformer to obtain 6 V. If the primary coil has 840 turns and the secondary coil has 42 turns, what is the primary voltage?   (19.6)
17. Choose the condition that best prevents heat loss when transmitting electrical energy over long distances:
    (a) low voltage, high current
    (b) low current, high voltage
    (c) high current, high voltage   (19.7)
18. Determine at least two electric devices in your home or school that use a transformer. In each case state whether the transformer increases or decreases the voltage.
19. What is the origin of 60 Hz electromagnetic waves?   (19.8)

19.10 Answers to Selected Problems

PRACTICE QUESTIONS

1. The magnetic field around the conductor must be changing.
2. (a) left and right
   (b) up and down
   (c) up and down
6. (i) step-up, 1000 V
   (ii) step-up, 1500 V
   (iii) step-down, 100 V

7. (a) $N_s = N_p \cdot \dfrac{V_s}{V_p}$

   (b) $N_p = N_s \cdot \dfrac{V_p}{V_s}$

   (c) $V_p = V_s \cdot \dfrac{N_p}{N_s}$

8. (a) 6 V
   (b) 240 V
   (c) 15 000 windings
   (d) 4800 windings

REVIEW ASSIGNMENT

3. (a) yes
   (b) no
   (c) yes
4. (a) The current increases.
   (b) The current increases.
   (c) The current increases.
   (d) The current stops.

8. (a) The right end becomes N.
13. 30 V
14. 60 windings
15. 200 windings
16. 120 V
17. (b)

VII. Light and Colour

20

The Nature and Reflection of Light

GOALS: After completing this chapter you should be able to:
1. Give a basic definition of light.
2. Describe the importance of light energy.
3. List sources of light energy.
4. Name the parts of a shadow.
5. Draw diagrams of solar and lunar eclipses.
6. List characteristics of light as it is transmitted from one place to another.
7. Define and give examples of objects that are transparent, translucent and opaque.
8. Explain the difference between absorbing, reflecting and transmitting light energy.
9. State the law of reflection for mirrors.
10. In a diagram locate the image of an object in a plane mirror.
11. State the meaning of the following terms for curved mirrors: concave, convex, centre of curvature, radius of curvature, principal axis, focal point and focal length.
12. Know and apply the ratio f/r for curved mirrors.
13. State three rules used to draw ray diagrams for curved mirrors.
14. For each image in a mirror, state its attitude, size and type.
15. Draw a ray diagram to find the image of an object located at various distances from a curved mirror.
16. State applications of all types of mirrors.

Knowing the information in this chapter will be especially useful if you plan a career in:

• optical dispensing
• fashion design
• interior decorating
• photography
• audio-visual technology
• live theatre
• TV (lighting crews)

20.1 The Importance of Light

Light is the form of energy that is visible to the eye. Light energy radiates outward as a sort of wave from a source. That is why it is one example of radiation. (Other examples of radiation are radio waves, heat waves and X rays.)

Most of our light energy comes from the sun. Without energy

from the sun, life on the earth would be impossible. Light provides the energy needed for plants to grow, and it helps keep the temperature on the earth suitable for life.

Light energy makes it possible for us to see. If you are in a totally dark room, you see nothing. As soon as a light is turned on, you can see because the light bounces off objects to your eyes. When the light is turned off again, you see nothing. Figure 20-1 shows a photograph of a moving hand. The hand is seen only when a strobe light in the room is flashed on. The strobe light used to take the picture flashed on every 0.1 s.

Figure 20-1 A hand is visible only when light bounces off it to the observer

In the chapters on light we will study how light behaves. This will help us understand many things that go on around us.

PRACTICE
1. List three reasons that light is important to us.

20.2 Sources of Light Energy

An object that gives off its own light is called **luminous**.

Objects that are luminous because they have a high temperature are called **incandescent**. They are the most common sources of light. Examples of incandescent objects are fires, incandescent light bulbs and the sun.

Substances that are luminous only when they are struck by high-energy waves or electrons are called **fluorescent**. Gases used in fluorescent and neon lights react this way.

Phosphorescent materials become luminous when struck by high-energy waves or electrons and remain luminous for a time. Luminous dials on watches give off light for several hours after they absorb energy. An oscilloscope screen gives off light for a short time after the oscilloscope is turned off.

Chemicals that react to produce light at cool temperatures are called **chemiluminescent**. Fireflies, certain fish and some safety lights give off light energy in this manner.

PRACTICE

2. Listed below are examples of luminous objects. State whether each should be called incandescent, fluorescent, phosphorescent or chemiluminescent.
   (a) white-hot molten iron
   (b) material on a Halloween outfit that gives off light
   (c) television screen
   (d) liquid light (two chemicals are added together to produce a faint glow in the dark)
   (e) headlight of a car

20.3 Experiment 44: Shadows

INTRODUCTION

If light from a source is blocked by an object, a shadow is created behind the object. If the light comes from a source that appears small, the shadow is entirely dark and is called an **umbra**. See Figure 20-2(a).

If light comes from a large source or from two sources, the shadow has more than one part. The umbra is the part where no light falls. The **penumbra** is the part where some light falls. Refer to Figure 20-2(b).

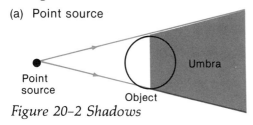
(a) Point source

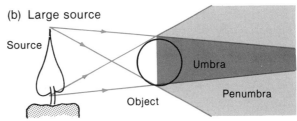
(b) Large source

Figure 20-2 Shadows

In Figure 20-2 the lines coming from the source represent paths of light. We call these lines **light rays**. Notice that the direction of each ray is shown.

To obtain light rays in experiments, a device called a **ray box** is used. One type of ray box is shown in Figure 20-3. One end of the box has mirrors so it acts as three light sources. The other end acts as a single source of light rays.

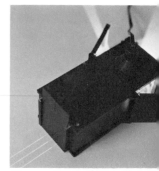

Figure 20-3 A ray box

PURPOSE: To study the formation of shadows.

APPARATUS: ray box; power supply; small object (such as a 100 g mass)

PROCEDURE
1. Place the ray box and object on a piece of paper. Aim a light beam from one source toward the object. Draw and label the shadow formed **behind** the object.
2. Aim light beams from two directions toward the object. (Use either the mirrored end of the ray box or two ray boxes.) Draw a labelled diagram of the shadow behind the object.
3. Repeat # 2 with light coming from three directions. In this case you should have an umbra, two dark penumbras and three light penumbras.

QUESTION
1. Which statement best explains why shadows are formed?
   (a) Light bends easily around corners.
   (b) Light is a form of energy.
   (c) Light travels in straight lines.
   (d) Light may come from many types of sources.

20.4 The Transmission of Light

As light travels from one place to another it displays several properties or characteristics. Three of those properties are discussed in this section.

The fact that **light travels in straight lines** was shown in the shadow experiment. It can also be observed in the beam of light of a movie projector or laser. See Figure 20-4.

Another illustration of light travelling in straight lines is the for-

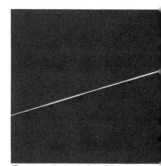

Figure 20-4 Light travels in a straight line

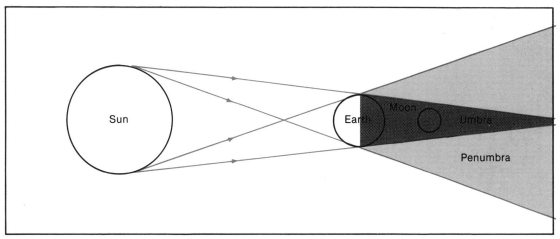

Figure 20–5 A lunar eclipse

mation of **eclipses.** Eclipses of the sun or moon occur when light
from the sun is blocked by the moon or the earth. Figure 20–5
shows how an eclipse of the moon occurs. This is called a **lunar
eclipse.** An eclipse of the sun, a **solar eclipse,** can only occur when
the moon is between the earth and the sun.

Another property of light energy is that **it does not need a
material in which to travel.** This is different from sound energy,
which must travel through particles of air or other material. In
fact, light travels fastest when there are no particles at all, in other
words, in a vacuum. That is why light from the sun and stars can
reach us through the vacuum of outer space. (If light could not
travel in a vacuum, imagine what would happen to the bell-in-a-
vacuum demonstration in Figure 13–1, Section 13.2. The bell
would disappear!)

One other property of light is its very fast speed. All radiation,
including light, travels at a **speed of 3.0 × 10⁸ m/s** in a vacuum.
That speed is only slightly less in air, so we use the same value for
calculations. At a speed of 3.0 × 10⁸ m/s, light can travel around
the world more than seven times in one second!

PRACTICE

3. The diagram illustrates how a solar eclipse occurs. Redraw the
   diagram exactly in your notebook. Complete the diagram to
   show the formation of shadows. Label the place on the earth
   where the observers would see a total eclipse of the sun.

4. Assume light takes 500 s to reach the earth from the sun. How
   far is the earth from the sun?   ($d = vt$)

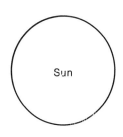

5. The average distance from the earth to the moon is 4.0×10^8 m. How long does it take light to travel from the earth to the moon? $(t = \frac{d}{v})$

20.5 How Light Behaves When It Strikes Objects

First we defined light. Then we studied sources of light and how light is transmitted. Now we will study what happens when light strikes various objects or materials.

Figure 20-6 shows a photograph of materials that treat light in three different ways.

The glass and the water are **transparent**. They allow light to pass through easily.

The ice is **translucent**. It allows some light to pass through, but it does not allow a clear image to be seen through it.

The spoon is **opaque**. No light passes through it. All the light is either absorbed or reflected.

Light energy is easily absorbed by dark, dull surfaces. It is easily reflected by shiny, smooth surfaces. An apparatus that demonstrates this is the Crookes' radiometer, shown in Figure 20-7.

When bright light is aimed toward the radiometer, the vanes spin rapidly. The shiny surfaces reflect light. The dark surfaces absorb light energy, which changes to heat energy. This heat energy in turn warms nearby air particles. The particles jump off the dark surfaces, forcing the vanes to spin in one direction.

Finally, the way light reflects off a shiny surface depends on how smooth the surface is. Two possible situations of reflection are shown in Figure 20-8.

Figure 20-6
Transparent, translucent and opaque objects

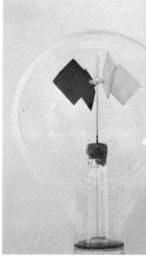

Figure 20-7 The radiometer

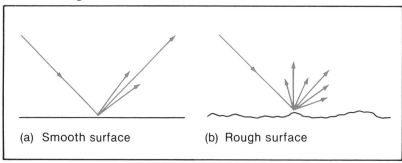

(a) Smooth surface   (b) Rough surface

Figure 20-8 Reflection of light off shiny surfaces

PRACTICE
6. Name two other objects that are:
   (a) transparent
   (b) translucent
   (c) opaque
7. Why is it better to wear white clothing than black clothing on a hot, sunny day?

20.6 Experiment 45: Reflection of Light in a Plane Mirror

INTRODUCTION
Several new terms are introduced in this experiment. The following definitions are illustrated in Figure 20–9.
(a) An **incident ray** is a ray of light travelling from the source to some object, such as a mirror.
(b) A **reflected ray** is a ray of light that bounces off an object.
(c) A **normal** (N) is a line at an angle of 90° to the surface where the incident ray strikes. (A normal is not a light ray.)
(d) The **angle of incidence** ($\angle i$) is the angle between the incident ray and the normal.
(e) The **angle of reflection** ($\angle r$) is the angle between the reflected ray and the normal.

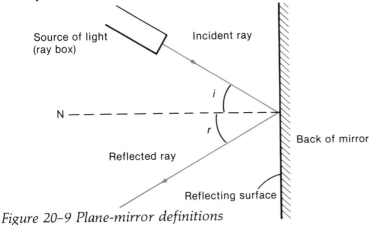

Figure 20–9 Plane-mirror definitions

PURPOSE
(a) To study the law of reflection for plane mirrors.
(b) To learn how to locate an image in a plane mirror.

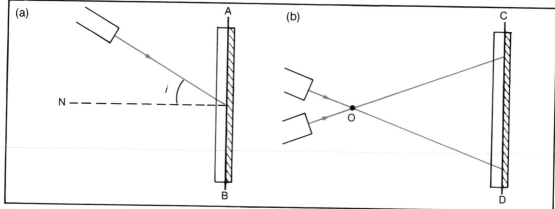

Figure 20–10

APPARATUS: ray box with a single-slit window; plane mirror; protractor

PROCEDURE
1. Look at the mirror to see how it is made. The reflecting surface is probably at the back of the glass.
2. Draw a line AB on a piece of paper. Place the mirror so its reflecting surface is along AB, as shown in Figure 20–10(a).
3. Aim an incident ray from the ray box to the mirror. Use small dots to mark the incident and reflected rays. Remove the ray box and mirror and use a straight edge to draw the rays.
4. Draw a normal from the point where the rays meet at the mirror. Label and measure the angles of incidence and reflection. Compare those angles.
5. Repeat #2 to #4 using two new diagrams and distinctly different angles. If the angle of incidence equals the angle of reflection in each case, you are ready to state the **law of reflection** for plane mirrors.
6. Draw a line CD and a point O on a piece of paper as shown in Figure 20–10(b). Locate the image (I) of the object (O) in the mirror in the following way:
   (a) Obtain two sets of incident and reflected rays. (The incident rays must go through O.)
   (b) Remove the mirror and extend the reflected rays straight back behind the mirror, because that is where they appear to come from.
   (c) Find the point where the extended lines meet. This is the image (I).
7. Look into a plane mirror. Is the image upright or inverted? Is the image smaller than the object, or larger or the same size?

QUESTIONS
1. State the angle of reflection if the angle of incidence is:
   (a) 22°   (b) 75°   (c) 0°
2. In a plane mirror, how does the distance of the object to the mirror compare to the distance of the image to the mirror? (Refer to your diagram for procedure #6.)

20.7 Ray Diagrams for Plane Mirrors

A ray diagram can be used to locate the image of an object seen in a mirror. The rays are drawn so that the law of reflection ($\angle i = \angle r$) is obeyed.

Consider object OB in Figure 20–11(a). From point O two incident rays are drawn. The normals and reflected rays are then drawn. To find the image (I), the reflected rays must meet, so they are extended behind the mirror. Then all the rays from point O reflect off the mirror and appear to come from point I.

The same procedure is used to find the image of B. That image is labelled M. Then the total image, IM, can be drawn.

In Figure 20–11(a), the distance from the object to the mirror equals the distance from the image to the mirror. This provides a short cut for finding an image in a plane mirror. An example of this short cut is shown in Figure 20–11(b).

After an image has been located in a ray diagram, we can describe the image. Three characteristics are generally needed to describe each image.

(a) **Attitude**   Is the image upright or inverted?

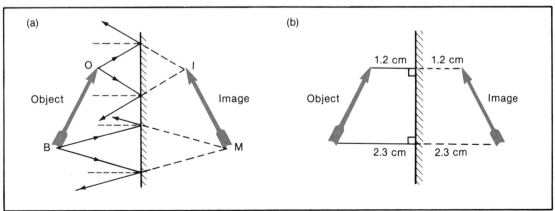

Figure 20–11 Locating an image in a plane mirror

382

(b) **Size**   Is the image larger than the object, smaller or the same size?

(c) **Type**   Is the image real or imaginary? A **real image** can be placed onto a screen, as you will observe later in the chapter. The light rays that create the real image actually meet each other. An **imaginary image** cannot be placed on a screen. The light rays that create such an image only appear to meet each other. To see an imaginary image in a mirror, you have to look into the mirror. In a single mirror an imaginary image is upright.

PRACTICE
8. Describe the image in a plane mirror by stating its:
   (a) attitude   (b) size   (c) type
9. Copy each diagram into your notebook. Locate the image of the object.

20.8 Applications of Plane Mirrors

Plane mirrors are commonly used as looking glasses. But they also have other interesting applications.

For example, in an arcade shooting gallery a plane mirror is used in some machines. Figure 20–12 shows the type of machine in which a rifle shoots light that bounces off a mirror to a target below. A direct hit is recorded if the light strikes a sensitive part of the target. The image of the target appears straight ahead and at the distance illustrated in the diagram.

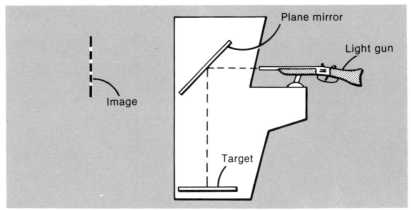

Figure 20–12 Changing the direction of light in a shooting gallery

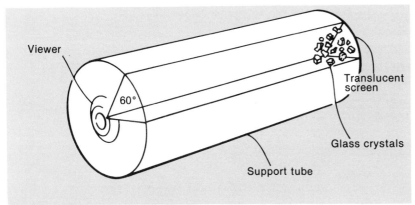

Figure 20–13 The kaleidoscope

Another application of plane mirrors comes from the fact that when two mirrors are placed at an angle to each other, multiple images are seen. The smaller the angle, the greater the number of images. At 90° between the mirrors, three images are formed, and at 60°, five images are formed. (Reference: Appendix F.) A kaleidoscope, shown in Figure 20–13, has two plane mirrors at an angle of 60° to each other. Coloured glass crystals are placed between the mirrors. Thus, five identical images of the crystal pattern can be seen at any one time. As the kaleidoscope is turned, new patterns are formed.

Figure 20–14 Curved mirrors

(a) Concave

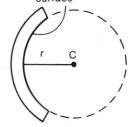

PRACTICE

10. **Mini-experiment:** Obtain two plane mirrors. Hold them at angles to each other of 90° and less. Observe how multiple images can be obtained.

20.9 Curved Mirrors

(b) Convex

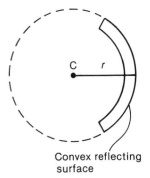

Although plane mirrors are very common, they are not the only type of mirror used. Curved mirrors, both concave and convex, also have many applications.

A **concave mirror** has a reflecting surface that curves inwards. A **convex mirror** has a reflecting surface that bulges outwards. Refer to Figure 20–14.

In Figure 20–14 there are two symbols, C and r, that should be defined. The point C is the **centre of curvature** of the circle from which the mirror is made. The distance r is the **radius of curvature** of a curved mirror. It is the distance from the centre of curvature to the reflecting surface.

Other definitions for curved mirrors are illustrated in Figure 20-15. The **principal axis** (P.A.) of a curved mirror is the line drawn through C that strikes the middle of the mirror. The **focal point** (F) is the position where parallel incident rays meet when they reflect. The **focal length** (f) is the distance from F to the reflecting surface.

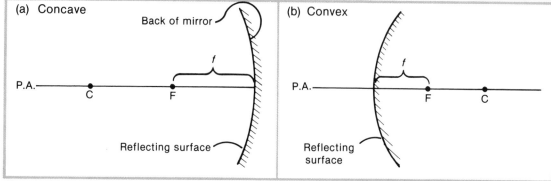

Figure 20-15 Curved-mirror definitions

11. The diagrams represent curved mirrors. For each mirror:
    (a) state whether it is concave or convex
    (b) measure its radius of curvature

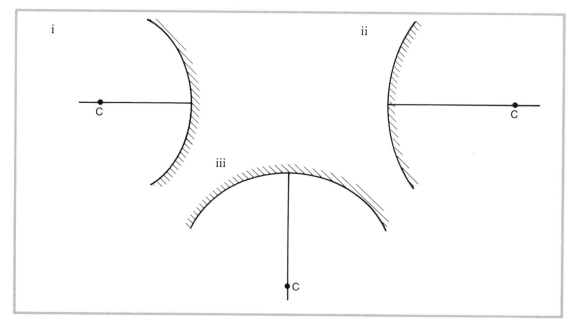

20.10 Experiment 46: Reflection in a Circular Concave Mirror

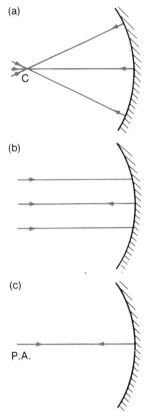

INTRODUCTION

If you are asked to find the radius of curvature of a curved mirror, you can do so by first finding C, the centre of curvature. One way to find C is to use the knowledge that a light ray through C strikes the surface of the mirror at an angle of 90°. Thus, the ray reflects onto itself. Two or three such rays will help you locate C.

PURPOSE: To find the following for a circular concave mirror:
   (a) the ratio of the focal length to the radius of curvature
   (b) three rules for light rays that strike the mirror

APPARATUS: ray box with single-slit and triple-slit windows; circular concave mirror

Figure 20–16

PROCEDURE

1. Set the mirror flat on a piece of paper and draw its shape. Aim a single light ray so it reflects onto itself. Draw that ray. Repeat this from two other directions, as shown in Figure 20–16(a). The point where the rays meet is C. Measure the radius of curvature, r.
2. Starting a new diagram, aim three **parallel** rays of light toward the mirror so that the middle ray reflects onto itself. See Figure 20–16(b). Draw all incident and reflected rays. The point where the reflected rays meet is F. Label it and measure the focal length, f.
3. Compare the focal length with the radius of curvature and find the ratio f/r.
4. Starting a new diagram, aim a single ray, P.A., toward the middle of the mirror so it reflects onto itself. See Figure 20–16(c). Find and label points C and F. Use a single ray to determine each of three important **rules** for concave-mirror reflection. Draw all the rays on your diagram. The rays used to determine these rules are:
   (a) an incident ray parallel to the principal axis
   (b) an incident ray through F
   (c) an incident ray through C (but not along the principal axis)

QUESTIONS

1. What is the focal length of a concave mirror whose radius of curvature is:
   (a) 12 cm   (b) 60 mm   (c) 4.6 cm?
2. What is the radius of curvature of a concave mirror whose focal length is:
   (a) 2.1 cm   (b) 24 mm   (c) 5.8 cm?
3. Describe where each ray will reflect:

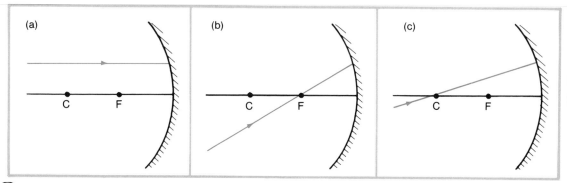

4. The word converge means to come together. Explain why a concave mirror may be called a converging mirror.

20.11 Images in Concave Mirrors

An image can be described as having three main characteristics: attitude, size and type. These were defined in Section 20.7. The characteristics of the image in a concave mirror depend on where the object is located.

Figure 20–17(a) shows an image in a concave mirror when the object is close to the mirror. The image is upright, larger than the object and imaginary.

Diagram (b) shows an image when the object is just beyond the focal point. In this case, the image is inverted, larger than the object and real. It is real because it can be placed onto a screen. You can look at the screen to see the image; you do not have to look at the mirror. When a single mirror is used, a real image is inverted.

PRACTICE

12. **Mini-experiment:** Hold a glass concave mirror at arm's length and view an object far away in the room. Describe the image by stating its attitude, size and type.

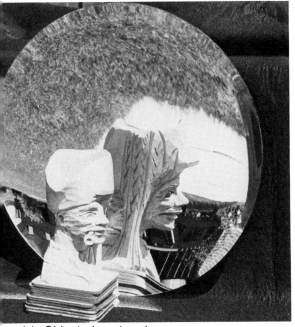

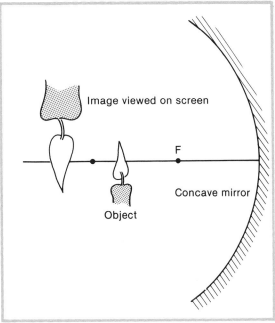

Image viewed on screen

F

Concave mirror

Object

(a) Object close to mirror   (b) Object just beyond the focal point

Figure 20–17 Images in concave mirrors

20.12 Experiment 47: Reflection in a Circular Convex Mirror

INTRODUCTION

In the previous experiment you learned how light behaves when it reflects from a concave mirror. This experiment will be almost identical to that one in the procedure, except that a convex mirror will be used.

As you will learn in the experiment, light rays that reflect off a convex mirror spread apart so they do not meet each other. In order to make such rays meet, you must extend the rays behind the mirror.

PURPOSE: To find the following for a circular convex mirror:
(a) the ratio of the focal length to the radius of curvature
(b) three rules for light rays that strike the mirror

APPARATUS: ray box with single-slit and triple-slit windows; convex mirror (circular); convex mirror (glass)

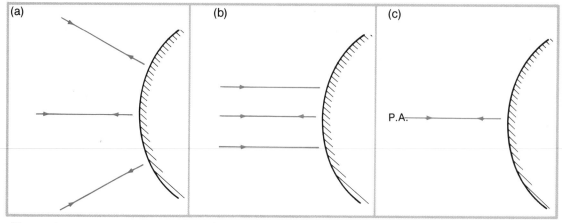

Figure 20–18

PROCEDURE

1, 2 and 3: Refer to steps 1, 2 and 3 in Experiment 46 and repeat them using a convex mirror. Use Figure 20–18(a) and (b) as a guide.

4. Determine three important rules for convex-mirror reflection and draw all the rules on a new diagram. To determine these rules use:
   (a) an incident ray that is parallel to the principal axis
   (b) an incident ray that is aimed toward F
   (c) an incident ray that is aimed toward C

5. View the image of an object seen in a convex mirror made of glass. Describe the image by stating its attitude, size and type.

QUESTIONS

1. How does the ratio f/r for convex mirrors compare to the ratio f/r for concave mirrors?
2. Describe where each ray will reflect:

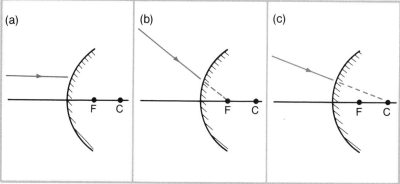

3. The word diverge means to spread apart. Explain why a convex mirror may be called a diverging mirror.

20.13 Ray Diagrams for Curved Mirrors

A ray diagram can be used to locate the image of an object in a curved mirror. The method we will use here involves the rules for curved-mirror reflection that you learned in the previous two experiments. (Those rules are actually based on the fact that $\measuredangle i = \measuredangle r$ for all mirrors, although we did not prove it in the experiments.)

In the sample problems that follow notice that the reflected rays must meet in order to locate an image. In the second case the reflected rays do not meet unless they are extended behind the mirror. Thus, the image is imaginary.

Sample problem 1: Use a ray diagram to find the image of a 2.0 cm high object located 10 cm from a concave mirror that has a focal length of 2.5 cm. State the attitude, size and type of the image.

Solution: If $f = 2.5$ cm, then $r = 5.0$ cm. Draw a concave mirror with $r = 5.0$ cm. Label C and F. Draw the object sitting on the principal axis. (This saves a lot of time.) Use three incident rays and reflected rays, according to the rules, to locate the image.

The image is inverted, smaller than the object and real.

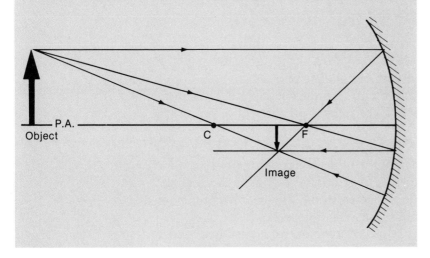

Sample problem 2: Repeat sample problem 1 using a convex mirror and an object distance of 6 cm.
Solution: The image is upright, smaller than the object and imaginary.

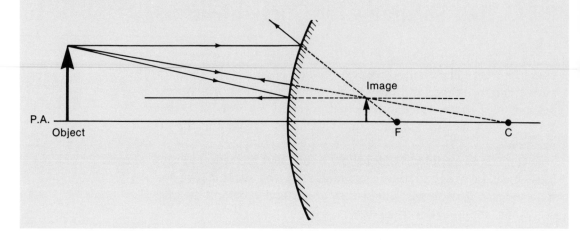

PRACTICE

13. A 1.5 cm object is located 8.0 cm in front of a concave mirror that has a radius of curvature of 6.0 cm. Draw a ray diagram to locate the image. State the image's attitude, size and type.
14. Repeat # 13 for a convex mirror but place the object 3.0 cm from the mirror.

20.14 Applications of Curved Mirrors

Concave Mirrors
Figure 20-19 shows four uses of concave mirrors.

A car headlight, in diagram (a), has the source of light near the focal point of the mirror. Light reflects off the mirror to form a directed beam.

Concave mirrors are used to concentrate sunlight. A solar cooker, illustrated in diagram (b), has the cooking pot located at the focal point of the mirror. Diagram (c) shows the largest solar collector in the world. It is located in the Pyrenees Mountains in Southern France and is used for scientific research.

Diagram (d) illustrates how a concave mirror is used in a reflecting telescope. Huge telescopes (up to 6 m in diameter) collect much

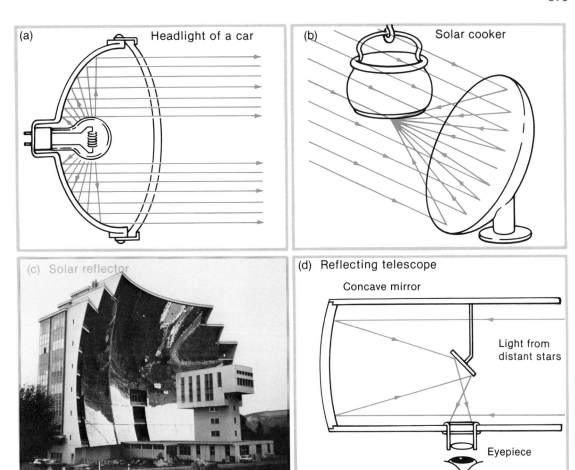

Figure 20–19 Applications of concave mirrors

light, so astronomers can view distant stars which are not visible through small telescopes.

Convex Mirrors

Images in convex mirrors are always upright and smaller than the object. Therefore convex mirrors allow the viewer to see over a wide angle. Such mirrors are used in stores to discourage shoplifting. They are also used as rear-view mirrors on trucks, buses and motorcycles. (See Figure 20–20.) When using a convex mirror, a driver must be careful because a vehicle is much closer than it appears in the mirror.

PRACTICE

15. List uses of concave mirrors other than those mentioned in this section.

Figure 20–20 A convex mirror gives a wide angle view

20.15 Review Assignment

1. What is light?   (20.1)
2. (a) Define incandescence.
   (b) State four examples of incandescent objects.   (20.2)
3. What property of light is illustrated by the formation of shadows?   (20.3, 20.4)
4. Name the type of eclipse in which the sun's light is blocked off by the moon.   (20.4)
5. Describe evidence that light travels in a straight line.   (20.4)
6. How fast does light travel in a vacuum?   (20.4)
7. The distance light travels in one year is called a **light year**.
   (a) Calculate that distance, assuming there are 3.2×10^7 s in one year.
   (b) The nearest star to our sun is about 4 light years away. Use your answer in (a) above to find the distance to the nearest star.   (20.4)
8. How long does it take light to travel across Canada, a distance of 6.0×10^6 m?   (20.4)
9. From each pair of materials listed, choose the material that absorbs light more easily.   (20.5)
   (a) a black surface; a white surface
   (b) a transparent material; an opaque material
   (c) a smooth surface; a rough surface
   (d) a dull surface; a shiny surface
10. State the law of reflection for mirrors.   (20.6)
11. Describe the differences between a real image and an imaginary image.   (20.7)
12. If you stand 60 cm in front of a plane mirror, how far are you from your image in the mirror?   (20.7)
13. Each diagram shows an object in front of a plane mirror. Draw the diagrams in your notebook and locate the images. (20.7)

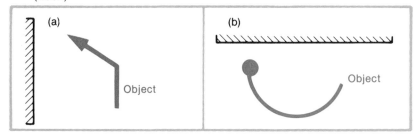

(a)

Object

(b)

Object

14. For circular curved mirrors, state the ratio of:
    (a) f/r (b) r/f (20.10, 20.12)

15. Draw a fully labelled ray diagram to locate the image of the object for each situation described in the chart below. In each diagram use two rules. For each image, state its attitude, size and type.   (20.9 to 20.13)

Type of mirror	Focal length (cm)	Height of object (cm)	Distance of object to mirror (cm)
(a) concave	3.0	1.5	12.0
(b) concave	4.0	2.0	8.0
(c) concave	4.0	1.5	6.0
(d) concave	4.0	1.0	4.0
(e) concave	4.0	1.5	2.0
(f) convex	3.0	2.0	10.0
(g) convex	3.0	1.5	3.0

16. Assume you have a shiny tablespoon. How would you hold the spoon to obtain an image of yourself that is:
    (a) small and imaginary?
    (b) small and real?
    (c) large and imaginary?   (20.9 to 20.13)

17. The diagram shows one way of heating water using solar energy.
    (a) Describe at least two ways in which this is an application of topics in this chapter.
    (b) How would you determine where to position the black hose for the best heating effects?

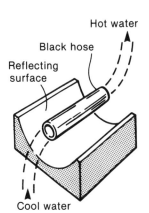

Hot water

Black hose

Reflecting surface

Cool water

20.16 Answers to Selected Problems

PRACTICE QUESTIONS
4. 1.5×10^{11} m
5. 1.3 s
8. The image is upright, the same size as the object and imaginary.
11. (i) concave, 2.5 cm
    (ii) convex, 3.5 cm
    (iii) concave, 3.0 cm
12. The image is inverted, smaller than the object and real.

REVIEW ASSIGNMENT
7. (a) 9.6×10^{15} m
   (b) 3.84×10^{16} m
8. 0.02 s or 2.0×10^{-2} s
12. 120 cm
14. (a) 1/2
    (b) 2/1

21

Refraction and Lenses

GOALS: After completing this chapter you should be able to:
1. Define refraction and explain why it occurs.
2. State the laws of refraction.
3. Measure the angles of incidence, refraction and emergence in a diagram of refraction of light through a prism.
4. Explain demonstrations of refraction.
5. Describe the conditions needed for total internal reflection.
6. Define critical angle and be able to find it experimentally.
7. Explain applications of total internal reflection.
8. Draw the shapes of convex and concave lenses.
9. State what is meant by primary focal point and secondary focal point for lenses.
10. State three rules for drawing ray diagrams for both convex and concave lenses.
11. State the attitude, size and type of image seen in a lens.
12. Draw a ray diagram to find the image of an object located at various distances from a lens.

Knowing the information in this chapter will be especially useful if you plan a career in:

• optical dispensing
• fashion design
• interior decorating
• photography
• audio-visual technology
• live theatre
• TV (lighting crews)

21.1 Refraction of Light

Have you ever noticed that the sun appears to change colour as it is setting? Or, looking down, have you ever noticed the distortion as you walk in clear, waist-deep water? Such effects occur because of the bending of light.

In the previous chapter you learned that light travels in straight lines. That is true as long as the light is travelling in one uniform material. However, when light travels from one material into another, it can bend. This bending is called **refraction**.

Figure 21–1(a) shows a beam of light travelling from air into a block of glass. The light refracts upon entering the glass. (Some of the light also reflects off the side of the block.) The reason the light

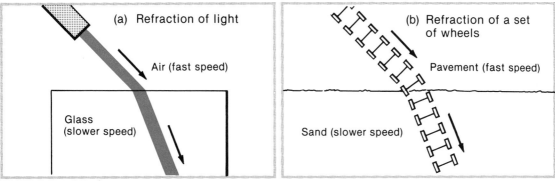

(a) Refraction of light

Air (fast speed)

Glass
(slower speed)

(b) Refraction of a set
of wheels

Pavement (fast speed)

Sand (slower speed)

Figure 21-1 Refraction

refracts is that its speed changes. Light travels more slowly in the glass than in the air. This results in a change in direction. Another example of how a change in speed causes a change in direction is illustrated in Figure 21-1(b). The wheels of a car change direction when they travel from pavement to sand, as shown in the diagram.

21.2 Experiment 48: Refraction in Rectangular Prisms

INTRODUCTION

Certain symbols are used to label diagrams of refraction of light. Figure 21-2 shows a typical refraction diagram with several of these symbols. The normals are drawn at 90° to the surfaces where the rays enter and leave the prism. The angles of incidence ($\angle i$), refraction ($\angle R$) and emergence ($\angle e$) are measured between the normal and the ray.

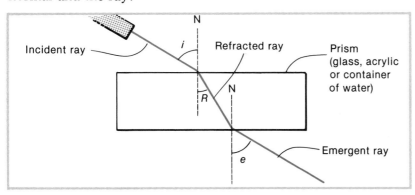

Incident ray

N

i

Refracted ray

R   N

e

Emergent ray

Prism
(glass, acrylic
or container
of water)

Figure 21-2 Symbols used in refraction diagrams

PURPOSE: To study the refraction of light in plastic and water and to learn the laws of refraction of light.

APPARATUS: ray box with single-slit window; rectangular acrylic prism; rectangular plastic box to hold water; protractor

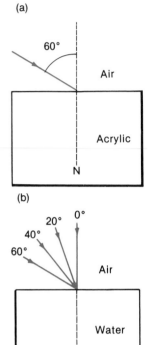

PROCEDURE

1. Place the acrylic prism on a page and draw its outline. Use a broken line to draw a normal as shown in Figure 21–3(a). Draw an incident ray so that ∡ i = 60°. Repeat this procedure using three more diagrams and angles of 40°, 20° and 0° from the normal.
2. Aim a single ray along the incident ray in the first diagram. Draw the ray that emerges on the opposite side of the prism. Remove the prism and draw the entire path of the light. Label and measure the angles of refraction (∡ R) and emergence (∡ e).
3. Repeat #2, using the other angles of incidence. What have you discovered?
4. Repeat #1 to #3, using water in a plastic container. You can draw all the rays on one diagram if you place them near one corner, as shown in Figure 21–3(b).
5. Tabulate the measurements in a chart similar to the one below.

Figure 21–3

Material	∡ i	∡ R	∡ e
Acrylic	60°		
	40°		

6. State the **laws of refraction** by answering these questions:
   (a) When light travels at an angle from a material of low density (such as air) to one of higher density (such as water), does it bend away from or toward the normal?
   (b) When light travels at an angle from a material of high density to one of lower density, does it bend away from or toward the normal?

QUESTIONS

1. When light travels from air into acrylic, at what angle of incidence is there no refraction?
2. How does the direction of the incident ray toward a rectangular prism compare to the direction of the emergent ray?
3. How does refraction in water compare to refraction in acrylic?
4. In which substance, water or acrylic, do you think light travels faster? What evidence do you have for your answer?

21.3 Demonstrations of Refraction

When light travels from water to air, it bends away from the normal. This fact is easily demonstrated by placing a ruler part way into water. The ruler appears to be bent, as shown in Figure 21-4.

Another easy demonstration of refraction can be done using a coin in a cup and some water. Place the coin in the middle of the empty cup. Position your eyes at a level so you just miss seeing the coin. [Refer to Figure 21-5(a).] Slowly add water to the cup without moving the coin. Observe the results. The ray diagram in Figure 21-5(b) explains what happens.

(a) Before adding water

(b) The light refracts as it leaves the water allowing the observer to see the coin.

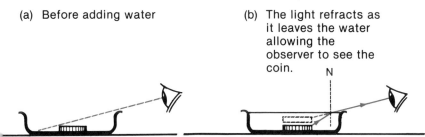

Figure 21-4 Demonstrating an effect of refraction

Figure 21-5 Another demonstration of refraction

PRACTICE

1. Two swimmers, B and G, stand at the edge of a clear lake. B bets he can stand on the rock he sees and the water will be at his waist. G bets he is wrong because she thinks the rock is too far below the surface.
   (a) According to the diagram, who is right?
   (b) Why was the other swimmer wrong? (A ray diagram will help you explain your answer.)

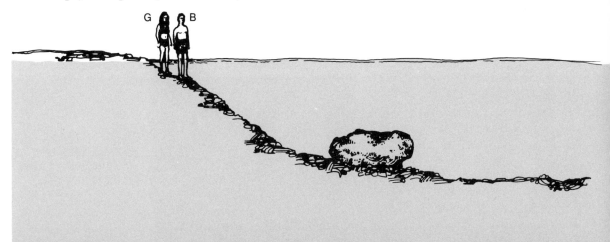

21.4 Experiment 49: Total Internal Reflection

INTRODUCTION

You have learned that when light travels from a material of high density (such as acrylic) into one of lower density (such as air), it refracts away from the normal. This means that the angle of emergence in air is greater than the angle in the acrylic. This is shown in diagram (a) of Figure 21-6.

In diagram (b) the angle in the acrylic has increased, and the angle of emergence is almost 90°. When this happens, the white light actually starts splitting up into rainbow colours. Also, some light is internally reflected; in other words, some light reflects off the inside surface of the acrylic.

When the light just disappears along the surface of the plastic, as in diagram (c), the angle in the acrylic is called the **critical angle**, $\angle c$. This is the angle at which **total internal reflection** begins.

For any angle in the acrylic greater than the critical angle, total internal reflection occurs. This is shown in diagram (d). Total internal reflection can occur in any transparent material. Figure 21-6(e) shows laser light internally reflecting in water.

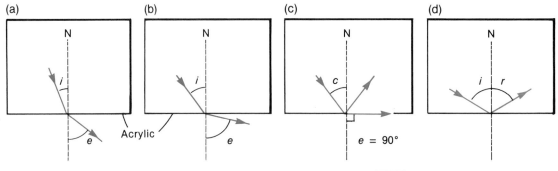

Figure 21-6
Explaining total internal reflection

(e) Total internal reflection of laser light in water

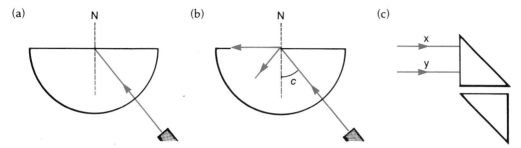

Figure 21-7

PURPOSE: To find the critical angle for acrylic and to study an application of total internal reflection.

APPARATUS: ray box with single-slit and double-slit windows; semicircular acrylic prism; 2 triangular acrylic prisms (45°, 90°, 45°)

PROCEDURE

1. Place the semicircular prism on your page and draw its outline. Aim a single ray from the curved side directly toward the **middle** of the flat edge as shown in Figure 21-7(a). Complete the diagram.
2. Slowly move the ray box around until you can see the rainbow colours near the flat edge. Move the box slightly further until the light in the air just disappears. See Figure 21-7(b). Mark the rays. Remove the prism, draw the normal and measure the critical angle, ∢c, for acrylic. Have the teacher check your value.
3. Start a new diagram with the same prism. Aim a single ray from the curved side so that the angle inside the acrylic is greater than ∢c. Complete the diagram of total internal reflection. Determine whether the law of reflection (∢i = ∢r) is valid.
4. A periscope is an interesting application of total internal reflection. To see how light travels in a periscope, set up the triangular prisms as shown in Figure 21-7(c). Aim two rays, X and Y, as shown and draw the paths followed by those rays. Determine whether the final emergent rays are upright or inverted when compared to the incident rays.

QUESTION

1. The critical angle of ordinary glass is 42°. Which of the following angles of light rays in glass would result in total internal reflection?
   (a) 35°   (b) 50°   (c) 43°   (d) 3°

21.5 Applications of Total Internal Reflection

Plane mirrors reflect light just like prisms that allow total internal reflection. However, plane mirrors have disadvantages. They tarnish easily so they do not last as long as prisms. Also they absorb more light than prisms. Thus, internal reflection in prisms is used where reflection in mirrors might not be good enough.

The principle of the periscope was described in the previous experiment. A more detailed diagram of a periscope is shown in Figure 21–8(a).

Bicycle reflectors also use total internal reflection. Light from a car behind the bicycle strikes the reflector and bounces back toward the car. This is illustrated in Figure 21–8(c).

Prisms are also used in binoculars, shown in Figure 21–8(b). Without prisms, the binoculars would have to be made longer to give the same enlargement of images.

Diagram (d) shows solid plastic tubing in which a laser beam is internally reflected every time it strikes an inside surface. This idea can be applied to the transmission of telephone and television messages in thin fibres.

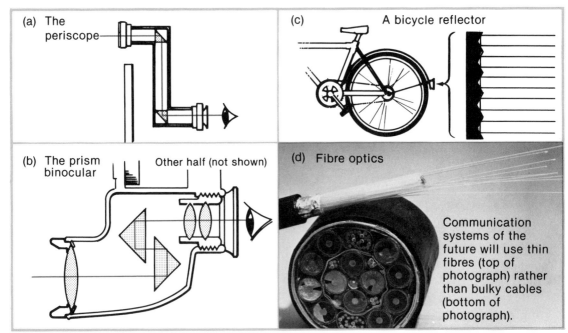

(a) The periscope

(c) A bicycle reflector

(b) The prism binocular   Other half (not shown)

(d) Fibre optics

Communication systems of the future will use thin fibres (top of photograph) rather than bulky cables (bottom of photograph).

Figure 21–8 Applications of total internal reflection

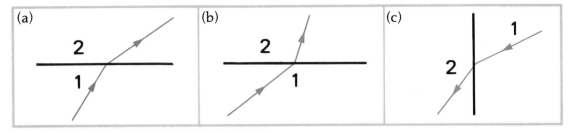

PRACTICE

2. Each diagram shows a light ray going from one material into another. In each case decide whether or not total internal reflection could occur in the first material.

21.6 Lenses

Anyone who has used a magnifying glass to light a campfire has made use of refraction of light in a lens. Anyone who wears contact lenses or ordinary eye glasses is taking advantage of the refraction of light in lenses.

A lens is a transparent device with at least one curved edge. Whether the edge is flat or curved, the laws of refraction are obeyed. One example of how light refracts in a lens is shown in Figure 21–9.

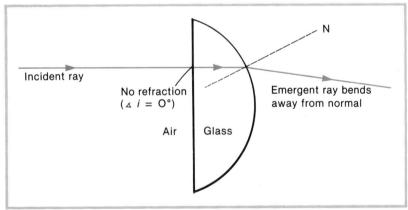

Figure 21-9 An example of refraction of light in a lens

Lenses may be either convex or concave. A **convex lens** is thicker in the middle than at the outside edge. A **concave lens** is thicker at the outside edge than in the middle. Refer to Figure 21–10.

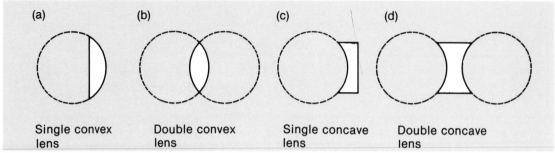

Figure 21-10 The design of lenses

PRACTICE

3. Determine the approximate direction of the light ray as it emerges from the lens shown. (Hint: Draw a normal from the curved edge and apply the laws of refraction.)

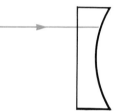

21.7 Experiment 50: Double Convex Lenses

INTRODUCTION

A lens has a principal axis (P.A.) just as a curved mirror does. However, a lens has two focal points and no centre of curvature. For a convex lens the **primary focal point** (P.F.) is located on the side of the lens opposite the source of light. The **secondary focal point** (S.F.) is located on the same side as the source. See Figure 21-11. The focal length (f) is measured to the middle of the lens, as indicated.

PURPOSE: To study how light behaves as it passes through a convex lens and to learn three rules for light rays striking such a lens.

APPARATUS: ray box with single-slit and triple-slit windows; 2 double convex lenses (one acrylic and one glass)

PROCEDURE

1. Place the acrylic lens at the middle of the page and draw its outline and principal axis. Aim three parallel rays toward the lens so that the centre ray falls along the principal axis. The point where the emergent rays meet is the primary focal point (P.F.). Label it and measure the focal length (f).

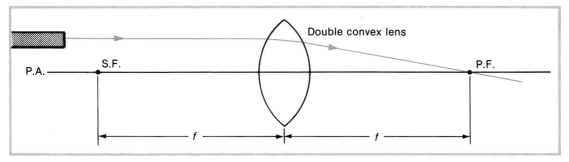

Figure 21–11 A double convex lens

2. Remove the lens and draw in the secondary focal point (S.F.) a distance *f* from the lens on the opposite side of the primary focal point.
3. Using your diagram, perform the procedure needed to complete these rules for drawing ray diagrams for convex lenses:
    (a) An incident ray parallel to the principal axis . . .
    (b) An incident ray through the secondary focal point . . .
    (c) An incident ray through the middle of the lens at a small angle to the principal axis . . .
4. Hold the glass convex lens at arm's length and view objects around you. Describe the image (attitude, size and type) when the object viewed is:
    (a) close to the lens
    (b) far from the lens

QUESTIONS

1. Describe where each ray will refract:

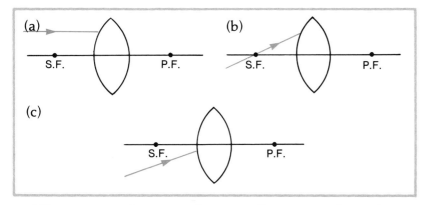

2. When a convex lens is used as an ordinary magnifying glass, is the image real or imaginary?
3. Is a convex lens a converging or diverging lens? Explain your answer.

21.8 Experiment 51: Double Concave Lenses

INTRODUCTION

Figure 21–12 shows that for a concave lens the primary focal point is on the same side of the lens as the source of light. Compare this situation to that for a convex lens in Figure 21–11.

The light rays that are parallel to the principal axis determine where the primary focal point is. In Figure 21–12 notice that the primary focal point is found by extending the emergent ray back to the principal axis.

Again the focal length (f) is measured to the middle of the lens.

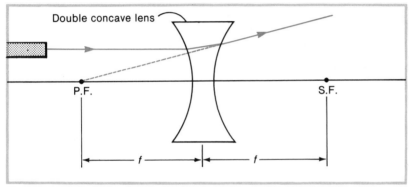

Figure 21–12 A double concave lens

PURPOSE: To study how light behaves as it passes through a concave lens and to learn three rules for light rays striking such a lens.

APPARATUS: ray box with single-slit and triple-slit windows; 2 double convex lenses (one acrylic and one glass)

PROCEDURE

1. Set the acrylic lens in the middle of the page and draw its outline and principal axis. Aim three parallel rays toward the lens so that the middle ray falls along the principal axis. Draw all the rays. Remove the lens and extend the emergent rays straight back until they meet. The point where they meet is the primary focal point (P.F.). Label it and measure the focal length (f).
2. Draw in the S.F. a distance f from the lens on the side of the lens opposite the P.F.

3. Using your diagram, perform the procedure needed to complete these rules for drawing ray diagrams for concave lenses:
   (a) An incident ray parallel to the principal axis . . .
   (b) An incident ray toward the secondary focal point . . . (Hint: Remove the lens and aim an incident ray toward the S.F. Then replace the lens.)
   (c) An incident ray through the middle of the lens at a small angle to the principal axis . . .
4. Use the glass concave lens to view objects around you. Describe the images by stating their attitude, size and type.

QUESTIONS
1. Describe where each ray will refract:

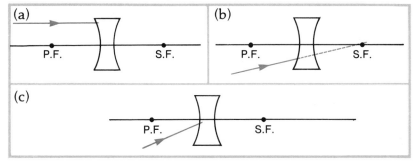

2. Can an image in a concave lens be placed onto a screen? Why or why not?
3. Is a concave lens a converging or diverging lens? Explain your answer.

21.9 Ray Diagrams for Lenses

A ray diagram can be used to locate the image of an object situated near a lens. Ray diagrams for lenses resemble those for mirrors. Any two of the three rules learned in the experiments may be used to locate an image. After the image is located, its attitude, size and type can be stated.

Light that passes through a lens can refract twice, first when entering the lens and again when leaving. This double refraction is not easy to draw, so a short cut can be used. A straight line is drawn to represent the lens, so the light ray in the diagram refracts only once. The type of lens is indicated in the centre of the diagram.

406

Sample problem 1: The focal length of a double convex lens is 2.5 cm. A 1.5 cm high object is located 8.0 cm from the lens. Use a ray diagram to locate the image of the object. Describe the image.

Solution: Draw the diagram according to the instructions. Place the object on the principal axis. Use two or three rules to find where the refracted rays meet.

The image is inverted, smaller than the object and real.

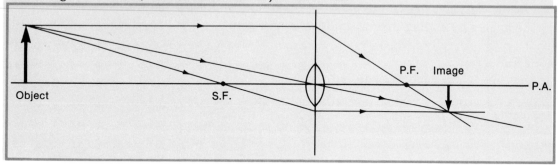

Sample problem 2: Repeat sample problem 1, using a double concave lens.

Solution: The refracted rays must be extended back to where they meet, which is on the side of the lens where the object is located.

The image is upright, smaller than the object and imaginary.

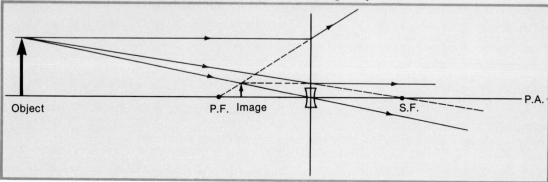

Several optical instruments, including the camera and the eye, use lenses. These applications are discussed in the next chapter.

PRACTICE

4. A 2.0 cm high object is located 6.0 cm from a double convex lens of focal length 2.5 cm. Locate and describe the image.
5. Repeat # 4, using a double concave lens.

21.10 Review Assignment

1. When light travels from air into water at an angle, why does it refract? (21.1)
2. In which material, A or B, is light travelling more slowly? (21.1)

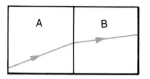

3. State the laws of refraction. (21.2)
4. Draw a diagram of a 3.0 cm × 5.0 cm glass prism that has a ray of light striking the long edge such that $\angle i = 50°$, $\angle R = 30°$ and $\angle e = 50°$. (21.2)
5. The angle of incidence of a ray into a rectangular prism is 0°. What is the size of the angle of:
   (a) refraction   (b) emergence? (21.2)
6. What is meant by the term critical angle? (21.4)
7. State two conditions necessary for total internal reflection. (21.4)
8. A right-angled periscope can be used to see around corners. Assume you are given one triangular prism and two cardboard tubes. Draw a diagram of how you would arrange those pieces to make such a periscope. (21.4, 21.5)
9. The diagram represents a block of diamond. If the critical angle in diamond is only 24°, which of the light rays shown will reflect internally? (21.4, 21.5)

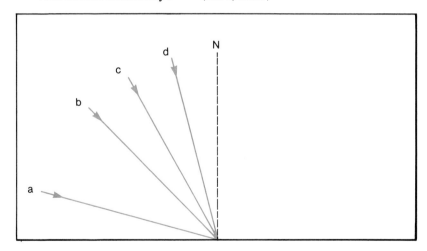

10. A double convex lens can be used to start a fire in bright sunlight. Draw a diagram of a lens ($f = 5.0$ cm) to explain why the lens can start a fire. (Hint: Light rays from a distant source are parallel to each other.) (21.6, 21.7)

11. Draw a fully labelled ray diagram to locate the image of the object for each situation described in the chart below. In each diagram use two rules. For each image that results, state its attitude, size and type.   (21.7, 21.8, 21.9)

	Type of lens	Focal length (cm)	Height of object (cm)	Distance of object to lens (cm)
(a)	convex	3.0	2.0	6.0
(b)	convex	3.0	1.5	4.5
(c)	convex	4.0	1.5	4.0
(d)	convex	4.0	1.5	2.0
(e)	concave	4.0	2.0	4.0

21.11 Answers to Selected Problems

PRACTICE QUESTIONS
1. (a) G is right
2. (a) yes
   (b) no
   (c) yes

REVIEW ASSIGNMENT
2. B
5. (a) 0°
   (b) 0°
9. a, b and c will internally reflect

22

Optical Instruments

GOALS: After completing this chapter you should be able to:
1. Describe ways in which optical instruments help us see.
2. Draw a ray diagram to locate the image in a pinhole camera and describe that image.
3. Compare a lens camera with a pinhole camera.
4. Describe the functions of the main parts of a camera.
5. State the names and functions of the main parts of the human eye.
6. Describe defects of the human eye and corrections for those defects.
7. Draw a diagram to show how the final enlarged image is created in a two-lens optical instrument.
8. Make a basic refracting telescope and microscope given two lenses.

Knowing the information in this chapter will be especially useful if you plan a career in:

- optical dispensing
- fashion design
- interior decorating
- photography
- audio-visual technology
- live theatre
- TV (lighting crews)

22.1 The Importance of Optical Instruments

Optical instruments are devices that help us see. The most important optical instrument is the eye. Of course, if we did not have eyes, the topic of light would have little meaning.

Optical instruments have the following important functions:

(1) *They record images.* The eye records images temporarily. A camera records images permanently.

(2) *They improve weak vision.* Eyeglasses and contact lenses help people with weak vision to see normally.

(3) *They create large images.* A microscope can make an image look hundreds of times larger than the original object. A projector places a large image onto a screen from a small photograph.

(4) *They help to view distant objects.* Binoculars and telescopes make faraway objects, such as the moon, appear much clearer than can be seen with the ordinary eye.

The experiments and ideas in this chapter will give you a basic understanding of a few optical instruments. Each instrument studied uses at least one lens. Thus, the information is based on what you learned about lenses in Chapter 21.

PRACTICE

1. Which of the two types of image does a projector project onto a screen? (If you cannot recall the difference between a real image and an imaginary image, check back to Section 20.7.)

22.2 Experiment 52: The Pinhole Camera

INTRODUCTION

The first photograph ever taken with a camera was in 1826 in France. The type of camera used at that time was similar to the type you will use in this experiment.

The image seen in a pinhole camera results from the fact that light travels in a straight line. This fact will be applied when you draw ray diagrams to explain your observations.

PURPOSE: To study images formed using a pinhole camera.

APPARATUS: pinhole camera; small piece of black paper; candle; pin; glass convex lens with a focal length greater than the length of the camera
Caution: Do not allow the camera to come too close to the flame.

PROCEDURE

1. Tape the black paper over the end opening of the camera and poke a pinhole in the middle of it. Aim the pinhole toward the candle's flame. Describe the image seen on the translucent screen by stating its attitude, size and type.
2. Determine what happens to the image on the screen when the camera is moved closer to and farther from the flame.
3. Add a second pinhole about 5 mm above the first. View the flame and describe the images.
4. Add several more pinholes in the shape of a large "R" and describe the effect. (A diagram of the image would help your description.)
5. Stand about one metre from the candle and hold the lens between the camera and the flame. Adjust the position of the lens

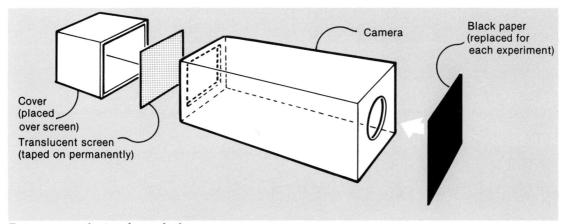

Figure 22-1 A simple pinhole camera

until you obtain a single bright image of the flame on the screen. Describe what occurs.

6. Remove the black paper and try to produce a clear image of the room or a scene outside using the lens in front of the camera. (In this case the room lights may have to be on.)

QUESTIONS

1. When an image is formed in a pinhole camera, what property of light is illustrated?
2. Draw a ray diagram to show each of the following:
   (a) A flame is located close to a pinhole camera with a single pinhole. See Figure 22-2.
   (b) A flame is located far from the camera in (a) above.
   (c) A flame is located far from a pinhole camera with two pinholes.
3. In procedure #6 you obtained an image similar to one that you found in Question #11, Section 21.10. Which diagram in #11 relates to procedure #6?

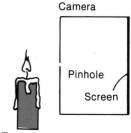

Figure 22-2

22.3 Modern Cameras

A camera is an optical device that can make a permanent record of a scene. The image is recorded on a film that is sensitive to light.

A pinhole camera can be used to take a picture. However, the pinhole does not let in much light, so a long time is needed to expose the film. To overcome this problem, a convex lens is used.

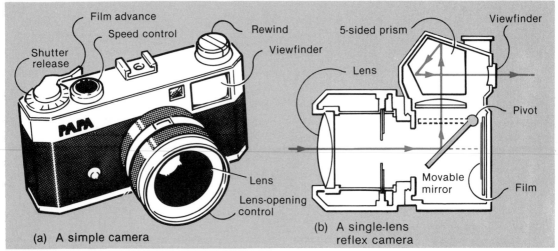

Film advance
Speed control
Shutter
release
Rewind
Viewfinder
5-sided prism
Viewfinder
Lens
Pivot
Lens
Lens-opening
control
Movable
mirror
Film

(a) A simple camera

(b) A single-lens
reflex camera

Figure 22–3 Modern 35 mm cameras

In the mid-1800s cameras used a lens and a single photographic plate. In 1889 roll films were invented, so cameras became more practical. It was not until 1935 that an efficient way of taking colour photographs was invented.

The modern camera box contains the following main parts:

(1) A **convex** lens gathers light and focusses it onto a film.

(2) A **shutter** controls the amount of light striking the film when the picture is taken. The shutter's speed and size of opening can be adjusted.

(3) A **film** records the image and is later developed to give a permanent picture.

(4) A **viewfinder** allows the photographer to see what he or she is photographing.

Figure 22–3 shows two common types of modern cameras. They are called 35 mm cameras because 35 mm of film is exposed each time a picture is taken.

In both cameras shown in Figure 22–3, the image on the film is inverted. The photographer, however, wants to see an upright image in the viewfinder, so this is done in one of two ways. In a simple camera a small separate viewfinder is used. In a single-lens reflex camera a plane mirror and a five-sided prism causing total internal reflection are used. The plane mirror flips up for an instant when the picture is taken.

PRACTICE

2. State the attitude and type of image that occurs on the film of a modern camera.

22.4 The Human Eye

Of all the known optical instruments, a healthy human eye comes closest to being perfect. Although the details of how we see are complex, we can compare the seeing process to a colour television system.

Figure 22–4(a) shows the basic setup of a television system. Diagram (b) illustrates how a similar system allows us to see.

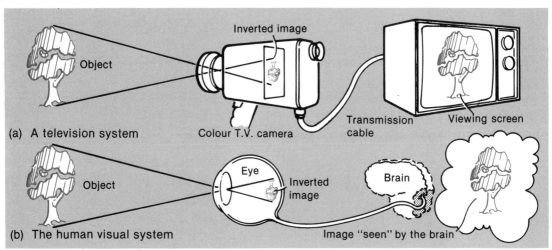

(a) A television system

(b) The human visual system

Figure 22–4 Systems that create images

The human eye, with an average diameter ranging from 20 mm to 25 mm, has many features to help us see. Figure 22–5 shows the basic structure of the eye.

Light rays that enter the eye are focussed by both the **cornea** and the **lens**. The cornea is a fixed transparent layer at the front of the eye. Light refracts a large amount as it passes through the cornea.

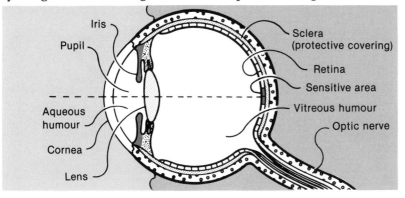

Figure 22–5 The structure of the right eye as viewed from above

Then the light refracts slightly as it passes through the convex lens. The image is focussed on the back of the eye, just as the image in a camera is focussed on the film.

The lens is somewhat flexible so that its shape can be controlled by muscles. When you look at faraway objects, the lens has a normal shape, and the muscles are relaxed. When you look at a close object, the muscles force the lens to become thicker so the image remains in focus. That is why eyestrain results if you look at nearby objects too long.

The **pupil** of the eye is the "window" through which light enters the lens. It appears black because most of the light that enters a human eye is absorbed inside. The pupil is surrounded by the **iris**, the coloured portion of the eye. The iris controls the size of the pupil. In bright light the pupil becomes small and in dark light the pupil becomes large to let in more light. You have likely experienced walking into a dark room or theatre. You are unable to see until your pupils have become larger. One way to overcome this problem is to cover one eye for a while before you enter the theatre.

Two important liquids, the **aqueous humour** and **vitreous humour**, help the eyeball maintain its shape. The aqueous humour has the added function of supplying cells to repair damage to the cornea or lens.

At the rear of the eye is the light-sensitive **retina**. It consists of two types of nerve endings. There are an estimated 120 million **rods** in each eye that are sensitive to black and white. There are also more than 6 million **cones** that are sensitive to colours. (Colour vision is discussed in the next chapter.)

When you look straight at an object, the clearest part of the image is located in the region marked "sensitive area" in Figure 22-5. The image is inverted and real. The information received by the rods and cones is sent through the **optic nerve** to the brain. Then the brain interprets what you see.

1

2

Hold the book at arm's length. Cover your left eye and stare at the number 1 with your right eye. Move the book toward you until the 2 disappears. This shows the location of your blind spot. If you move the book closer to or farther from you, the 2 will reappear.

Figure 22-6 Determining your blind spot

At the point where the retinal nerves join the optic nerve, a **blind spot** occurs. The blind spot is noticed only when one eye is closed. Figure 22-6 illustrates how you can find your own blind spot.

PRACTICE

3. Name the part of the eye that performs each function listed below:
   (a) controls the amount of light entering the pupil
   (b) keeps the shape of the inside of the eye
   (c) causes most of the refraction of light entering the eye
   (d) reacts to black and white light
   (e) carries signals from the eye to the brain
   (f) allows the focussing of nearby and faraway objects

22.5 Vision Defects and Their Corrections

Normal, healthy eyes allow the formation of a clear image on the retina of the eye. A person with normal eyes is said to have 6/6 vision. This means that the eye sees clearly at a distance of 6.0 m. Someone with 6/12 vision must be 6.0 m from an object in order to see it as clearly as someone with normal eyes at a distance of 12.0 m.

If the image in the eye comes to a focus in front of the retina, the defect is called **near-sightedness**. Objects far away appear out of focus or blurry. A concave lens corrects this fault, as shown in Figure 22-7.

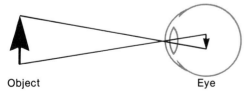

Object   Eye

(a) The image comes to a focus in front of the retina

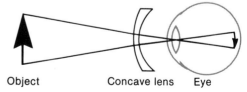

Object   Concave lens   Eye

(b) A concave lens corrects the defect

Figure 22-7 Near-sightedness

416

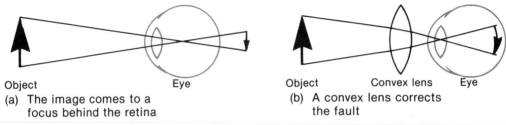

Object   Eye
(a) The image comes to a
     focus behind the retina

Object   Convex lens   Eye
(b) A convex lens corrects
     the fault

Figure 22–8 Far-sightedness

Far-sightedness is caused when the image comes to an imaginary focus behind the retina. Objects close to the eye are blurred. This defect can be corrected using a convex lens, as illustrated in Figure 22–8.

Another defect, called **astigmatism**, results when the cornea has an uneven surface. Equal-sized lines in different directions appear to have different thicknesses. Figure 22–9 gives a simple test for astigmatism. The defect can be corrected using a lens that is shaped like a cylinder. It can also be corrected by wearing hard contact lenses. A layer of water between the contact lens and the cornea automatically corrects the astigmatism.

Hold the book at arm's length. View the diagram with one eye at a time. If all the lines appear equally bright, you do not have astigmatism.

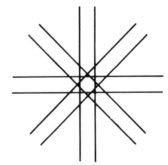

Figure 22–9 A test for astigmatism

Colour blindness is a defect in which certain shades of colour are not clear. Some cones in the retina do not respond to the light energy received. About 8% of all males and 0.5% of all females have some form of colour blindness.

PRACTICE
4. Which eye has better vision, one with 6/8 vision or one with 6/16 vision?
5. State the type of lens that can be used to correct for:
   (a) near-sightedness   (b) far-sightedness   (c) astigmatism

22.6 Using More than One Lens in an Optical Instrument

Lenses have four general functions as described in Section 22.1. We have discussed how lenses record images and improve weak vision. In the last parts of this chapter we will discuss how they create large images and help us view distant objects.

A convex lens can be used as a magnifying glass. The image may appear two or three times larger than the object. But this is not sufficient if tiny objects, such as skin cells, are to be viewed. To obtain larger images, two or more lenses can be combined to make an instrument called a **microscope**. (See Figure 22–10.)

Similarly, one lens by itself does not help the normal eye to view distant objects. Two or more lenses must be used together to enlarge the image of faraway objects such as stars and the moon. One optical instrument that makes distant objects appear larger and closer is called a **telescope**. See Figure 22–11. (A telescope that uses lenses is called a **refracting telescope**. A reflecting telescope, which uses a concave mirror, was described in Section 20.14.)

To make a basic microscope or telescope, two convex lenses can be used. The lens that is closest to the object is called the **objective** lens. The lens that the eye looks into is called the **eyepiece**.

To see how an image is created in a two-lens instrument, refer to the three ray diagrams in Figure 22–12.

Diagram (a) shows an object located beyond the focal point of the objective lens. A real image is formed. The real image then becomes the "object" for the eyepiece in diagram (b). This object is located between the eyepiece and the secondary focal point of the eyepiece. The usual rules for drawing ray diagrams are used, even though the object is below the principal axis.

Diagrams (a) and (b) are combined to obtain (c). The result is a complete ray diagram of a two-lens instrument. Notice that the final image is larger than the original object.

Many microscopes and refracting telescopes have three or more lenses. Extra lenses are added to get an even larger image or to create an upright image.

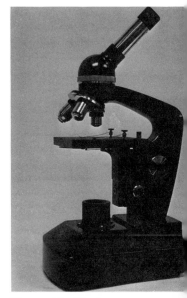

Figure 22–10 A student microscope

Figure 22–11 A simple refracting telescope

PRACTICE

6. In Figure 22–12(c), state the attitude, size and type of the final image compared to the original object.

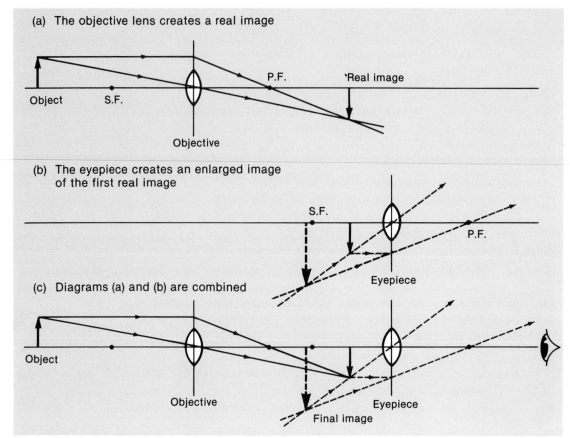

(a) The objective lens creates a real image

Object S.F. P.F. 'Real image

Objective

(b) The eyepiece creates an enlarged image of the first real image

S.F. P.F.

Eyepiece

(c) Diagrams (a) and (b) are combined

Object

Objective Eyepiece

Final image

Figure 22–12 A two-lens optical instrument

22.7 Experiment 53: The Microscope and Refracting Telescope

INTRODUCTION

In any two-lens optical instrument, the objective lens must create a real image. Then the eyepiece must enlarge that real image. In order for this to happen, the eyepiece should be fairly close to the first real image.

Convex lenses of different focal lengths are needed in this experiment. A simple method to find the focal length of a glass convex lens is illustrated in Figure 22–13.

PURPOSE: To make and compare microscopes and telescopes that have two lenses.

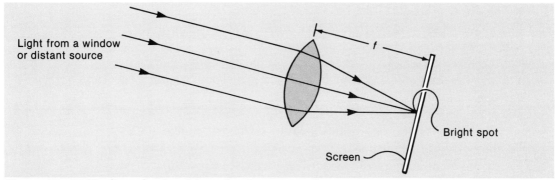

Figure 22–13 Finding the focal length of a convex lens

APPARATUS: several glass convex lenses (with focal lengths ranging from 5 cm to 20 cm); optical bench apparatus

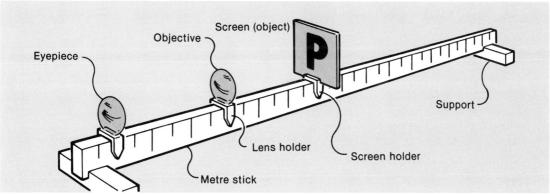

Figure 22–14 An optical-bench setup

PROCEDURE A: THE MICROSCOPE

1. Obtain two lenses and determine their focal lengths.
2. Place the lenses in the holders on the optical bench so they are separated by the sum of their focal lengths. (See Figure 22–14.) Place the screen holder almost twice the focal length of the objective away from the objective. Insert a piece of paper with small printing into the screen holder.
3. Look through the eyepiece and move the lenses and your eye back and forth until you find the clearest and largest image. The image should be inverted and larger than the object if the system is acting like a microscope.
4. Repeat #1 to #3 using various combinations of lenses; for example, objective $f = 5.0$ cm and eyepiece $f = 5.0$ cm; objective $f = 5.0$ cm and eyepiece $f = 10$ cm. Try to discover which combination gives the largest and clearest image.

PROCEDURE B: THE REFRACTING TELESCOPE
1. Obtain two lenses and determine their focal lengths.
2. Place the eyepiece at one end of the optical bench. Place the objective a distance equal to the sum of the focal lengths away from the eyepiece. (The screen holder is not needed for the telescope.)
3. Look through the eyepiece at a distant object. Move the lenses and your eye until you obtain the clearest and largest image of the object. Describe that image.
4. Try various combinations of lenses to discover which combination will give the largest image of the same object. Describe your observations.

22.8 Review Assignment

1. Name two optical instruments that are used to:
   (a) record images
   (b) improve weak vision
   (c) create large images
   (d) help view distant objects   (22.1)
2. What happens to the size of the image in a pinhole camera when the object is moved farther from the camera?   (22.2)
3. In what ways does a lens camera differ from a pinhole camera? (22.2, 22.3)
4. State the type of image (real or imaginary) in each of the following:
   (a) pinhole camera
   (b) lens camera
   (c) television camera
   (d) human eye   (22.2 to 22.4)
5. State the main function of these parts of the human eye:
   (a) cornea
   (b) lens
   (c) iris
   (d) retina   (22.4)
6. Set up an experiment to determine the angle from your line of vision (∡ *a* in the diagram) at which your blind spot occurs. Compare your value with that of other students.   (22.4)
7. (a) What is near-sightedness and how can it be corrected?
   (b) What is far-sightedness and how can it be corrected?
       (22.5)

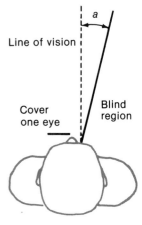

Line of vision

a

Blind region

Cover one eye

8. Draw a ray diagram to locate the final image in a two-lens instrument using the diagram below as a guide. Label your diagram.   (22.6)

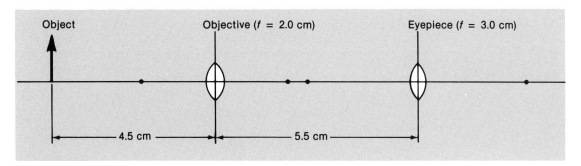

9. Describe how a microscope and refracting telescope are:
   (a) similar
   (b) different   (22.6, 22.7)

22.9 Answers to Selected Problems

PRACTICE QUESTIONS
1. real
2. inverted and real
3. (a) iris
   (b) vitreous humour
   (c) cornea
   (d) rods
   (e) optic nerve
   (f) lens

4. 6/8
5. (a) concave lens
   (b) convex lens
   (c) cylindrical lens
      or hard contact lens
6. inverted, larger than the
   object and imaginary

REVIEW ASSIGNMENT
4. They are all real.
6. For most people the angle is between 16° and 22°.

23

Colour

Knowing the information in this chapter will be especially useful if you plan a career in:

- optical dispensing
- fashion design
- interior decorating
- photography
- audio-visual technology
- live theatre
- TV (lighting crews)

23.1 The Beauty of Colour

One thing that humans have in common with bees, apes and goldfish is that they are able to see colours. Most animals are not so lucky. Their eyes are sensitive to black and white only. Their world would seem very dull to us.

It would be hard to imagine a world without colour. Most of us take for granted the beauty of sunsets, flower blossoms, rainbows and colourful underwater life.

Scientists admit they do not fully understand how we see colours. They do, however, agree on the important fact that colours could not exist without light.

In this chapter you will study the properties of light that cause

different colours. The experience should help you answer questions such as:
—What is white light?
—Why does a red rose appear red in white light?
—Why does a red rose appear black in blue light?
—How are the colours created on a colour television screen?
—What causes rainbows?

23.2 Experiment 54: Dispersion of White Light

INTRODUCTION

More than 300 a ago, in 1666, Sir Isaac Newton discovered that white light can be split up into many colours. This splitting up of white light into its colours is called **dispersion**. The resulting band of colours is called the **visible spectrum**.

The speed of light in a vacuum is 3.0×10^8 m/s. (See Section 20.4.) When light enters a transparent material, such as glass, it refracts because its speed decreases. The amount of decrease in speed depends on the colour of the light. Violet is the colour that slows down the most and red slows down the least. These colours refract different amounts, as you will learn in the experiment.

PURPOSE: To observe and describe the formation of a visible spectrum.

APPARATUS: ray box with a single-slit window; equilateral triangular prism (acrylic or glass)

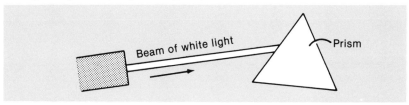

Figure 23–1

PROCEDURE

1. Place the prism on white paper and aim a beam of light toward it, as shown in Figure 23–1. Adjust the position of the light beam to obtain the best spectrum.
2. Draw a diagram showing how the white light is dispersed. Label the positions of the six main colours (red, orange, yellow, green, blue and violet).

QUESTIONS

1. (a) Which colour slows down the most when entering a prism?

   (b) Which colour refracts the most when entering a prism?

2. Sunlight consists of more radiant energies than our eyes can see. It also has invisible radiations beyond the visible spectrum. Two such radiations are called **infrared** and **ultraviolet**. Include these invisible radiations where you think they belong in your diagram of observations.

23.3 Experiment 55: Adding Light Colours and Viewing Colour Shadows

INTRODUCTION

In the previous experiment white light was split up into its spectral colours using a single prism. It is possible to use a second prism or a convex lens to bring the colours back together to obtain white light. However, not all colours of the spectrum are needed to produce white light. Only three, called the **primary light colours**, are required.

Two colours should be defined for this experiment. **Cyan** (pronounced si'an) is a greenish-blue colour. **Magenta** (with a soft "g") is a pinkish-purple colour.

When observing colours in this experiment you may notice that what should be white appears to be yellowish. This occurs because better results can only be obtained with high-quality equipment not always available to schools.

PURPOSE

(a) To study the addition of light colours.

(b) To observe and describe colour shadows.

APPARATUS: ray box with 2 mirrors at one end (or 3 ray boxes); set of 6 plastic or glass filters (red, green, blue, yellow, cyan and magenta)

PROCEDURE

1. Place a green filter and a red filter in the ray box as shown in Figure 23–2(a). Adjust the mirror so that the two colours of light overlap on a nearby white screen. Record the colour observed.

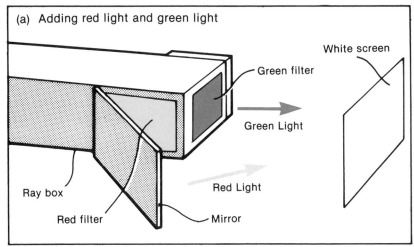

(a) Adding red light and green light

White screen

Green filter

Green Light

Ray box

Red Light

Red filter

Mirror

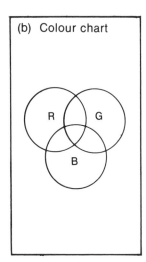

(b) Colour chart

R G

B

Figure 23–2

2. Repeat #1 for the following colour combinations:
   (a) green and blue
   (b) blue and red
   (c) red, green and blue
   (d) blue and yellow
   (e) red and cyan
   (f) green and magenta
3. Draw a colour chart like the one shown in Figure 23–2(b). Use the first letter of each colour (R, G, B, Y, C, M and W) to complete it. Have your teacher check the chart.
4. With the red, green and blue filters placed in separate positions at the end of the ray box, aim the colours so they overlap on the screen. Place your finger or fingers between the ray box and the screen to obtain various colour penumbras and a black umbra. Explain how the shadows occur.

QUESTIONS
1. List the three primary light colours.
2. List the three **secondary light colours**, which are colours that result from the mixing of pairs of primary colours.

23.4 Additive Colour Mixing

Additive colour mixing is an expression that refers to the process of adding light to light. It was observed in the previous experiment. When separate colours, such as red, green and blue, are added together, the light becomes lighter and brighter.

It is important to distinguish between mixing light colours (an adding process) and mixing pigment colours (a subtracting process). The next experiment deals with the subtraction of light energies.

PRACTICE
1. The lighting crew in a live theatre uses lights with red, green and blue filters. Assume a director wants a white object to appear yellow. What filters should the lighting crew use?

23.5 Experiment 56: Subtracting Light Colours

INTRODUCTION
In the previous experiment colours of light were added together to yield lighter colours. In this experiment colour filters will be used to absorb or subtract certain colours. For instance, a blue filter is a piece of plastic or glass that transmits blue light but absorbs red and green light.

If your colour chart in Figure 23-2(b) is correct, it can be used to predict which light energies are reflected by coloured opaque objects. For example, a blue object is blue because it reflects blue light. A yellow object is yellow because it reflects both red and green light, and so on.

Figure 23-3 illustrates how to predict the colour transmitted through a filter when light reflects from an opaque object. The prediction does not always come true because the colours of opaque objects and filters are not pure.

Figure 23-3 Predicting the colour transmitted through a filter

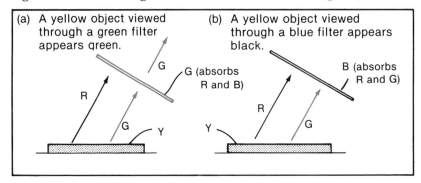

(a) A yellow object viewed through a green filter appears green.

(b) A yellow object viewed through a blue filter appears black.

PURPOSE: To learn how filters subtract light colours.

APPARATUS: set of colour filters (or coloured lights); set of coloured opaque objects (Both sets should have these colours: red, green, blue, yellow, cyan and magenta.)

PROCEDURE
1. In your notebook set up a chart like the one shown.

Colour of opaque object in white light

		Red	Green	Blue	Yellow	Cyan	Magenta
Colour of filter or light	Red						
	Green						
	Blue						
	Yellow						
	Cyan						
	Magenta						

Colour seen by observer

2. View a red object through each of the colour filters, one at a time. Record the colours seen in your chart.
3. Repeat #2 for an object that is green, then one that is blue and so on. Complete your chart.

QUESTIONS
1. State the primary light colours reflected by an opaque object that is:
   (a) red   (b) yellow   (c) cyan
2. Explain why a red object viewed in green light or through a green filter should appear black.
3. Explain why a magenta object viewed in red light or through a red filter should appear red.

23.6 Subtractive Colour Mixing

You have likely had the experience of mixing two paints or colour pigments together to get a third colour. For example, if you mix yellow and cyan pigments, you get green pigment. The process of

These diagrams illustrate the colours that can be transmitted by pure filters.

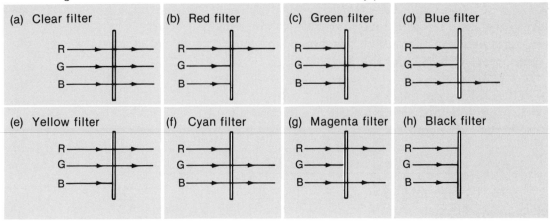

Figure 23-4 White light striking filters

mixing colour pigments together to obtain new colours is called **subtractive colour mixing**.

To understand why colours are subtracted when pigments are mixed together, we will summarize the facts based on the previous experiment.

Filters and opaque objects are similar in that they absorb (subtract) certain colours. Figures 23-4 and 23-5 show what happens to the primary light colors that strike filters and opaque objects.

Figure 23-5 White light striking opaque objects

These diagrams illustrate the colours that can be reflected by opaque objects.

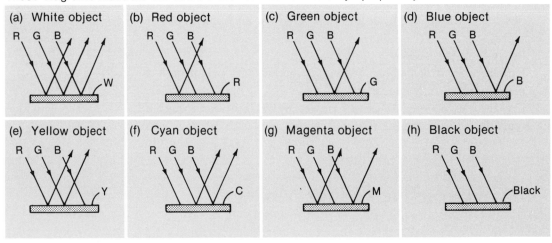

Sample problem 1: What colour is transmitted when cyan light (B, G) strikes a filter that is:
(a) blue   (b) red   (c) yellow?
Solution:
(a) blue (A blue filter transmits blue. Cyan is made up of blue and green.)
(b) no colour, or black (A red filter absorbs both green and blue.)
(c) green (A yellow filter absorbs blue.)

Sample problem 2: What colours are reflected when magenta light strikes a yellow object?
Solution: Magenta light is made up of red and blue. A yellow object absorbs blue, so only red can reflect. The object appears red.

Colour pigments absorb light energies just as filters and opaque objects do. Colour pigments, however, can be mixed together to absorb more light energies. Figure 23–6 gives two examples of mixing colour pigments.

Three colour pigments can be mixed to obtain black pigment. Those three pigments, yellow, cyan and magenta, are called the **primary pigment colours**. When any two primary pigments are

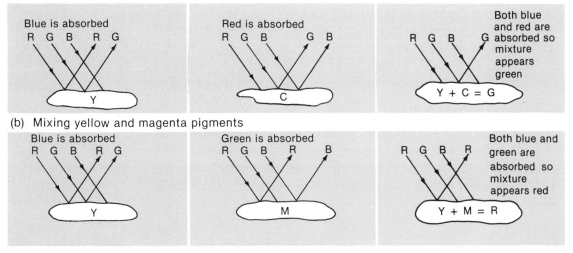

(a) Mixing yellow and cyan pigments

Blue is absorbed
R G B   R G
Y

Red is absorbed
R G B   G B
C

Both blue and red are absorbed so mixture appears green
R G B   G
Y + C = G

(b) Mixing yellow and magenta pigments

Blue is absorbed
R G B   R G
Y

Green is absorbed
R G B   R   B
M

Both blue and green are absorbed so mixture appears red
R G B   R
Y + M = R

Figure 23–6 Mixing colour pigments (in white light)

mixed, a **secondary pigment colour** results. The secondary pigment colours are red, green and blue.

Try not to confuse light colours with pigment colours. The primary light colours (R, G and B) are the secondary pigment colours. The secondary light colours (Y, C and M) are the primary pigment colours.

PRACTICE
2. State the colour(s) transmitted in each case:
   (a) white light (R, G, B) strikes a red filter
   (b) white light (R, G, B) strikes a cyan filter
   (c) red light strikes a green filter
   (d) yellow light (R, G) strikes a red filter
3. State the colour(s) reflected off each opaque object:
   (a) white light (R, G, B) strikes a magenta object
   (b) blue light strikes a green object
   (c) blue light strikes a magenta object
   (d) cyan light strikes a yellow object
4. What colour results when the following pigments are mixed?
   (a) Y + M   (b) Y + C   (c) C + M   (d) Y + C + M

23.7 Applications of Colour

(1) The Rainbow
To see a rainbow, the observer must be in a position between the raindrops and the sun, as shown in Figure 23–7(a). The sun's rays travel to the water droplets where some of them internally reflect and travel to the observer.

Figure 23–7(b) shows a beam of light that enters a single droplet of water. The light refracts and splits into its spectral colours. The various colours reflect off the inside surface of the droplet (total internal reflection). When the light leaves the droplet, it remains split into the same colours seen in the dispersion experiment.

In diagram (b) violet light is entering the observer's eye. The observer would have to look slightly higher in the sky to see red light from other droplets. The resulting arrangement of colours is illustrated in diagram (c).

(2) Colour Vision
The human eye is well adapted to see colours of light. The retina at the back of the eye has both rods and cones. (This was discussed in

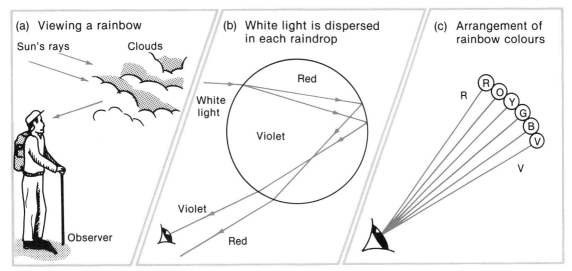

Figure 23–7 How a rainbow is produced

Section 22.4.) The rods react to black and white. The cones, which are important for this discussion, react to colours.

It is generally believed that there are three types of cones. Each type of cone reacts to one primary light colour (red, green or blue). The cones can react in various combinations to give all the shades of colour we see. For instance, if the cones sensitive to green and blue are stimulated equally, we see cyan.

An interesting way to demonstrate the theory of three types of cones is to stare at a coloured object until your eyes become tired or fatigued. To observe this "retinal fatigue", place a small, bright red object on a white background. Stare at the centre of the object for about 40 s. Then stare intently at one spot on a piece of white paper. Try to explain what you observe.

When you stare at a red object, the red cones in your retina become fatigued. Then when you stare at a white surface, which reflects red, green and blue light, only your green and blue cones react. Thus, you see the colour cyan.

Colour vision is very complex. Scientists are researching to try to prove (or disprove) the theory of three types of cones.

(3) Colour Television

If you look at a colour television screen through a magnifying glass, you will notice thousands of tiny dots or bars. (**Caution:** If you try this, do not stare at the screen for more than a minute. Harmful radiation may be emitted when the television is turned on.)

The dots or bars on the screen are arranged in groups of three. You can likely guess the colours of the dots. They are the primary light colours, red, green and blue.

Three electron guns, one for each colour of dot, are located at the back of the television. At selected instants they send electrons toward the dots on the screen. Each time a dot is struck by electrons it gives off light energy. To produce blue light, only blue ones are struck. To produce yellow light, both green and red dots are struck. Black results when none are struck. In fact, a great number of colours can be created by properly aiming the electrons at the coloured dots.

PRACTICE

5. Which cones in the human eye must be activated in order to see the colour magenta?
6. If you stare at a bright yellow circle and then stare at a white piece of paper, you see a blue circle. Explain why this happens.
7. In a colour television, which dots must be struck by electrons in order to create:
   (a) cyan   (b) white   (c) black?

23.8 Light Theory and the Electromagnetic Spectrum

Light is a form of energy. In some ways it acts the way particles act. In other ways it acts more like waves. Scientists have a special name for a package of energy that acts like a particle and a wave: they all it a **photon**.

In the study of waves (Chapter 12) you learned that waves have a wavelength and a frequency. Light waves or photons also have wavelengths and frequencies. Each colour in the visible spectrum has a different wavelength and frequency, as shown in Figure 23–8. Notice that as the frequency increases, the energy also increases. (Notice also that as the frequency increases the wavelength decreases. This can be verified by considering the universal wave equation, $v = f\lambda$, in which v is the speed of light, 3.0×10^8 m/s.)

Photons are created by the changing motion of charged particles. (This was discussed in Section 19.8.) Both electric and magnetic forces are involved in changing the motion of these particles. Thus, the resulting photons are called **electromagnetic waves**.

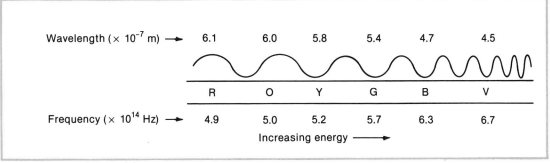

Figure 23-8 The visible spectrum

Visible light is only a small portion of the entire set of electro-magnetic waves. The entire set, called the **electromagnetic spec-trum**, is illustrated in Figure 23-9. Once again notice that the high-frequency waves have high energies. Also notice that infrared light has lower frequencies than visible light. ("Infra" means below.) It is ultraviolet light that has higher frequencies than visible light. ("Ultra" means above.)

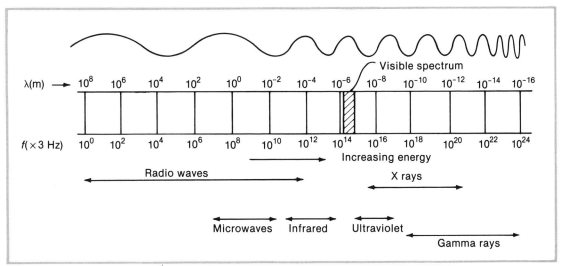

Figure 23-9 The electromagnetic spectrum

PRACTICE

8. Direct sunlight contains large amounts of ultraviolet light, which can cause sunburn. Why is ultraviolet light more likely to cause sunburn than visible light? (Hint: Compare the energies of the photons.)

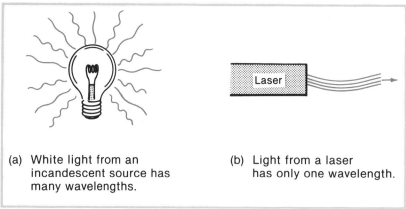(crop at top)

23.9 The Laser: An Application of the Entire Topic of Light

A laser is a source of light energy that is different from other sources. Most light sources, such as incandescent lamps, give off many wavelengths of light, as illustrated in Figure 23–10(a). A laser, however, usually gives off light of only one wavelength. That is why laser light has a pure colour. Not only are the waves the same wavelength, but they are also in step with each other, as shown in Figure 23–10(b).

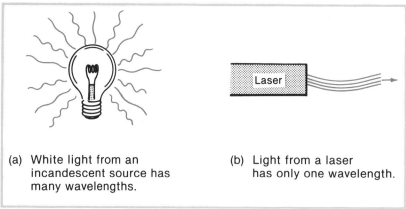

(a) White light from an incandescent source has many wavelengths.

(b) Light from a laser has only one wavelength.

Figure 23–10 Comparing white light with laser light

Since the invention of the first laser in 1960, there have been many types of lasers developed. They all work on the idea that energy can be given to tiny particles, which then give off light energy of one wavelength.

One example of this process is illustrated in Figure 23–11. It shows a basic laser with a long thin rod made of ruby. The ruby is surrounded by a flash lamp. Electrical energy is given to the flash lamp. Then the lamp gives energy to the ruby crystals. The crystals give off their extra energy in the form of light of one wavelength. Some waves escape through the sides of the ruby, but those that do not escape bounce back and forth between the mirrored ends of the laser. The mirror at the front end of the laser reflects about 90% of the light striking it. The remaining 10% is emitted as a laser beam.

The advantages of laser beams are that they:
(1) do not spread as they travel
(2) are easily controlled
(3) can have very high energies

Many applications have been developed to make use of these advantages.

Laser beams that are continuous are used in constructing tunnels, highways and skyscrapers. They are also used to transmit telephone and other communication messages.

Laser beams that pulse for short periods of time are used to weld tiny pieces of metal together and to drill precise holes in metal. They are also used to mend damaged blood vessels in the retina of the human eye.

Lasers are used for advanced scientific research and for entertainment purposes. A laserium show, held in planetariums

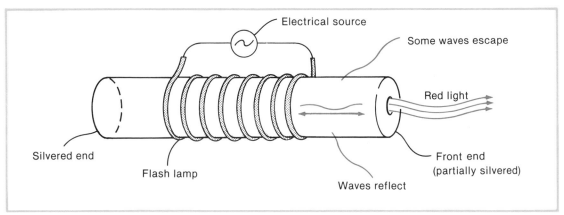

Figure 23–11 A basic ruby laser

throughout North America, uses a laser as well as prisms and lenses to create patterns of light that "dance" to music. One such pattern, produced by a laser having four distinct wavelengths, is shown on the cover of this book.

23.10 Review Assignment

1. List the colours of the visible spectrum.　(23.2)
2. Explain why blue light refracts more than yellow light when the two colours enter a prism from the same direction. (23.2)
3. What light colour results when the following pairs of colours are added together?
   (a) red and green　(b) blue and yellow
   (c) magenta and green　(23.3, 23.4)

4. Bright white fabrics gradually become yellowish. The whiteness can be restored by using a bleach that adds a certain colour to the yellowed fabrics.
   (a) What primary light colours are reflected by yellow?
   (b) What other primary light colour should be added to create white?
   (c) Why do bleaches have "bluing agents"?   (23.2 to 23.5)
5. Why does a red rose appear red in white light?   (23.5)
6. Why does a red rose appear black in blue light?   (23.5)
7. If black objects absorb all light energies, how can we see them?
8. Fluorescent lighting in stores has a lot of blue light and very little red. Is this an advantage or disadvantage when customers are choosing an appropriate shade of red lipstick?   (23.6)
9. A man who is designing his own restaurant wants it to have a colourful interior. He decides to use blue lighting in a room having no other source of light.
   (a) State what colour a customer would see when looking at:
      (i) a red rose
      (ii) white plates
      (iii) green lettuce
      (iv) a rare steak (reddish centre)
      (v) yellow beans
   (b) What would you suggest that the man should do to improve the situation?   (23.6)
10. What two paints can be mixed together to obtain a paint that is:
    (a) red   (b) green   (c) black?   (23.6)
11. How do the colours of the rainbow compare to the colours produced when white light passes through a triangular prism? (23.2, 23.7)
12. When seen from the ground, a rainbow has the shape of an arc. What do you think its shape would be when viewed from an airplane high above the raindrops?
13. Describe the similarities between colour vision and colour television.   (23.7)
14. **Class demonstration:** Set up a pendulum about 1 m in length. Stand about 5 m or more from the pendulum.
    (a) Hold a red filter over your left eye. Keeping **both** eyes open, watch the pendulum as it swings perpendicular to your line of sight. Describe the motion observed.
    (b) Repeat (a) with the filter over your right eye.
    (c) Repeat the experiment using filters of different colours.

15. Describe the ways in which visible light differs from the rest of the electromagnetic spectrum.   (23.8)
16. A certain laser produces electromagnetic waves having a wavelength of 2.0×10^{-10} m. Are the waves visible?   (23.8, 23.9)
17. In what ways does laser light differ from other light? (23.9)

23.11 Answers to Selected Problems

PRACTICE QUESTIONS
1. red and green
2. (a) red
   (b) cyan, blue and green
   (c) no colour (black)
   (d) red
3. (a) red + blue (magenta)
   (b) no colour (black)
   (c) blue
   (d) green

4. (a) red
   (b) green
   (c) blue
   (d) black
5. red and blue
7. (a) blue and green
   (b) red, green and blue
   (c) none

REVIEW ASSIGNMENT
3. (a) yellow
   (b) white
   (c) white
4. (a) red and green
   (b) blue
10. (a) yellow and magenta
    (b) yellow and cyan
    (c) There are three possible answers. One is "cyan and red".
16. no

VIII. Atomic Physics

24

Atoms and Radioactivity

GOALS: After completing this chapter you should be able to:
1. Compare the time periods called classical physics and modern physics.
2. State the uses and dangers of X rays.
3. Define radioactivity and state the three types of emissions from a radioactive substance.
4. Describe the shapes of tracks in a cloud chamber left by the three types of radioactive emissions.
5. Define background radiation.
6. Define and use the unit of radiation activity.
7. Name materials that absorb radioactive emissions.
8. Name and compare the three main types of particles that make up an atom.
9. Define element and write the symbols of common elements.
10. Draw a model of an atom, given its atomic mass and atomic number. (This is required only for the first eighteen elements.)
11. Define isotope.
12. Define the half-life of a radioactive substance and plot a half-life curve on a graph.
13. Given the activity of a radioactive sample, find the activity after one, two or three half-lives.
14. Describe uses of radioactivity.

Knowing the information in this chapter will be especially useful if you plan a career in:

• chemistry
• radiology
• nursing
• nuclear power generation
• medicine

24.1 From Classical to Modern Physics

Toward the end of the 1800s, scientists thought they knew almost everything about matter and energy. They believed that matter was made up of tiny atoms, which could not be divided into smaller parts. They thought they understood motion, forces and such energies as heat and light. The physics up to that time is called **classical physics**.

Then in 1895 an important discovery forced scientists to change their minds. That year marked the beginning of what we call **modern physics**. Listed below are some of the important events of modern physics.

   (1) 1895–discovery of X rays
   (2) 1896–discovery of radioactivity
   (3) 1897–discovery of electrons
   (4) 1911–discovery of the nucleus
   (5) 1919–observation of the first nuclear reaction
   (6) 1920–naming of the proton
   (7) 1932–discovery of the neutron
   (8) 1939–discovery of nuclear fission
   (9) 1942–first sustained reaction in a nuclear reactor
(10) 1948–invention of the transistor
(11) 1960–invention of the laser

Today's physicists know there are many tiny particles besides protons, neutrons and electrons. They no longer assume they know everything. On the contrary, they admit there is much to learn. They continually search for new particles and ideas.

PRACTICE
1. Review the following in class:
   (a) the structure of an atom (Section 15.2)
   (b) the parts of the electromagnetic spectrum that have high-energy waves (Section 23.8)

24.2 The Discovery and Use of X Rays

In 1895 a German physicist named Wilhelm Roentgen (1845-1923) was experimenting in his darkened laboratory. He had a glass tube with invisible particles travelling in it and black paper surrounding it. On a table near the tube were some crystals that he was not even paying attention to. Suddenly he noticed that the crystals were glowing in the dark. He decided to search for the cause of the glow.

After experimenting, Roentgen discovered that the particles travelling in the glass tube struck the end of the tube, giving up much energy. That energy was given off in the form of invisible rays that could travel through black paper as well as cause certain crystals to glow. The rays he had accidentally discovered had no name. He decided to call them **X rays**.

After more experimenting, Roentgen found that the X rays from his tube could pass through flesh and reveal bone structure on a film. He used his wife's hand to make the world's first X-ray photograph.

Later it was learned that X rays travel in straight lines at the speed of light. They belong to the high-energy end of the electromagnetic spectrum. (That spectrum, discussed in Section 23.8, contains radio waves, infrared light, visible light, ultraviolet light and gamma rays.)

X rays are very useful for medical purposes. They are used to take X-ray photographs of teeth, bones and internal organs. Figure 24–1 shows a typical X-ray photograph.

Figure 24–1 This X-ray photograph shows a distinct fracture of a child's leg. The ring shape near the top is a metal splint used to support the leg.

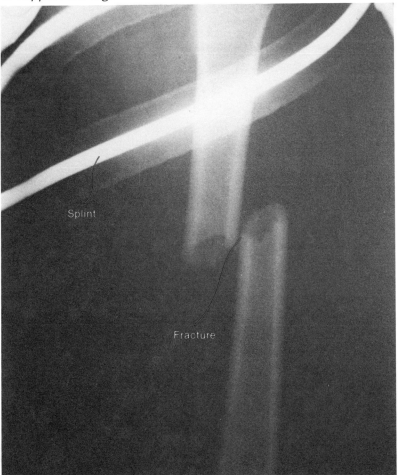

442

X-ray cameras are used to check the contents of baggage at airports. The rays can pass through leather, plastic and cloth, but they are absorbed by metal objects such as guns and knives.

X rays are not always useful. In fact, they can be very harmful, especially when too many of them strike certain cells in the body. X-ray photographs of your teeth or body should only be taken when necessary. If possible, a lead shield, which absorbs X rays, should be used to protect areas where unwanted rays may do damage.

PRACTICE

2. When a person needs several X-ray photographs of his or her teeth, a lead shield is placed around the neck just under the chin. Why is this wise?

24.3 The Discovery of Radioactivity

Just a few months after Wilhelm Roentgen discovered X rays, a French scientist named Antoine Becquerel made another accidental discovery. Becquerel, who lived from 1852 to 1908, was experimenting with minerals that he thought might give off X rays when struck by sunlight.

One cloudy day in 1896 he stored some mineral samples on an unexposed, covered film in a drawer. Four days later, just for interest, he developed the film, expecting to see nothing. To his amazement, the film had been exposed. Some invisible energy had passed through the opaque cover of the film and had exposed it. Becquerel had discovered **radioactivity**, which is the emission of particles and energy from the nucleus of an atom.

Antoine Becquerel (1852–1908)

Becquerel's experiment can be repeated in the laboratory. If a radioactive substance is placed on an unexposed film for three or four days, it will cause the film to be exposed. An example of this is shown in Figure 24–2.

After radioactivity was discovered, many experiments were performed to find out more about radiation. We now know that when a radioactive substance gives out emissions, it turns into a

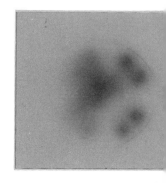

Figure 24–2 The photograph shows the faint outline of a radioactive watch, which was built in the 1950s. The watch was left on an unexposed film for three days before the film was developed.

different substance. We also know that there are three types of emissions given off by radioactive substances:

(1) **Alpha** (α) **particles** have a positive charge and are fairly slow moving.
(2) **Beta** (β) **particles** have a negative charge, are much lighter than alpha particles and are fast moving.
(3) **Gamma** (γ) **rays** are high-energy radiations, not particles; thus, they have no mass. They belong to the electromagnetic spectrum.

Methods of detecting and using radioactive emissions will be discussed later in the chapter. As you study radioactivity, remember that these emissions come from the nucleus of an atom.

PRACTICE

3. Which waves have higher energies, X rays or γ rays? (Clue: Refer to Figure 23–9 in Section 23.8.)

24.4 Experiment 57: Evidence of Radio-activity in a Cloud Chamber

INTRODUCTION

Since radioactive emissions (α, β and γ) cannot be seen, what evidence do we have that they exist? One piece of evidence is that they expose film. This is how Antoine Becquerel discovered radioactivity. Other evidence can be observed in a device called a **cloud chamber**.

A cloud chamber is a container that has a layer of alcohol resting on a cold surface. See Figure 24–3(a). The cold temperature, provided by solid carbon dioxide (dry ice), is needed so that the alcohol does not evaporate too quickly. As the alcohol evaporates, alcohol vapour (a "cloud") forms in the chamber. A radioactive source provided with the chamber is placed in the vapour.

Although you cannot see the radioactive emissions in a cloud chamber, you can see where they have travelled. The action is like the vapour trail of a jet aircraft high in a clear sky. You may not be able to see the jet, but you can see the trail it leaves behind.

In a cloud chamber the radioactive emissions cause the alcohol vapour to condense and form tracks of visible droplets. That allows you to see where the emissions have travelled. Figure

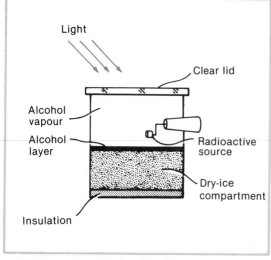

(a) Design of a simple chamber

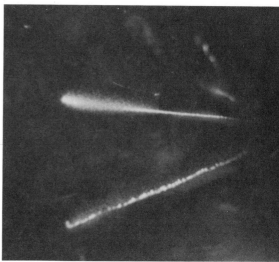

(b) Tracks seen in the chamber

Figure 24-3 The cloud chamber

24-3(b) shows a photograph of some tracks in a cloud chamber.

The vapour tracks have three different shapes, depending on what caused them:

(1) α particles are heavy, so their tracks are short, fat and straight.

(2) β particles are light and fast moving, so their tracks are longer, thinner and often curved.

(3) γ rays are high-energy waves that create tracks that are thin and often in a zigzag pattern.

PURPOSE: To view and describe tracks left by emissions from a radioactive source in a cloud chamber.

APPARATUS: cloud chamber; dry ice; alcohol; radioactive source(s); piece of cloth; masking tape

Caution:

(a) Do not handle the dry ice with your bare hands.

(b) Do not aim a radioactive source toward any person.

PROCEDURE

1. Place the dry ice in the lower compartment of the cloud chamber. Add a layer of alcohol to the chamber and check to be sure that the chamber is level. Place the lid on the chamber.

2. Put the radioactive source in the appropriate position. Allow the vapour to settle for three or four minutes.

3. Observe and describe the tracks. Draw diagrams of the shapes

and sizes of the tracks. (If very few tracks are seen, try rubbing the lid of the container with a cloth.)

4. If a different type of source is available, use it to repeat #1, #2 and #3. Compare the two sources.

5. Remove the radioactive source. Cover the hole where it was positioned with masking tape. Wait for a few minutes, and then determine if there are any tracks visible in the vapour. Such tracks are caused by **background radiations,** which come from the sun, the walls and even from yourself.

24.5 Experiment 58: Radiation Detection and Absorption

INTRODUCTION

In the cloud chamber it was possible to view the effects of radioactive emissions. A more mathematical way of detecting radioactivity is to use the electronic equipment available for this experiment. The detection apparatus consists of a Geiger-Müller tube (G.M. tube), which is connected to an electronic counter.

The unit used to measure radioactivity is the becquerel (Bq), named after Antoine Becquerel who discovered radioactivity.

$$1 \text{ Bq} = 1 \text{ emission/s}$$

Thus, an activity of 120 emissions/min is the same as 2 emissions/s or 2.0 Bq.

PURPOSE: To detect radioactive emissions and learn how well various materials can absorb them.

APPARATUS: G.M. tube; electronic scaler; watch; source of β particles (e.g., strontium 90); source of γ rays (e.g., cesium 157); holder for the sources; various thicknesses of paper, aluminum and lead

Caution: Only the teacher should handle the radioactive sources listed here.

PROCEDURE

1. Connect the G.M. tube to the scaler. Turn on the scaler and set the counter to zero. Record the background radiation for 1.0

min. Repeat this measurement five times to obtain an average value. Express your answer in becquerels (Bq).

2. Have your teacher place the β source in a holder about 30 cm from the G.M. tube. Find the count for 1.0 min and subtract the background radiation. Record the answer in becquerels.

3. Place a piece of paper about 5 cm from the source and repeat #2. Try varying the thickness of the paper.

4. Repeat #3 using pieces of aluminum and lead. Which substance appears to be the best absorber of β particles?

5. Repeat #2 to #4 using the γ ray source.

QUESTIONS

1. How would you design an experiment to measure the total activity of each source in this experiment?

2. A certain source has an activity of 5 Bq. How many emissions come from the source in one minute?

3. Calculate the activity (in Bq) of a source that:
   (a) has 500 emissions in 10 s
   (b) has 2500 emissions in 20 s
   (c) has 4200 emissions in 1.0 min

24.6 Developing a Model of the Atom

After the discovery of X rays and radioactivity, scientists searched for the answer to the question, "What is an atom made of?"

In 1897 an Englishman, Sir Joseph Thomson (1856-1940), discovered that atoms contain electrons with negative charges. Then he reasoned that because the atoms were neutral, they must also contain some positively charged material to neutralize the

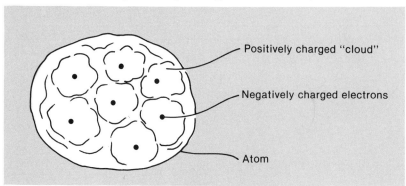

Positively charged "cloud"

Negatively charged electrons

Atom

Figure 24-4 J. J. Thomson's model of the atom

negative electrons. He developed a model or description of an atom as shown in Figure 24–4.

Then in 1911 in England, Ernest Rutherford (1871-1937) performed an important experiment to check Thomson's model of the atom. Rutherford aimed α particles at a thin sheet of gold foil. All around the foil was a screen that could detect the positively charged α particles. See Figure 24–5(a).

Rutherford discovered that almost all the α particles went straight through the gold foil. A few particles were deflected, as illustrated in Figure 24–5(b). He concluded that the positively charged part of each atom must be concentrated in a small nucleus of the atom. His model of the atom is shown in Figure 24–6.

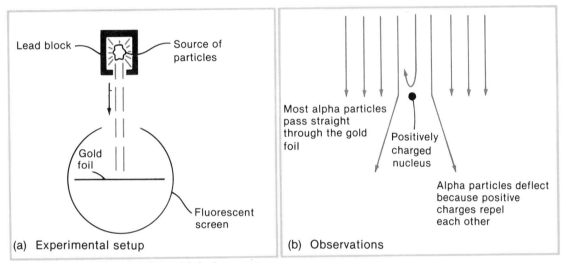

Figure 24–5 Rutherford's gold-foil experiment

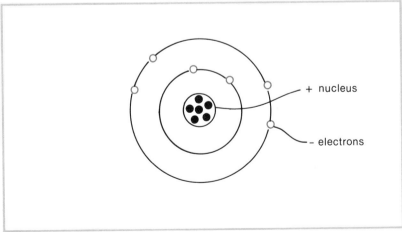

Figure 24–6 Rutherford's model of the atom

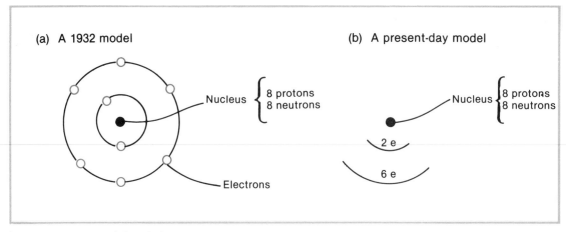

(a) A 1932 model

(b) A present-day model

Nucleus { 8 protons
8 neutrons

Electrons

Nucleus { 8 protons
8 neutrons

2 e

6 e

Figure 24–7 Models of the oxygen atom

By 1920 scientists knew that the atom contained electrons travelling around a nucleus. The nucleus contained positively charged **protons** as well as some other particles. Those particles were finally discovered in 1932 by James Chadwick (1891-1974), an English physicist. The particles had a neutral charge and were named **neutrons**. Now the model of the atom was more complete. Figure 24–7(a) shows a 1932 model of an oxygen atom.

Today's scientists believe that electrons travel in regions, not in set orbits as shown in Figure 24–7(a). Thus, they use a simplified model of the atom, as shown in diagram (b).

PRACTICE
4. Name the type of charge on:
    (a) an electron   (c) a neutron
    (b) a proton   (d) a nucleus

24.7 Atoms, Elements and Isotopes

You probably know that water is made up of hydrogen and oxygen (H_2O). Hydrogen and oxygen are examples of elements. An **element** is a substance that cannot ordinarily be divided into other substances.

All atoms of one element have the same number of protons. For example, every hydrogen atom has one proton and every oxygen atom has eight protons. The number of protons in each atom is called the element's **atomic number**.

Protons and neutrons have nearly equal masses. An element's **atomic mass** is the sum of the masses of the protons and neutrons in each atom. For instance, an atom of oxygen has 8 protons and 8 neutrons, so its atomic mass is 16.

Electrons, which have hardly any mass, travel in regions around the nucleus. Each region can hold a certain maximum number of electrons. For example, the region closest to the nucleus can hold up to two electrons and the next region up to eight. In a neutral atom the total number of electrons is the same as the total number of protons.

There are more than 100 known elements. The first 18 elements are listed in Table 24–1 in the order of their atomic numbers. Each box in Table 24–1 contains the following information:

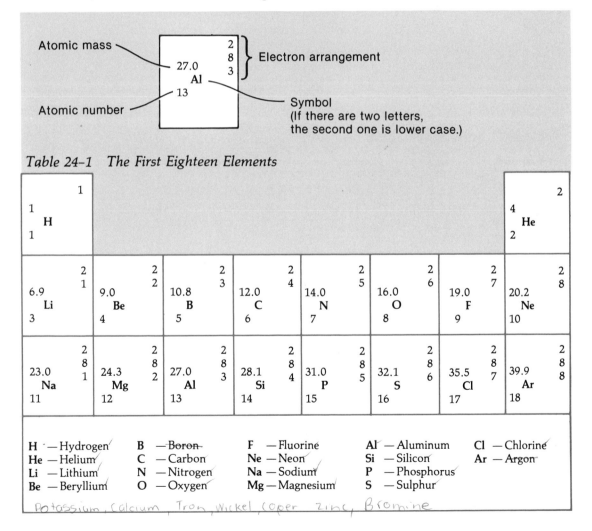

Atomic mass
27.0
Al
13
Atomic number

2
8
3 } Electron arrangement

Symbol
(If there are two letters,
the second one is lower case.)

Table 24–1   The First Eighteen Elements

1									2
1 **H** 1									4 **He** 2
2 1 6.9 **Li** 3	2 2 9.0 **Be** 4	2 3 10.8 **B** 5	2 4 12.0 **C** 6	2 5 14.0 **N** 7	2 6 16.0 **O** 8	2 7 19.0 **F** 9	2 8 20.2 **Ne** 10		
2 8 1 23.0 **Na** 11	2 8 2 24.3 **Mg** 12	2 8 3 27.0 **Al** 13	2 8 4 28.1 **Si** 14	2 8 5 31.0 **P** 15	2 8 6 32.1 **S** 16	2 8 7 35.5 **Cl** 17	2 8 8 39.9 **Ar** 18		

H — Hydrogen   **B** — Boron   **F** — Fluorine   **Al** — Aluminum   **Cl** — Chlorine
He — Helium   **C** — Carbon   **Ne** — Neon   **Si** — Silicon   **Ar** — Argon
Li — Lithium   **N** — Nitrogen   **Na** — Sodium   **P** — Phosphorus
Be — Beryllium   **O** — Oxygen   **Mg** — Magnesium   **S** — Sulphur

Potassium, Calcium, Iron, Nickel, Coper, Zinc, Bromine

Sample problem 1: For an atom of the element sodium (Na), state the:
(a) number of protons in the nucleus
(b) number of neutrons in the nucleus
(c) number of electrons in the third region
Solution:
(a) 11   (b) 23 – 11 = 12   (c) 1

Notice in Table 24–1 that some of the atomic masses are not whole numbers. Sulphur, for instance, has an atomic mass of 32.1. This results from the fact that many elements have isotopes. An **isotope** is a substance whose atoms have the same number of protons but a different number of neutrons than other atoms of the same element. Most sulphur atoms have 16 protons and 16 neutrons (atomic mass of 32). A few sulphur atoms have 16 protons and 19 neutrons (atomic mass of 35). The average atomic mass works out to 32.1.

Refer to Table 24–2 for a list of some of the isotopes of these elements.

Table 24–2 Some Common Isotopes

Element	Name	Symbol	Comment
Hydrogen	hydrogen	$_1^1 H$	most abundant form of H
	deuterium	$_1^2 H$	
	tritium	$_1^3 H$	radioactive
Carbon	carbon 12	$_6^{12} C$	most abundant form of C
	carbon 13	$_6^{13} C$	
	carbon 14	$_6^{14} C$	radioactive
Uranium	uranium 232	$_{92}^{232} U$	radioactive
	uranium 235	$_{92}^{235} U$	radioactive
	uranium 238	$_{92}^{238} U$	radioactive; most abundant form of U

PRACTICE
5. For an atom of the element fluorine (F), state the:
   (a) atomic number
   (b) number of neutrons in the nucleus
6. Find the number of neutrons in an atom of:
   (a) nitrogen   (b) hydrogen   (c) phosphorus

7. Draw a model similar to that shown in Figure 24–7(b) of an atom of:
   (a) beryllium
   (b) helium
   (c) phosphorus
8. Calculate the number of neutrons in each isotope of uranium listed in Table 24–2.

24.8 Experiment 59: The Concept of Half-Life

INTRODUCTION

As the nuclei (plural of nucleus) of a sample of a radioactive isotope emit particles, they become nuclei of a different element. The average length of time for half the original nuclei to become other nuclei is called the **half-life** of the radioactive isotope. After one half-life has elapsed, the rate of emissions is half of the original rate.

As an example, consider an isotope that has an average half-life of eight days. Assume you have a sample that emits at a rate of 1000 Bq. Eight days later the activity will be half the original, or 500 Bq. After another eight days, the activity will be 250 Bq, and

Figure 24–8 Graphing half-life curves

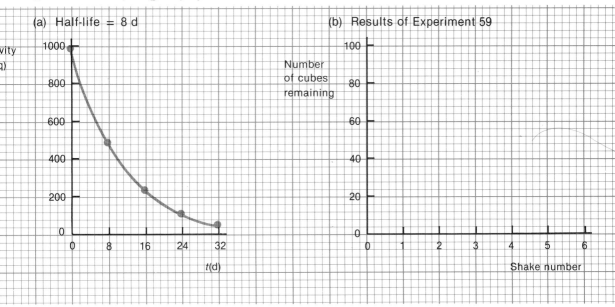

so on. When plotted on a graph, this half-life example yields a smooth curve, as shown in Figure 24–8(a).

Half-lives of radioactive substances range from a small fraction of a second to billions of years.

PURPOSE: To study the concept of half-life.

APPARATUS: large cardboard box; 100 sugar cubes, each with a coloured dot on three sides

PROCEDURE

1. Assume that the sugar cubes represent a radioactive substance and each coloured dot represents an emission. Place the cubes in the box and shake them thoroughly.
2. Remove and count all the cubes with dots facing upwards. Record the number in a chart like this one:

Shake number	Number of cubes removed	Number of cubes remaining
		100
1		
2		
3		

3. Again shake the cubes thoroughly. Remove and count the cubes with the dots facing upwards. Continue this procedure until there are only one or two cubes left. Complete your chart.
4. Plot a graph of the number of cubes remaining as a function of the shake number, as shown in Figure 24–8(b). Draw a curve of best fit.
5. Relate the graph to the concept of the half-life of a radioactive substance.

QUESTIONS

1. What is the half-life (in shakes) of the sugar cubes in this experiment?
2. How do you think your answer to #1 would change if each cube had only one coloured dot?
3. A certain sample of radioactive substance has an activity of 100 Bq at time 0.0 h. The half-life of the substance is 2.0 h. Calculate the activity when the time is:
   (a) 2.0 h   (b) 4.0 h   (c) 6.0 h

24.9 Uses of Radioactivity

(1) Radioactive dating

The process of using the half-life of a radioactive substance to find the age of some object is called **radioactive dating**. It is a valuable aid to archaeologists, geologists and historians.

Consider, as an example, the problem of determining the age of a tree that died centuries ago. During its lifetime, the tree absorbed both carbon 12 ($^{12}_{6}C$) and radioactive carbon 14 ($^{14}_{6}C$). After the tree died its carbon 12 content remained the same because it was no longer taking in carbon dioxide. However its carbon 14 content gradually decreased due to radioactive emissions. After about 5700 a, the carbon 14 emissions would be cut in half. (The half-life of carbon 14 is 5700 a.) Thus, scientists can determine the tree's age by calculating the amount of carbon 14 remaining in a sample of the tree.

Carbon-14 dating can be used on any object that was alive at some time. However, the process is inaccurate for ages greater than about 30 000 a.

Radioactive dating can also be used to determine the age of rocks and mineral deposits. Uranium 238 ($^{238}_{92}U$), with a half-life of 4.5×10^9 a, has been used to determine that some rocks on earth are about 4×10^9 a old!

(2) Tracers

Very small amounts of a radioactive substance can be injected into the liquid of a system. As the isotope travels in the liquid through the system, it continually gives off emissions. These emissions help a detector trace the path of the isotope and thus analyse the system.

Sodium 24 ($^{24}_{11}Na$), with a half-life of 15 h, is often used in biological systems, such as the human body or plants, to trace the flow of blood, food or water. Iron 59 ($^{59}_{26}Fe$) can be used as a tracer in mechanical systems, such as an engine, to trace the flow of lubricating oil.

(3) Radiation therapy

Cobalt 60 ($^{60}_{27}Co$), a radioactive source of γ rays, is used in radiation therapy or treatment. Gamma rays, like X rays, travel through some cells but are absorbed by others. In radiation therapy γ rays are aimed from several directions at a cancerous growth. If the radiation is successful, the cancer cells absorb the γ rays and are then destroyed by them. See Figure 24–9.

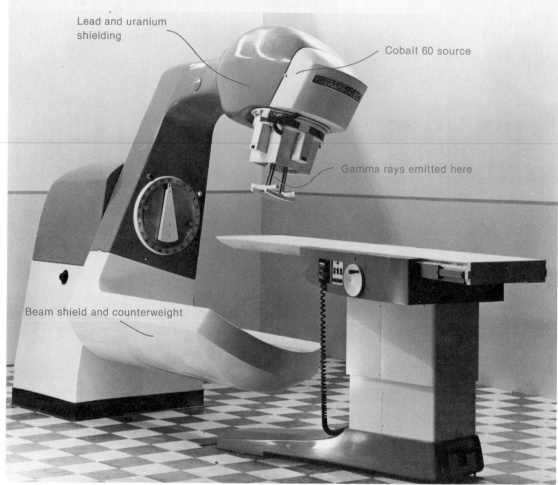

Lead and uranium shielding

Cobalt 60 source

Gamma rays emitted here

Beam shield and counterweight

Figure 24–9 The cobalt 60 source in this machine emits gamma rays that kill cancerous cells.

(4) Industrial applications

Radioactive sources can be used to control the thickness of manufactured products such as paper, steel and aluminum foil. If the product becomes too thick, it absorbs more radiation and adjustments can then be made.

Sources of γ rays can be used to detect flaws inside metal parts. Gamma rays can penetrate where visible light and even X rays cannot.

PRACTICE

9. Assume that a 2000 Bq source of sodium 24 (half-life of 15 h) is injected into a body. How long will it take the activity to decrease to 250 Bq?

24.10 Review Assignment

1. What discovery marked the end of classical physics and the beginning of modern physics?   (24.1)
2. The X rays used to view luggage at airports can damage high-speed film. What type of material may be used to protect a film from exposure?   (24.2)
3. In past centuries one of the substances in artists' paints was lead. Today, little or no lead is used in such paints. How could X rays be used to distinguish a genuine old masterpiece from a modern fake of that masterpiece?   (24.2)
4. From what part of the atom does radioactivity originate?   (24.3)
5. Name the three types of emissions from radioactive sources and the kind of charge on each emission.   (24.3)
6. List three methods of detecting radioactivity.   (24.3 to 24.5)
7. What is meant by the term background radiation?   (24.4)
8. Assume you are asked to make a container to store a β source. What materials and design would you use?   (24.5)
9. At a rate of 10 Bq, calculate how many emissions a source would send out in:
   (a) 1.0 min
   (b) 3.0 min   (24.5)
10. Which of the following substances are elements?
    H, H_2O, He, Na, NaCl, H_2SO_4, Al   (24.7)
11. Draw a model of an atom of:
    (a) fluorine
    (b) carbon 12
    (c) carbon 14   (24.7)
12. An experiment was performed to measure the half-life of iodine 131 ($^{131}_{53}I$). The following activities were recorded at noon on each observation day.

Observation day	0	4	8	12	16
Activity (Bq)	10 000	7100	5000	3500	2500

(a) Plot a graph of the activity.
(b) According to the graph, what is the half-life of iodine 131?
(c) What would be the activity on day 24?   (24.8)

24.11 Answers to Selected Problems

PRACTICE QUESTIONS

3. γ rays have higher energies than X rays.
4. (a) negative
   (b) positive
   (c) neutral
   (d) positive
5. (a) 9
   (b) 10
6. (a) 7
   (b) 0
   (c) 16
8. 140, 143, 146 respectively
9. 45 h

REVIEW ASSIGNMENT

9. (a) 600 emissions
   (b) 1800 emissions
10. H, He, Na, Al
12. (b) 8 days
    (c) 1250 Bq

25

Using Nuclear Energy

GOALS: After completing this chapter you should be able to:
1. Name and describe the force that binds the particles of the nucleus together.
2. Define nuclear fission.
3. Calculate the amount of energy produced when a given amount of mass changes into energy ($E = mc^2$).
4. Describe how a chain reaction is set up.
5. Explain the basic operation of a nuclear generating station.
6. State advantages and disadvantages of using nuclear energy to generate electricity.
7. Name sources of electrical energy other than nuclear energy.
8. Define nuclear fusion.
9. State the contributions of physics to our society in the past, present and future.

Knowing the information in this chapter will be especially useful if you plan a career in:
- chemistry
- radiology
- nursing

25.1 The Force That Binds the Nucleus

In Chapter 24 you learned that the nucleus of an atom contains neutrons and protons. Neutrons have a neutral electric charge and protons have a positive electric charge.

Since protons are positively charged, they repel each other (like charges repel). This repelling must be overcome by some powerful force in order to hold the protons in the nucleus.

The force that binds the particles of the nucleus together is called the **strong nuclear force**. This force acts between all protons and neutrons. It results only when the particles are extremely close together.

Although scientists do not fully understand the strong nuclear force, they do know that it is sometimes not strong enough to hold the nucleus together. That is when the nucleus splits up into smaller parts, giving off radioactivity, as discussed in Chapter 24.

Whenever the strong nuclear force is weakened, and the nucleus splits, energy is released. It is the creation and use of this energy that you will study in this chapter.

PRACTICE

1. When a nucleus is held together, which force is stronger, the repelling electric force or the attracting nuclear force? How can you tell?

25.2 Nuclear Fission

In the type of nuclear activity (called "radioactivity") discussed in Chapter 24, the nucleus of an atom emitted alpha or beta particles, as well as energy. Such activity can occur for both light and heavy elements. However, another type of activity, called nuclear fission, occurs only for heavy elements.

Nuclear fission, discovered in 1939, is the splitting up of a large nucleus into smaller nuclei that are nearly equal in size. Fission is accompanied by the release of neutrons and a large amount of energy.

In a fission reaction the mass of the original nucleus is greater than the sum of the masses of the resulting nuclei. In other words, some mass disappears. That lost mass changes into energy.

It was Albert Einstein (1879-1955), the famous German scientist, who first predicted that matter and energy could interchange in a nuclear reaction. His well-known equation states that the amount of energy created when some mass disappears equals the product of the mass and the square of the speed of light.

$$E = mc^2 \text{ where } c = 3.0 \times 10^8 \text{ m/s}$$

Albert Einstein (1879–1955)

Sample problem 1: If 1.0 kg of a substance changes entirely into energy, how much energy is created?
Solution: $E = mc^2$
$$= 1.0 \text{ kg} \times (3.0 \times 10^8 \frac{\text{m}}{\text{s}})^2$$
$$= 9.0 \times 10^{16} \text{ J}$$

Until 1905 scientists believed that matter could not be created or destroyed (law of conservation of mass) and energy could not be created or destroyed (law of conservation of energy). Then Einstein proposed the relation $E = mc^2$, and the conservation laws had to be joined into one law. Now, the **law of conservation of mass-energy** states that:

for an enclosed system, the total amount of mass-energy remains constant

An element commonly used for nuclear fission is uranium. Most uranium nuclei do not fission easily. However, one isotope, uranium 235, can be made to fission by bombarding it with a neutron. (An alpha particle is not used because it would be repelled by the nucleus.) The extra neutron is absorbed by the uranium 235, creating an unstable isotope, which almost immediately splits into two smaller nuclei, releasing neutrons and energy.

A typical uranium fission reaction is:

uranium 235 + neutron → uranium 236 → strontium 90 + xenon 143 + 3 neutrons + energy

Using symbols, the equation for this reaction is:

$$^{235}_{92}U + {}^{1}_{0}n \rightarrow {}^{236}_{92}U \rightarrow {}^{90}_{38}Sr + {}^{143}_{54}Xe + 3{}^{1}_{0}n + energy$$

In the fission reaction above the three neutrons are capable of joining with other uranium 235 nuclei. Every time such a nucleus absorbs one neutron, it splits up and emits three more neutrons. Soon thousands and then millions of nuclei will be splitting up. This process is called a **chain reaction**. It is illustrated in Figure 25–1. A tremendous amount of energy is released in a chain reaction.

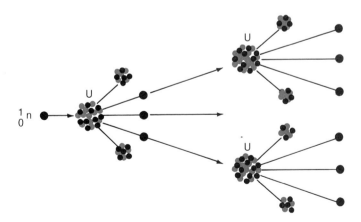

Figure 25–1 The beginning of a chain reaction

A fission bomb is designed on the principle of chain reactions. A bomb is made up of two separate masses of a material that can undergo fission. The masses are too small to start a chain reaction until they are joined together. When the barrier separating the masses is broken by a chemical explosion, a chain reaction begins, and the bomb explodes. (A more important and peaceful use of the energy released in chain reactions is the topic of the next section.)

PRACTICE
2. When a nucleus undergoes fission, what products result?
3. Calculate the amount of energy released if the following amounts of matter change entirely into energy:
   (a) 500 g (0.5 kg)   (b) 2.0 kg   (c) 10 kg

25.3 Using Nuclear Fission to Generate Electricity

One of the most important uses of nuclear fission is in nuclear reactors. In such devices a controlled fission process causes a substance such as uranium to split into new substances with the release of much energy.

There are at least eight different designs for nuclear reactors. We will study the details of the reactor designed in Canada. If you understand its operation, you will have little trouble understanding other types of reactors.

The nuclear reactor designed and built in Canada is called the **CANDU** reactor. This name indicates that the reactor is **CAN**-adian in design, uses **D**euterium ($^{2}_{1}H$) in its moderator and uses **U**ranium as its fuel.

(1) **Fuel for the reactor** Uranium oxide is the fuel used in a CANDU reactor. Most of its uranium atoms are $^{238}_{92}U$, but a small number (0.7%) are the fissionable $^{235}_{92}U$.

The fuel is pressed into pellets and placed in long metal tubes, each tube having a mass of about 300 kg. Hundreds of these tubes are placed horizontally into an assembly called a **calandria**. The horizontal arrangement has the advantage that the tubes can be replaced one at a time without shutting down the entire reactor. See Figure 25–2.

(2) **Controlling the fission reaction** When uranium 235 is bombarded with neutrons, it splits up. This creates new substances as well as high-speed neutrons and energy. The neutrons are needed

Figure 25–2 In the calandria of a CANDU reactor the fuel rods are horizontal. This allows the rods to be replaced by using the mechanical device shown in the photograph. The calandria shown is located at the Pickering Generating Station, Pickering, Ontario.

to continue the chain reaction, but they must be controlled so that the desired number are absorbed by the uranium nuclei.

The number of neutrons is regulated by using **control rods** that absorb neutrons. These rods, often made of cadmium, can be moved into or out of the calandria to adjust the number of neutrons striking the uranium 235.

In order for the neutrons to react with the uranium 235 nuclei, they must travel slowly enough to be absorbed. Therefore, they must be slowed down to increase the chances of reaction. A substance that causes neutrons to slow down is called a **moderator**.

The moderator used in a CANDU reactor is heavy water. Its chemical name is deuterium oxide (D_2O) because it contains deuterium and oxygen. Deuterium (D or $_1^2H$) is an isotope of hydrogen that was mentioned in Section 24.7. Each deuterium atom has one proton and one neutron. It is the extra neutron that helps prevent heavy water from absorbing neutrons from the fission reaction. That is why heavy water is an excellent moderator.

(3) **Producing steam in the reactor** The energy released from the

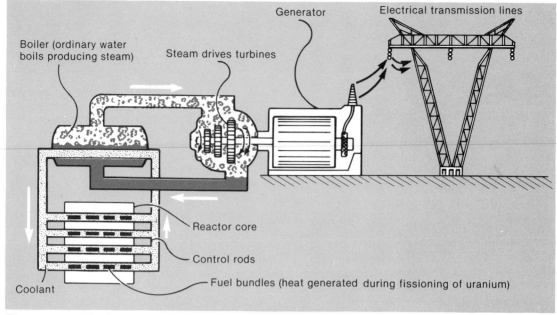

Figure 25–3 The basic operation of a nuclear generating station

fission reaction heats a substance called a **coolant**. The coolant, again heavy water, delivers its heat energy to ordinary water, which in turn changes to steam. Then the steam is directed through huge turbines to make them spin. Refer to Figures 25–3, 25–4 and 25–5.

(4) **Creating electricity** The ultimate purpose of a nuclear generating station is to produce electricity. The spinning turbines are connected to alternating current (AC) generators. The generators create electrical energy, which is delivered to consumers.

PRACTICE

4. Describe the main differences between a nuclear generating station and a hydro generating station (Section 19.4).

25.4 Nuclear-Reactor Safety

The design of nuclear generating stations takes into consideration the dangers associated with nuclear reactions. If an emergency occurs in a CANDU reactor, the moderator is immediately drained to a huge safety tank below the calandria. Without the moderator, the chain reaction would come to a halt.

Figure 25–4 Installing parts of a turbine

Figure 25–5 Huge turbines at a nuclear generating station

Unwanted radiations from the nuclear reactor are controlled by shielding (e.g., thick cement walls) and special filters. People who work in generating stations are checked daily for exposure to radiation.

Not all reactor problems have been solved. Thermal (heat) pollution occurs in the body of water near a generating station and upsets the balance of life for fish and underwater plants. Although efforts are made to control thermal pollution, the problem may become worse as our demand for electricity increases.

Probably the greatest problem facing the nuclear industry is what to do with the waste products from the fission reactions. Some of the waste products are highly radioactive and have long half-lives. Scientists have short-term solutions to this problem, but they are searching for a long-term solution.

25.5 Alternate Sources of Energy

All thermal generating stations use non-renewable fuels to change water into steam to drive turbines. Non-renewable fuels include uranium, oil, natural gas and coal. The main problems with these sources of energy are (a) we are running out of them, and (b) they create many environmental problems, including thermal, water and air pollution.

Renewable sources of energy are clean and will always be available. They include the sun, falling water, wind, waves, tides, ocean currents and biomass. The problems with these resources are that (a) they are expensive to develop, and (b) they might not be consistent day in and day out. See Figure 25-6.

Energy is one of the main problems facing our society. Scientists are trying to develop a new source of energy, as you will learn in the next section. In the meantime each one of us can help solve the problem by learning to use energy wisely.

Figure 25-6 Erosion shows that there is much energy available from ocean tides. The Bay of Fundy, which separates Nova Scotia and New Brunswick, has tides that are among the highest in the world.

25.6 Nuclear Fusion

Nuclear fusion is the joining together of two small nuclei to make a larger nucleus. The mass of the final nucleus is less than the sum of the masses of the original nuclei. In other words, some mass disappears. As in nuclear fission the lost mass changes into energy

$(E = mc^2)$. The energy from a fusion reaction is even greater per unit mass than the energy from a fission reaction.

One example of a fusion reaction is the collision of two fast-moving deuterium nuclei creating a helium nucleus, a neutron and energy. The equation for this reaction is:

$$^2_1H + , ^2_1H \rightarrow ^3_2He + ^1_0n + energy$$

Nuclear fusion is the basic source of energy in our universe. It goes on continually in the sun and stars where hydrogen and other light elements fuse into heavier elements. In fact, it is believed that nuclear fusion is the process by which all elements have formed from atoms of hydrogen.

Scientists have been able to create fusion reactions on earth. A hydrogen bomb is an example of a use of fusion that is neither peaceful nor controlled. However, scientists are searching for ways of controlling fusion reactions. They are attempting to use lasers to obtain the high temperatures and pressures needed for fusion. If they are successful, we may have an answer to our energy supply problems. An almost endless number of hydrogen nuclei obtained from ocean water may be available for energy through fusion. Furthermore, fusion does not create the problems of radioactive by-products as does fission.

PRACTICE
5. Nuclear fusion on the sun changes an estimated 4×10^9 kg of mass into energy each second.
   (a) Calculate the amount of energy emitted by the sun each second.
   (b) What is the power rating of the sun? $(P = \dfrac{E}{t})$

25.7 Physics——Past, Present and Future

In Section 1.4 studying physics was compared to building a house. Your study of physics has built the foundation, added the framework and completed the details step by step. Your study of physics is now complete. Or is it?

You know how physics developed from the ancient Greeks to the times of Galileo Galilei and Isaac Newton and then through to this century. You have seen the importance of measurement, experimentation and observation.

You have also filled in many of the details of the structure of physics. Some of those details include the study of motion, mechanical forces and energy, power, heat energy, sound energy, electrical energy, electric and magnetic forces, light energy and nuclear energy. A model of the structure of matter was developed, and you have learned how matter and energy interact. You have seen many applications of physics concepts and have learned the importance of conserving energy.

But this study has not given a complete view of physics. No one textbook could do that. In fact, the more we learn about the subject of physics, the more there is to learn.

Thus, the process called the scientific method continues. We hypothesize, we experiment, we arrive at conclusions, we predict the results of a new situation and we start all over again. The researchers of "practical physics" tell us there may be hundreds of subatomic particles that have no explanation. They believe that every particle has an antiparticle. When a particle collides with its antiparticle, the two destroy each other, producing energy.

Astronomers make us feel very small by telling us how many billions of stars there are. They believe that the chances are high that somewhere else life exists in advanced forms. Much remains to be learned from the study of the universe.

Physicists are trying to solve two basic questions about nature:
(1) Is there a fundamental particle that makes up all other particles?
(2) Is there a unified theory that can explain all of nature's forces?

Where will all this knowledge and these questions lead us in the future? The rapid development of computers will give us more leisure time. We may be able to watch three-dimensional television using laser technology. Communication systems will make our planet seem very small. Perhaps we will learn more about how humans think so we can make wise decisions about our future.

This brings us to the end of this study, but it represents the beginning of a huge frontier of knowledge in the subject of physics.

25.8 Review Assignment

1. Name and describe the force that binds the particles of the nucleus together. (25.1)
2. Compare nuclear fission and nuclear fusion. (25.2 and 25.6)

3. Calculate the amount of energy released when the following quantities of matter change entirely into energy:
   (a) 3.0 kg
   (b) 400 g
   (c) 600 kg   (25.2)
4. Describe how a chain reaction begins.   (25.2)
5. State the function of each of the following components of a CANDU nuclear reactor:
   (a) uranium
   (b) control rods
   (c) moderator
   (d) coolant
   (e) steam
   (f) generator   (25.3)
6. What are the advantages and disadvantages of using nulear energy to produce electricity?   (25.3 to 25.5)
7. List three ways that you think you could use to help conserve energy in the future.

25.9 Answers to Selected Problems

PRACTICE QUESTIONS
3. (a) 4.5×10^{16} J
   (b) 1.8×10^{17} J
   (c) 9.0×10^{17} J
5. (a) 3.6×10^{26} J
   (b) 3.6×10^{26} W or 3.6×10^{23} kW (This is approximately a billion billion times as much power as our largest generating station puts out!)

REVIEW ASSIGNMENT
3. (a) 2.7×10^{17} J
   (b) 3.6×10^{16} J
   (c) 5.4×10^{19} J

Appendix A
Standard Form
(Scientific Notation)

For certain measurements in physics, the numbers can be very awkward, even when using the metric system. Numbers that are either very big or very small are better written using **standard form** (also called **scientific notation**). In standard form the number is written with one digit other than zero before the decimal and the correct power of ten notation after the number. The best way to see the usefulness of standard form is to study examples, such as those shown in the table below.

Table A-1   Examples of Standard Form

Quantity measured	Measurement	Standard form
Distance to the Andromeda Galaxy	19 000 000 000 000 000 000 000 m	1.9×10^{22} m
Distance to nearest star	40 000 000 000 000 000 m	4×10^{16} m
Distance of earth to sun	150 000 000 000 m	1.5×10^{11} m
Diameter of earth	13 000 000 m	1.3×10^{7} m
Length of football field	100 m	1.0×10^{2} m
Width of a fingernail	0.01 m	1.0×10^{-2} m
Thickness of credit card	0.0005 m	5×10^{-4} m
Thickness of human hair	0.000 05 m	5×10^{-5} m
Thickness of spider web strand	0.000 005 m	5×10^{-6} m

PRACTICE
1. Change these measurements to standard form:
   (a) 32 000 000 s in one year
   (b) 6 250 000 000 000 000 000 electrons/s
   (c) 9 192 000 000 vibrations/s
   (d) 0.000 085 cm
   (e) 0.000 000 000 38 kg
2. Write these measurements using ordinary notation:
   (a) 4.2×10^5 m   (c) 8.2×10^1 kg
   (b) 3.1×10^{-7} g   (d) 9.564×10^8 km

Appendix B
Measurements and Calculations Using Significant Digits

The number of students in a classroom is an exact number. The length of the classroom is a measurement that might appear to be exact. However, it cannot be known exactly. The surface area of the floor of the classroom involves a calculation (length × width) that also cannot be known exactly. In experiments performed in physics, several such measurements and calculations must be made. The following rules should be applied when recording measurements or performing calculations:
(1) Each recorded measurement should consist of the digits known for certain and no more than one estimated digit.
(2) When calculating a number using measured quantities, the answer should be rounded off so it has no more than one estimated digit.

Consider, for example, the measurement of the length of a page in your notebook. You might find that its length lies somewhere between 27.3 cm and 27.5 cm, so you would record its length as

27.4 cm. There are three digits that have a meaning in this measurement. They are called **significant digits**. The first two digits are known for certain, and the last digit is estimated. To indicate the estimated digit, some symbol, such as an "x", may be used (e.g., 27.4^x cm).

Now consider calculations involving measured quantities. The answer should contain no more than one estimated digit. The sample problems that follow illustrate examples of addition and multiplication. Subtraction and division involve similar results.

Sample problem 1: Add 3.45^x cm and 8.342^x cm.
Solution: 3.45^x cm
                  +8.342^x cm
                 ─────────
                 11.79^{x}2^x cm
∴ The answer is 11.79^x cm (only one estimated digit).

Sample problem 2: Multiply 3.4^x cm × 2.1^x cm.
Solution: 3.4^x cm
                  2.1^x cm
                 ─────────
                 3^{x}4^x
              68^x
              ─────────
              7.1^{x}4^x cm^2 or 7.1^x cm^2

Sample problem 3: Calculate 2.6^x m × 4.8^x m × 5.1^x m.
Solution: 2.6^x m   12.5^x m^2
              4.8^x m   5.1^x m
             ─────────   ─────────
             208^x 1^{x}2^{x}5^x
          104^x 62 5^x
          ─────────   ─────────
          12.4^{x}8^x m^2 or 12.5^x m^2 63^x.7^{x}5^x m^3

∴ The final answer is 64^x m^3.

PRACTICE
1. Add:
   (a) 2.2^x m + 3.5^x m
   (b) 5.1^x cm + 6.23^x cm
   (c) 2.3^x m + 2.3^x m + 2.3^x m
2. Multiply:
   (a) 2.2^x m × 1.3^x m
   (b) 2.4^x m × 2.62^x m
   (c) 1.2^x cm × 1.2^x cm × 1.2^x cm

Appendix C
Rules for Rounding Off Numbers

When calculated answers must be rounded off to the correct number of significant digits, the following rules should apply. Each example given is rounded off to two significant digits.

(1) If the first digit to be dropped is 4 or less, the preceding digit is not changed.

e.g., 8.74 becomes 8.7

(2) If the first digit to be dropped is 6 or more, the preceding digit is raised by 1.

e.g., 6.36 becomes 6.4

(3) If the digits to be dropped are a 5 followed by digits other than zeros, the preceding digit is raised by 1.

e.g., 3.45123 becomes 3.5

(4) If the digit to be dropped is a 5 (or a 5 followed by zeros), the preceding digit is not changed if it is even, but it is raised by 1 if it is odd.

e.g., 2.65 becomes 2.6
2.75 becomes 2.8

PRACTICE

1. Round off each number to two significant digits:

(a) 6.43   (b) 3.54
(c) 8.49   (d) 5.07
(e) 2.7538 (f) 7.15223
(g) 8.25   (h) 2.85
(i) 4.55   (j) 3.15

Appendix D
Scalar and Vector Quantities

In certain situations in the subject of physics it is necessary to distinguish scalar and vector quantities.

Scalar quantities have size but no direction. Distance, speed and time are scalar quantities. Examples of scalar measurements are: 2.5 m, 6.8 m/s and 180 s.

Vector quantities have both size and direction. **Displacement** is a vector quantity. It is the distance an object is located from its starting position. **Velocity** is another vector quantity. It is the ratio of the final displacement to the time to reach that displacement. Examples of vector measurements are: 6.3 m W and 3.2 km/h N.

To distinguish vector quantities from scalar quantities, an arrow is usually written above the symbol for a vector quantity; e.g., $\vec{d}$ for displacement and $\vec{v}$ for velocity.

The equation $v = \dfrac{d}{t}$ applies to the scalar quantities of distance and speed. The corresponding equation for vector quantities is:

$$\textbf{average velocity} = \frac{\textbf{displacement}}{\textbf{time}} \quad \text{or} \quad \vec{v} = \frac{\vec{d}}{t}$$

Sample problem 1: A girl walks 20 m E, then 5.0 m W in 15 s. Calculate her:
(a) final displacement
(b) average velocity
Solution:
(a) The girl's displacement is 15 m E of her starting point.

(b) $\vec{v} = \dfrac{\vec{d}}{t}$

$\quad = \dfrac{15 \text{ m } E}{15 \text{ s}}$

$\quad = 1.0 \ \dfrac{\text{m}}{\text{s}} E$

PRACTICE

1. State which measurements are scalar and which are vector:
   (a) 12 s   (b) 3.2 m S (c) 13.5 km W
   (d) 4.0 km/h E (e) 120 kg   (f) 8.5 m *up*

2. A car travels 40 km N, then 120 km S in 4.0 h. Calculate the:
   (a) final displacement   (b) average velocity

Appendix E
Details of Musical Scales

Section 14.2 mentions two main musical scales, the scientific and musicians' scales. This appendix is written for teachers and musical students who are interested in the mathematical derivations of the frequencies of those scales.

In general, two frequencies sound harmonious if they are in a certain ratio. Pleasant, harmonious sounds have high **consonance**. Unpleasant, unharmonious sounds have high **dissonance** or low consonance. Table E-1 lists some ratios of frequencies in order of decreasing consonance.

Table E-1

	Interval	Ratio of frequencies
	Unison	1/1
	Octave	2/1
Decreasing consonance	Fifth	3/2
	Fourth	4/3
	Major third	5/4
	Major sixth	5/3
	Minor third	6/5
	Minor sixth	8/5

The Scientific Scale

This scale is based on the number 2^8, which is 256. Thus, the standard frequency is 256 Hz. That standard is multiplied by certain ratios, such as 3/2, to obtain the scale. The calculations of one octave are shown in Table E–2.

Table E–2   The Scientific Musical Scale

Note	Ratio of frequencies	Frequency (Hz)	Interval
Middle C	1/1	256.0	standard
D	9/8	288.0	major whole tone (second)
E	5/4	320.0	major third
F	4/3	341.3	fourth
G	3/2	384.0	fifth
A	5/3	426.7	major sixth
B	15/8	480.0	major seventh
High C	2/1	512.0	octave (eighth)

The notes G and C sound pleasant because their frequencies are in the ratio 3/2. Another pleasant-sounding pair is B and E because 480/320 is 3/2.

Musicians' (Equitempered) Musical Scale

This scale is based on two main facts:

(1) There are exactly 12 notes per octave. On a piano keyboard, for example, there are 5 black keys and 7 white keys per octave.

(2) A note one octave above another is double in frequency.

Thus, the equitempered scale uses a number (approximately 1.0595) that when multiplied by itself 12 times gives the number 2. In other words, $\sqrt[12]{2} = 1.0595$. We will use the symbol "a" for this special number.

Starting with the standard frequency of 440 Hz, we multiply and divide 440 by a, then by a^2, then by a^3 and so on. This procedure gives new notes. It is repeated until the entire musical scale is calculated. Calculations for one octave are shown in Table E–3.

Table E–3   The Musicians' Musical Scale

Note	Calculation (using a = 1.0595)	Frequency (Hz)	App. ratio	Interval
A_4	standard	440.00	1/1	standard
A# , Bb	$440 \times a$	466.16	16/15	semi-tone
B_4	$440 \times a^2$	493.88	9/8	major whole tone minor whole tone
C_5	$440 \times a^3$	523.25	6/5	minor third
C# , Db	$440 \times a^4$	554.37	5/4	major third
D_5	$440 \times a^5$	587.33	4/3	fourth
D# , Eb	$440 \times a^6$	622.25	45/32	augmented fourth diminished fifth
E_5	$440 \times a^7$	659.26	3/2	fifth
F_5	$440 \times a^8$	698.46	8/5	minor sixth
F# , Gb	$440 \times a^9$	739.99	5/3	major sixth
G_5	$440 \times a^{10}$	783.99	9/5	minor seventh
G# , Ab	$440 \times a^{11}$	830.61	15/8	major seventh
A_5	$440 \times a^{12}$	880.00	2/1	octave (eighth)

One example of a pair of notes that sounds harmonious is E_5 and A_4. The ratio of their frequencies is 3/2.

Appendix F
Derivations of Equations

In the text certain equations are used without a complete explanation regarding their origin. The derivations of those equations are beyond the goals of the text but are included here for reference.

(1) **Section 3.4** "For every centimetre difference between the water levels there is a pressure difference of 100 Pa or 0.1 kPa."

The force of gravity pulling down on a 1.0 kg mass is about 10 N. A force scale can be used to show this fact.

The density of water is 1000 g/L, which is the same as 1.0 g/mL or 1.0 g/cm³. This means that the force of gravity pulling down on 1.0 cm³ of water is 10 N ÷ 1000 or 0.01 N.

Now imagine a cube of water 1.0 cm on each side. The bottom of the cube has a surface area of 1.0 cm². If the cube is resting on a surface, it exerts a pressure of:

$$p = \frac{F}{A}$$
$$= \frac{0.01 \text{ N}}{0.0001 \text{ m}^2} \quad (1.0 \text{ cm}^2 = 0.0001 \text{ m}^2)$$
$$= 100 \text{ N/m}^2, \text{ which is 100 Pa or 0.1 kPa}$$

This proves the original statement.

(2) **Section 3.8** "The force on 1.0 mL of water is 0.01 N."
This fact is explained in #1 above.

(3) **Section 8.5** $E_K = \frac{mv^2}{2}$

Assume that an object, starting from rest, has an amount of work, E, done on it. By the law of conservation of energy, that work is transformed into kinetic energy possessed by the object. Thus,

$$E_K = E$$
$$= Fd$$

where F is the unbalanced force $(F = ma)$ that causes the object to accelerate for a distance d given by the relation $d = v_{ave}t$. (v_{ave} is the average or half-time speed.)

Therefore, $E_K = (ma)(v_{ave}t)$

Now, $a = \dfrac{\Delta v}{t}$

$\quad\quad = \dfrac{v_f}{t}$ (The object started from rest and reached a final speed of v_f.)

Thus, $E_K = \dfrac{m(v_f)(v_{ave}t)}{t}$

$\quad\quad = mv_f v_{ave}$

Since $v_{ave} = \text{½ } v_f$, we conclude that $E_K = \dfrac{mv_f^2}{2}$

or $E_K = \dfrac{mv^2}{2}$

(4) **Section 16.12** "For a series circuit, $R_T = R_1 + R_2$."

Experimentally, it was found that for a series circuit having resistors R_1 and R_2, the current (I) remains constant in the circuit and the total voltage (V_T) equals the sum of the individual voltages $(V_1 + V_2)$.

Using Ohm's law, $R_T = \dfrac{V_T}{I}$

$\quad\quad R_T = \dfrac{V_1 + V_2}{I}$

$\quad\quad R_T = \dfrac{V_1}{I} + \dfrac{V_2}{I}$

$\quad\quad \therefore R_T = R_1 + R_2$

(5) **Section 16.12** "For a parallel circuit, $\dfrac{1}{R_T} = \dfrac{1}{R_1} + \dfrac{1}{R_2}$."

Experimentally it was found that for a parallel circuit having resistors R_1 and R_2, the voltage (V) remains constant and the total current (I_T) equals the sum of the individual currents $(I_1 + I_2)$.

Using Ohm's law, $R_T = \dfrac{V}{I_T}$ or $\dfrac{1}{R_T} = \dfrac{I_T}{V}$

$\quad\quad \dfrac{1}{R_T} = \dfrac{I_1 + I_2}{V}$

$\quad\quad \dfrac{1}{R_T} = \dfrac{I_1}{V} + \dfrac{I_2}{V}$

$\quad\quad \therefore \quad \dfrac{1}{R_T} = \dfrac{1}{R_1} + \dfrac{1}{R_2}$

(6) **Section 17.4** $P = VI$

For this derivation we will introduce a new symbol, Q, representing charge.

Current (I) is defined as the amount of charge (Q) that

passes through a circuit each second (t) (Section 16.3). Thus,

$$I = \frac{Q}{t} \text{ or } Q = It.$$

Voltage (V) is defined as the amount of energy (E) given to a certain amount of charge (Q) (Section 16.4). Thus,

$$V = \frac{E}{Q} \text{ or } E = VQ.$$

Now power is $P = \frac{E}{t}$ (where $E = VQ$)

$$P = \frac{VQ}{t} \text{ (where } Q = It)$$

$$P = \frac{V(It)}{t} \text{ (divide the } t)$$

$$\therefore P = VI$$

(7) **Section 20.8** "At 90° between the mirrors, three images are formed."

The number of images (n) formed by two mirrors at various angles to each other is given by the equation

$$n = \frac{360°}{a} - 1$$

where a is the angle between the mirrors.

Index

Photo Credits

Cover photos: Laser Images, Inc., California
Figure 3-3: The Toronto Star
Figure 4-3(c): General Motors Corporation, Warren, Mich.
Figure 8-1: The Toronto Star
Figure 12-15(a), (b) and (c): University of Washington College of Engineering/Library Services.
Figure 14-29: The National Arts Centre, Ottawa, Ont.
Figure 14-30(a): Bell-Northern Research Ltd., Ottawa
Figure 15-19(a): The Ontario Science Centre, Toronto
Figure 16-3(a): National Aeronautics and Space Administration
Figure 16-3(d): Pacific Gas and Energy, San Francisco, Cal.
Figure 21-8(d):Bell-Northern Research Ltd., Ottawa
Figure 24-9:Atomic Energy of Canada Ltd.
Figures 25-2, 25-4 and 25-5: Ontario Hydro
Figure 8-8: Miller Services.